D0687666

MODERN INSTRUMENTATION FOR SCIENTISTS AND ENGINEERS

Springer
New York
Berlin
Heidelberg
Barcelona
Hong Kong
London
Milan
Paris
Singapore
Tokyo

MODERN INSTRUMENTATION FOR SCIENTISTS AND ENGINEERS

James A. Blackburn

With 254 Figures

Springer

James A. Blackburn
Department of Physics and Computing
Wilfrid Laurier University
Waterloo, Ontario N2L 3C5
Canada

Cover illustration: Conceptual representation of a shear mode piezoelectric accelerometer. (Figure 15.14, page 234)

Library of Congress Cataloging-in-Publication Data
Blackburn, James A.
Modern instrumentation for scientists and engineers / James A. Blackburn.
p. cm.
Includes bibliographical references and index.
ISBN 0-387-95056-7 (alk. paper)
1. Scientific apparatus and instruments. I. Title.
Q185.B563 2001
502′.8—dc21 00-040040

Printed on acid-free paper.

Production managed by Allan Abrams; manufacturing supervised by Erica Bresler.
Typeset by TechBooks, Fairfax, VA.
Printed and bound by Hamilton Printing Co., Rensselaer, NY.
Printed in the United States of America.

9 8 7 6 5 4 3 2 1

ISBN 0-387-95056-7 SPIN 10770110

Springer-Verlag New York Berlin Heidelberg
A member of BertelsmannSpringer Science+Business Media GmbH

To Helena

Preface

Knowledge of instrumentation is for experimentalists a kind of fluency in the language of measurement. But it is a fluency not so commonly possessed, and without which much of the experimental process remains hidden and mysterious. The basic goal in writing this book is to provide a treatment of useful depth of the basic elements of the instrumentation "language," namely electronics, sensors, and measurement.

The present epoch is arguably a golden age for instrumentation. The crucial ingredient has been the exceptional development of semiconductor fabrication technology, and this has led to the present richness in both analog and digital integrated circuits. The former provide relatively inexpensive but high-performance electronic modules (such as the operational amplifier) which can serve as building blocks for more complex circuits, whereas the latter have culminated in the desktop computer, which has permeated modern life generally and revolutionized the instrumentation world with its capacity to act as a measurement controller and data storage center. Finally, silicon micromachining is creating a host of new sensors for such quantities as acceleration and pressure.

The sources of this book are threefold. First, it arises from the process of teaching the subject for more than a decade. Second, it is driven by the repeated experience of acting as a consultant on instrumentation to colleagues in physics, biology, chemistry, psychology, and geography. Finally, instrumentation has played a fundamental role in my own research. These distinctly different threads are reflected in the particular objectives that I wished the book to fulfill: to serve as a reference and handbook for researchers and to serve as a possible textbook for senior undergraduate or graduate students preparing for careers in experimental science or engineering.

Personal experience has taught me that when a book is consulted as a handbook or reference, it is read in a very localized manner. Only a particular block of material may be referenced at one time. For this reason, each chapter has been written in a reasonably self-contained style, with its own citations. The requisites of a textbook are somewhat different. Ideally, chapters should link logically from one to another in sequence. Teachers of course like to see problems, so a modest selection of worked examples and end-of-chapter questions is provided. These are intended to suggest typical lines of questioning. In a course setting, simulation assignments and individual projects can be the best way to develop familiarity and understanding of the concepts.

As a subject, instrumentation presents particular challenges and difficulties because of the diversity of an audience that might include physicists, biologists, chemists, experimental psychologists, geophysicists, engineers, and medical researchers. Each group possesses a very different background, so no simple common baseline exists from which a text can proceed. To address this reality, a certain amount of review material has been included, especially in the first four chapters. For some, this will be quite familiar, but hopefully still welcome as a summary. For others, these topics will be an essential component in a reference on instrumentation.

Following the preliminaries just noted, the next four chapters cover topics related to signal conditioning. These include amplification and filtering. Waveform generation is also discussed at this point. Hence, the first eight chapters provide a review and/or development of essential concepts in electronics. Many practical circuit examples based on tested PSpice simulations are included.

The logical flow of the remainder of the book is structured around the notion that physical quantities (such as heat and light) can be transformed by sensors into electrical quantities, which then can be measured. In choosing sensors for inclusion, the guiding principle was to focus on real-world scenarios. This meant selecting those sensors that are typically employed for the most common ranges of temperature, light intensity, magnetic field strength, pressure, and so forth. Transducers intended for use well outside these ranges (such as very high or low temperatures or exceptionally weak light levels) are considered here to be "special purpose" and thus are not covered.

The final chapter focuses on measurement systems. This represents, to an extent, the present state-of-the-art approach to complex instrumentation tasks. It is also an area of continuing rapid progress and change.

In writing a book such as this, an author is faced with basic choices in regard to what is to be included and what is to be excluded. Any manuscript could be expanded with additional material, or equally, compressed by trimming. On the one hand, an encyclopedic volume is usually impractical; yet the omission of key material must be avoided. In the end, however, only one book can be written. Experience, personal opinion, and taste play a role, which explains why no two authors would ultimately produce the same text. This is the book *I* wanted but could never find.

Finally, it is with pleasure that I acknowledge debts of various sorts to the following individuals: Dr. Robert G. Rosehart, President of Wilfrid Laurier University, for providing release time from my usual academic obligations in order to complete the final phase of writing; Dr. Thomas von Foerster and Jeannette Mallozzi of Springer-Verlag New York, Inc.; Manfred Gartner, for technical help in computer-related matters; Dr. Reinhard Neul of Robert Bosch, GmbH; Liz Searcy of Analog Devices; Michael F. Grimaldi of Kistler Instrument Corporation; Joseph C. Nowlan of Walker Scientific, Inc.; Bob Christensen of GMW Associates; Peter D. Stolpe of F.W. Bell; Carl Nybro, Vanessa Trujillo, and Don Clinchy of National Instruments; Brian Withnell and Bill Porter of Agilent Technologies; and especially, Professor John Smith of the University of Waterloo, for his lasting contributions, first as teacher and mentor, then as colleague, collaborator, and friend.

James A. Blackburn
Waterloo, Ontario, Canada
September 2000

Schematic files covering all of the PSpice simulations discussed in this text are available from the Springer website at:
http://www.springer-ny.com/detail.tpl?isbn=0387950567.
Instructions for downloading these files, and for obtaining the Evaluation Version of PSpice, may be found in Read_Me.

Contents

P A R T I

Electronics

C H A P T E R 1

Physical Quantities

At the heart of experimental science lies the issue of measurement, and each particular experiment entails the measurement of certain physical quantities. Before proceeding to issues relating to the means by which such quantities really are monitored, a brief review of definitions, units, and fundamental properties is called for.

1.1 CHARGE, POTENTIAL, AND CURRENT

In the MKS system, electrical charge (q) is expressed in units of coulombs (C). One coulomb is in fact an enormous quantity which would require an accumulation of gigantic numbers of elementary charged particles. The fundamental negative particle is the electron; it has a mass of 9.109558×10^{-31} Kg and a charge of $1.6021917 \times 10^{-19}$ C. The positive charge element, the proton, possesses the same amount of charge but is much more massive at 1.672614×10^{-27} Kg.

Electrons are normally located in the quantized orbital shells surrounding atomic nuclei, but they can be liberated from the parent atoms under a variety of circumstances, all of which involve the application of sufficient energy to activate an escape. The outermost electrons in an atom are liberated most easily because their orbits are most distant from the positive nucleus and the Coulomb binding forces are, consequently, weakest. For the simplest of all atoms, hydrogen, the solitary electron orbiting the nucleus (proton) has a binding energy of 13.605826 eV, where the electron volt (eV) is a unit of energy equal to $1.6021917 \times 10^{-19}$ joules. This means that about 13 eV is required to ionize hydrogen.

In a metal such as copper, at room temperature, there is enough thermal energy to free from each atom the outermost electron, which is then able to wander about in the lattice. There are approximately 10^{22} of these *conduction electrons* per cubic centimeter. The electron states in a semiconductor such as silicon are subject to somewhat complex quantization conditions, but the essential idea here is that liberation of the outermost atomic electron is considerably more difficult than in a metal. At room temperature, there might be only about 10^{10} or 10^{12} conduction electrons per cubic centimeter. All of the electrical properties of these metals or semiconductors are controlled by the response of the conduction electrons to various stimuli, such as electric and magnetic fields.

Electric fields ($\mathcal{E}$) are created in space by local concentrations or distributions of static charge. Electrical fields have MKS units of volts per meter (V/m). Between two points A and B, there will be a potential difference ϕ, measured in volts, defined by the integral $\phi = \oint \vec{\mathcal{E}} \bullet d\vec{\ell}$, where the integration is carried out along any path connecting the points. Properly speaking, only potential differences are defined, but if, as is done in many cases, one selects some reference point in a system and arbitrarily assigns to that point zero potential, then any other point can be spoken of as having a potential of so many volts.

When an electric field is present throughout some region in space, then all charged particles within that region experience forces equal to $\vec{F} = \vec{\mathcal{E}} q$. Free charges (conduction electrons in a solid material) move under the influence of such forces. In fact, the charged particles are accelerated by the force, but typically they are scattered by obstacles in their environment, such as lattice atoms or impurities, so that the actual motion of conduction electrons in a solid is a sequence of interrupted accelerations. Averaged over time, the free carriers appear to drift at a velocity determined by the smoothed-out actions of the collisions. Typical scattering times at room temperature might be about 10^{-6} sec, whereas drift velocities are approximately 0.003 m/sec per unit electric field (1 V/m) for copper and 0.15 m/sec for pure silicon.

Moving charge constitutes electrical current. One ampere (A) is defined as a flow rate of 1 C/sec. Current density is simply the current normalized to a unit of cross sectional area.

When they are scattered, conduction electrons lose energy to the host lattice within which they are traveling. This process gives rise to the heating associated with current flow in resistive media. For many materials, the ratio of the voltage across any particular sample to current flowing through it turns out very nearly to be a fixed quantity; that is, larger currents are matched by larger voltages, and conversely. This ratio is called the *resistance* (R), and it is measured in MKS units of ohms (Ω). Note that resistance is a geometry-dependent quantity—short, fat samples will exhibit a smaller resistance than long, thin samples. A constant (at least for a given fixed temperature) that reflects the resistive process inherent in any particular material is the *conductivity* (σ), which is defined as the ratio of current density to electric field. *Resistivity* (ρ) is just the reciprocal of conductivity. The resistivity of copper is approximately $2 \times 10^{-8}\ \Omega\,\mathrm{m}$, whereas for a semiconductor such as silicon $\rho \approx 2000\ \Omega\,\mathrm{m}$.

1.2 MAGNETIC FIELD

Just as static charges give rise to electric fields, moving charges (currents) are associated with magnetic fields, which extend out into the surrounding space. Three quantities are connected to descriptions of magnetic fields: flux (Φ: a scalar), flux density ($\vec{B}$: a vector), and field intensity ($\vec{H}$: a vector). The MKS unit of magnetic flux is the *weber* (Wb), which is equivalent to a volt-sec, flux density is then expressed in $\mathrm{Wb/m^2}$, and field intensity has dimensions of amp–turns/m. The magnitude of $\vec{B}$ is very often given in *gauss* (G), and the conversion is $1\ \mathrm{Wb/m^2} = 10^4$ G. The terrestrial magnetic field at the surface of Earth is typically about 0.5 G; within the gaps of fairly strong permanent magnets, $B \approx 10^3$ to 10^4 G.

Charged particles that move in a region occupied by a magnetic field will experience a force according to the Lorentz expression $\vec{F} = q\,\vec{v}\,\mathbf{x}\,\vec{B}$, where **x** denotes the vector cross product of the velocity and the magnetic field. The Lorentz force is thus orthogonal to the plane containing $\vec{v}$ and $\vec{B}$.

1.3 CAPACITANCE AND INDUCTANCE

A capacitor is a charge storage device, and in its simplest conceptual form can be thought of as just two flat metallic plates of area A held apart at a separation d. Such a structure can store electrical charge. The well-known relation between the amount of charge ($+Q$ on one plate and $-Q$ on the other) and the associated voltage difference between the plates is $Q = CV$, where the proportionality constant C is the capacitance, which for the parallel plate system is $C = 4\pi\epsilon_0 A/d$ with the permittivity of free space $\epsilon_0 = 8.85 \times 10^{-12}$ Farads/m. Clearly, the unit of capacitance is the Farad (F), but a one Farad capacitor would be a huge structure, and more common capacitances turn out to have values measured in microfarads ($\mu\mathrm{F} = 10^{-6}$ F), nanofarads ($\mathrm{nF} = 10^{-9}$ F), and picofarads ($\mathrm{pF} = 10^{-12}$ F). When the space between the plates is filled with a material substance such as wax or insulating plastic, then the capacitance will be enhanced by a factor k, called the dielectric constant of the filler. For a vacuum, $k = 1.0$ by definition, whereas for typical dielectrics such as polyethylene or polystyrene, k ranges up to about 3.

An inductor is, by analogy with the capacitor, a current storage device, and in its simplest conceptual form can be thought of as just a cylindrical coil of length ℓ made up of N turns of resistanceless wire. Faraday's law relates the rate of change of current flowing through the coil to an induced potential difference which consequently appears across its terminals; the expression is $V = -L\, dI/dt$, where L is the inductance measured in *henrys*. For the example of a simple coil, $L = \mu_0 N^2 \pi r^2/\ell$, with the permeability of free space given by $\mu_0 = 4\pi \times 10^{-7}$ H/m. If the coil is filled with a magnetizable medium, then the inductance will be enhanced because the permeability μ will be larger (for ferromagnetic materials) than μ_0. Steel and iron, for example, have permeabilities that exceed the free-space value by factors of several thousand, whereas cobalt and nickel are increased by a few hundred times.

CHAPTER 2

DC Circuits

A circuit consists of an arrangement of components and wires. Direct current (dc) circuits are understood to mean configurations of batteries and resistors in which steady currents are flowing. In other words, any switch-on transients that might have been present initially have died away. Capacitors will not be considered as components in this dc domain because no constant current can flow through the device—it looks like an open circuit. Similarly, inductors will not be included because under steady conditions ($dI/dt = 0$) no voltage will exist across an ideal inductor—it looks like a short circuit.

2.1 BRANCH AND NODE ANALYSIS

The method of branch and node analysis will be demonstrated through its application to a simple but generic circuit. Hence, consider the network shown in Fig. 2.1. The objective of our analysis will be to determine all currents flowing in all components. Of course, this will also effectively determine the voltages between any pair of nodes in the circuit.

This particular example is composed of resistors and batteries arranged in three cells. There are six distinct branches, and each one has been assigned a current ($I_1 \ldots I_6$). It turns out that the procedure to be followed does not require a correct guess as to the direction of these various currents. An incorrect direction will ultimately be revealed through a negative algebraic sign for the current.

To proceed, we invoke Kirchhoff's laws, which may be stated as follows:

(1) The algebraic sum of all voltages around any closed path in a circuit (i.e., returning to the original starting point) is zero. The choice of starting point is entirely arbitrary, and the path may be traversed in either a clockwise or

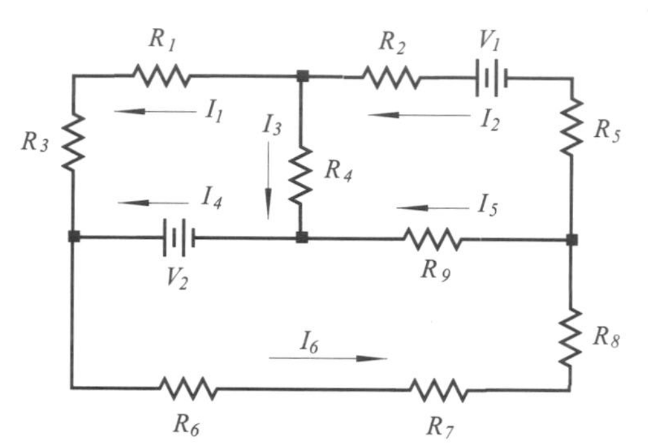

FIGURE 2.1. Branch current analysis of example circuit.

counterclockwise sense. Whichever choice is made, the following rules govern the algebraic sign of each voltage contribution to the sum. In moving through a resistor, the voltage is $+IR$ if the path sense is against the assumed current direction, and it is $-IR$ if the path sense is coincident with the assumed direction of current. In moving through a battery, the voltage is $+V$ if the path sense is from negative terminal to positive terminal and is $-V$ otherwise.

(2) The sum of all currents entering any node must equal the sum of all currents leaving that same node.

When applied to the present example, these laws can be seen to yield the following equations:

$$
\begin{aligned}
-I_3R_4 + V_2 + I_1R_3 + I_1R_1 &= 0, \\
I_2R_2 - V_1 + I_2R_5 - I_5R_9 + I_3R_4 &= 0, \\
I_6R_8 + I_6R_7 + I_6R_6 - V_2 + I_5R_9 &= 0, \\
I_2 - I_1 - I_3 &= 0, \\
I_5 + I_3 - I_4 &= 0, \\
I_6 - I_5 - I_2 &= 0.
\end{aligned}
$$

This system of six equations in six unknowns ($I_1 \ldots I_6$) is inhomogeneous and therefore is solvable. It should be noted that additional equations dictated by Kirchhoff's laws could be written, such as $-I_1R_1 - I_1R_3 - I_6R_6 - I_6R_7 - I_6R_8 - I_2R_5 + V_1 - I_2R_2 = 0$, which arises from a loop around the entire periphery of the figure, or $I_4 + I_1 - I_6 = 0$. In this example, at least three other voltage sum equations are possible as well, with paths that encompass pairs of primary loops. All such expressions are correct, but they do not contain new information. In other words, they are derivable from the first set of six equations. As a matter of fact, other groupings of six equations in the six unknowns can be selected from the total of eleven equations mentioned, but whatever group is chosen must represent an independent set, so that in the end there will always be just six independent equations to determine the six unknowns.

2.2 LOOP ANALYSIS

Loop analysis is an algebraic alternative to the previous method for obtaining component currents. To illustrate this approach, we return to the same example circuit, only now each of the three primary loops is labeled with a loop current I_A, I_B, I_C, as shown in Fig. 2.2.

Using Kirchhoff's law #1,

$$-I_A R_1 - I_A R_3 - V_2 - (I_A - I_B)R_4 = 0,$$
$$-I_B R_5 + V_1 - I_B R_2 - (I_B - I_A)R_4 - (I_B - I_C)R_9 = 0,$$
$$-(I_C - I_B)R_9 + V_2 - I_C R_6 - I_C R_7 - I_C R_8 = 0.$$

Notice that wherever a resistor is situated in a branch that is shared by two adjacent loops, such as R_4 and R_9 in this example, the net current flowing in that resistor will be the difference between the associated loop currents, since they pass through the resistor in opposite directions.

In this case, there are three equations in three unknowns (I_A, I_B, I_C). These are easily related to the six branch currents of the previous section:

$$I_1 = I_A,$$
$$I_2 = I_B,$$
$$I_3 = (I_B - I_A),$$
$$I_4 = (I_C - I_A),$$
$$I_5 = (I_C - I_B),$$
$$I_6 = I_C.$$

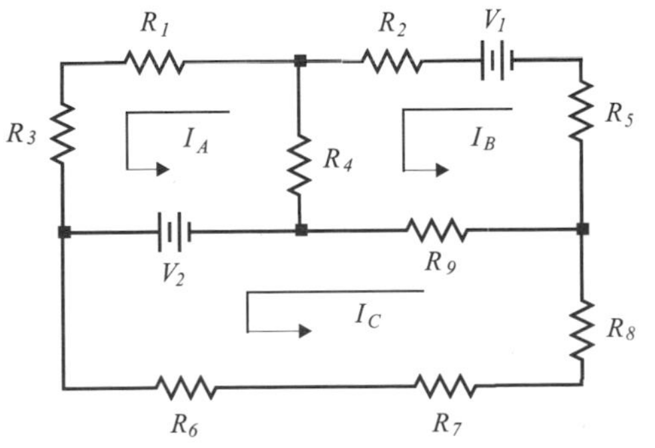

FIGURE 2.2. Analysis of example circuit using loop currents.

2.3 REMARKS

The simple network of Figs. 2.1 and 2.2 has served to illustrate two equivalent procedures for obtaining all the dc currents (and voltages). Because they are fully equivalent, there is no "best" choice between branch/node and loop formalisms. It would seem that the loop method offers some economy in that there are fewer linear equations to solve. However, there will be occasions when the branch/node scheme will be adopted because it relates directly to actual currents flowing in components.

PROBLEMS

Problem 2.1. Find the voltage at node A in the circuit shown in Fig. 2.3. [Ans. 1.75 V].

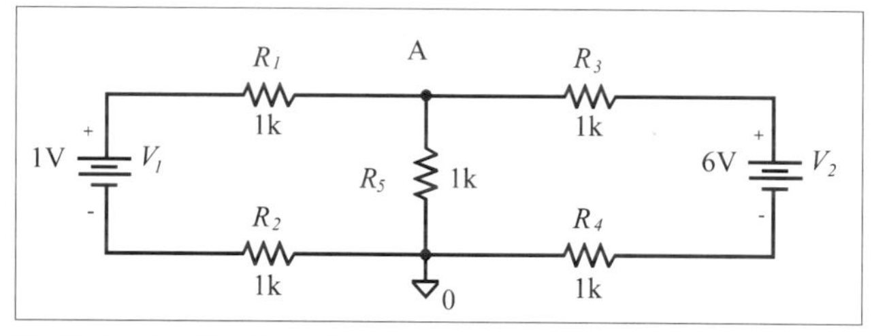

FIGURE 2.3. Problem 2.1.

Problem 2.2. Determine the magnitude and direction of the current that will flow through the 5 K resistor in Fig. 2.4. [Ans. 2 mA from left to right].

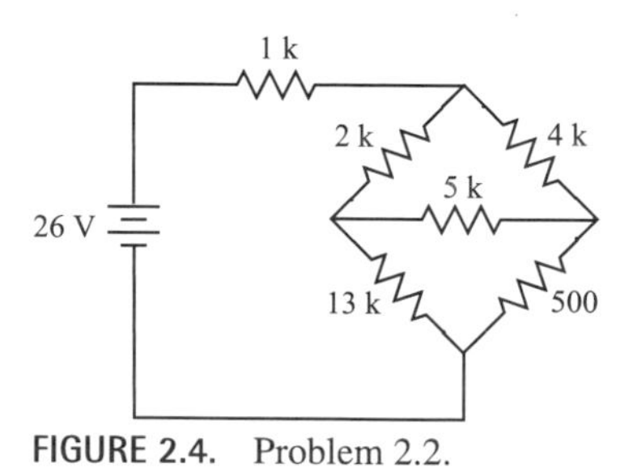

FIGURE 2.4. Problem 2.2.

C H A P T E R 3

AC Circuits

3.1 ALTERNATING VOLTAGE AND CURRENT

According to Fourier's theorem, any suitably well-behaved periodic waveform can be represented as a sum of harmonic functions. That is, suppose $F(t)$ is a time-dependent function satisfying $F(t+T) = F(t)$. The period of $F(t)$ is T. Then, $F(t)$ can be expanded in the series

$$F(t) = \frac{a_0}{2} + \sum_{n=1}^{\infty} (a_n \cos n\omega_0 t + b_n \sin n\omega_0 t) \tag{3.1}$$

with $\omega_0 = 2\pi/T$. The Fourier coefficients are given by

$$a_n = \frac{2}{T}\int_{-T/2}^{T/2} F(t)\cos(n\omega_0 t)\,dt, \quad n = 0, 1, 2, 3, \ldots, \tag{3.2}$$

$$b_n = \frac{2}{T}\int_{-T/2}^{T/2} F(t)\sin(n\omega_0 t)\,dt, \quad n = 1, 2, 3, \ldots, \tag{3.3}$$

and $a_0/2$ is seen to be the average value of $F(t)$ calculated over a complete period. A decomposition that is equivalent to Eqs. (3.1)–(3.3) can be phrased in terms of a series of cosines, as follows:

$$F(t) = C_0 + \sum_{n=1}^{\infty} C_n \cos(n\omega_0 t + \theta_n). \tag{3.4}$$

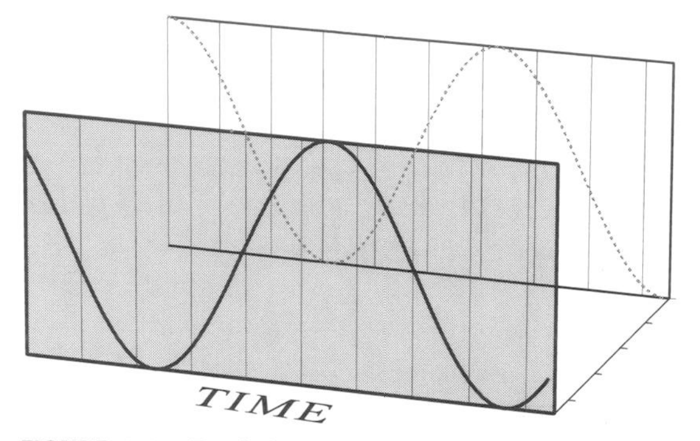

FIGURE 3.1. Depiction of two harmonic waveforms, $\cos(\omega t)$ at the rear and $\cos(\omega t+\theta)$ on the forward (shaded) plane.

Clearly, the net contribution of all terms in the series (3.1) at any given frequency must match exactly the corresponding term in series (3.4). This means that $a_n \cos n\omega_0 t + b_n \sin n\omega_0 t = C_n \cos(n\omega_0 t + \theta_n)$, from which one obtains

$$C_0 = \frac{a_0}{2}, \tag{3.5}$$

$$C_n = \sqrt{a_n^2 + b_n^2}, \quad n = 1, 2, 3, \ldots, \tag{3.6}$$

$$\theta_n = \tan^{-1}\left(\frac{-b_n}{a_n}\right), \quad n = 1, 2, 3, \ldots. \tag{3.7}$$

In the representation of Eq. (3.4), the series is specified by the amplitudes C_n and phases θ_n of the cosines. To understand the implication of the constants θ_n, consider the plots of $\cos(\omega t+\theta)$ and $\cos(\omega t)$ shown in Fig. 3.1. Because time increases to the right, right $\Longrightarrow$ later whereas left $\Longrightarrow$ earlier. Notice that $\cos(\omega t+\theta)$ appears shifted slightly to the left in Fig. 3.1. This means that any particular point on $\cos(\omega t+\theta)$ will happen sooner than the same point on $\cos(\omega t)$. In this context, it is said that the waveform $\cos(\omega t+\theta)$ *leads* the waveform $\cos(\omega t)$ by a phase angle θ. For similar reasons, $\cos(\omega t-\theta)$ would be said to *lag* $\cos(\omega t)$ by an amount θ.

Because periodic signals are essential ingredients in electronics, the properties of the constituent harmonic components are important, and the response of any electronic system to harmonic inputs becomes the fundamental question for analysis. Thus, we confine our attention largely to situations where the time-varying voltages and currents may be represented in forms such as $V(t) = V_0 \cos(\omega t+\theta)$ and $I(t) = I_0 \cos(\omega t+\phi)$. These oscillations run at an angular frequency of ω (radians/sec), or equivalently at a "conventional" frequency f, which is expressed in cycles/sec or Hertz (Hz). As noted earlier, the angular frequency and period are related through $\omega = 2\pi/T$, and because the frequency is simply the reciprocal of the period, $f = 1/T$, it is also true that $\omega = 2\pi f$.

The voltages and currents oscillate between the limits $(+V_0, -V_0)$ and $(+I_0, -I_0)$, respectively. V_0 and I_0 are the voltage and current amplitudes, but clearly the peak-to-peak excursions are $2V_0$ and $2I_0$. The waveforms have mean values of zero.

3.2 RESISTORS AS AC COMPONENTS

Consider a circuit composed only of a source of ac voltage $V_0 \cos(\omega t)$, which is applied to a resistor R, as shown in Fig. 3.2. Kirchhoff's law regarding the sum of voltages around any closed path must apply always, meaning at every moment in time. Hence, assuming an ac current I as indicated

$$V_0 \cos(\omega t) - I(t)\,R = 0. \tag{3.8}$$

Trivially, then,

$$\boxed{I(t) = \frac{V_0}{R}\cos(\omega t)}. \tag{3.9}$$

The ac current has the form $I(t) = I_0 \cos(\omega t)$ with amplitude $I_0 = V_0/R$, and there is no phase shift with respect to the applied voltage. Note that the connection between current amplitude, voltage amplitude, and resistance has the familiar form of Ohm's law.

3.3 CAPACITORS AS AC COMPONENTS

The voltage across a capacitor V_C is related to the charge Q residing on the plates: $Q = CV_C$. Taking the time derivative of both sides of this equation yields

$$\frac{dQ}{dt} = C\frac{dV_C}{dt}.$$

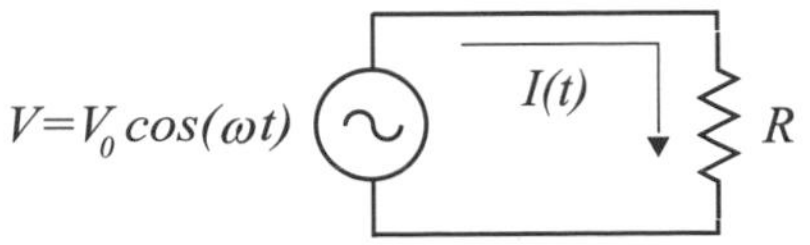

$V = V_0 \cos(\omega t)$ $I(t)$ R

FIGURE 3.2. Simple resistor driven by an ac source.

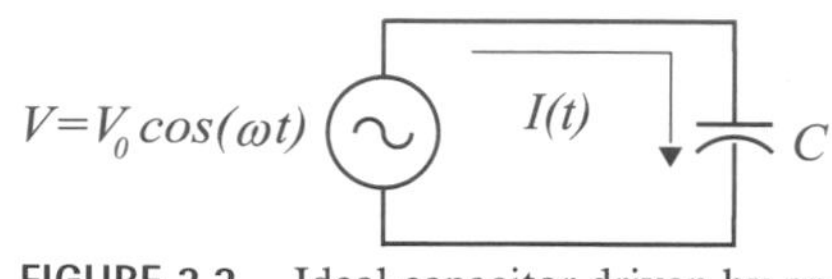

FIGURE 3.3. Ideal capacitor driven by an ac source.

But the rate of change of charge (the left-hand side) is just current flowing to and from the capacitor plates, so

$$I_C(t) = C\frac{dV_C}{dt}. \tag{3.10}$$

If the resistor in the previous example is replaced with a capacitor (Fig. 3.3), then Kirchhoff's voltage law requires

$$V_C = V_0 \cos(\omega t).$$

Differentiating this expression gives $-\omega V_0 \sin(\omega t) - \frac{dV_C}{dt} = 0$, from which with the help of Eq. (3.10), one obtains

$$I_C(t) = -\omega C V_0 \sin(\omega t).$$

Since $\sin(\omega t) = -\cos(\omega t + \frac{\pi}{2})$, this becomes

$$I_C(t) = \omega C V_0 \cos\left(\omega t + \frac{\pi}{2}\right), \tag{3.11}$$

which may be written

$$\boxed{I_C(t) = \left[\frac{V_0}{1/\omega C}\right] \cos\left(\omega t + \frac{\pi}{2}\right)}. \tag{3.12}$$

The ac current could be expressed in a standard form

$$I_C(t) = I_0 \cos(\omega t + \theta) \tag{3.13}$$

in which case $\theta = \pi/2$ and the amplitude is seen to satisfy

$$I_0 = \frac{V_0}{1/\omega C}. \tag{3.14}$$

Hence, the current through a capacitor leads the voltage across a capacitor by $\pi/2$. Equally correctly, it could be said that the voltage across a capacitor lags the current through a capacitor by $\pi/2$.

Equation (3.14) has the same form as Ohm's law—current equals voltage divided by resistance, only here the "resistance" is instead the *reactance* $1/\omega C$. This factor has units of ohms (Ω) and is clearly frequency dependent, becoming vanishingly small as $\omega \to \infty$.

3.4 INDUCTORS AS AC COMPONENTS

Faraday's law relates the voltage across an inductor to the time rate of change of current through it. Hence,

$$V_L = L\frac{dI_L}{dt}.$$

Integrating both sides of this equation,

$$I_L(t) = \frac{1}{L}\int V_L\,dt. \tag{3.15}$$

If the capacitor in the previous example is replaced with an inductor (Fig. 3.4), then Kirchhoff's voltage law requires

$$V_0\cos(\omega t) = V_L. \tag{3.16}$$

Substituting Eq. (3.16) for V_L into Eq. (3.15),

$$I_L(t) = \frac{V_0}{L}\int \cos(\omega t)\,dt$$

or

$$I_L(t) = \frac{V_0}{\omega L}\sin(\omega t),$$

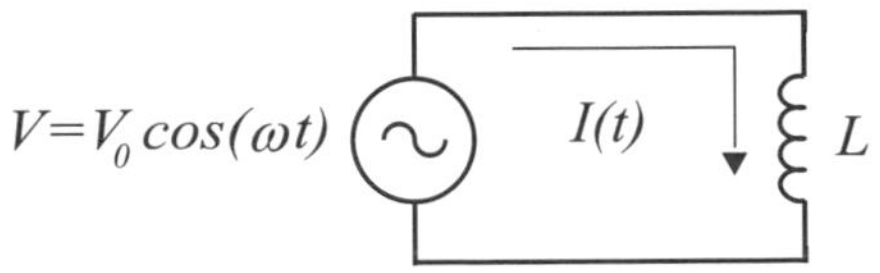

FIGURE 3.4. Ideal inductor driven by an ac source.

but $\sin(\omega t) = \cos(\omega t - \frac{\pi}{2})$, so

$$\boxed{I_L(t) = \left[\frac{V_0}{\omega L}\right] \cos\left(\omega t - \frac{\pi}{2}\right)}. \tag{3.17}$$

Hence, the ac current

$$I_L = I_0 \cos(\omega t + \theta) \tag{3.18}$$

has an amplitude that satisfies

$$I_0 = \frac{V_0}{\omega L}, \tag{3.19}$$

with $X_L = \omega L$ being the *inductive reactance,* which has units of ohms (Ω). Also apparent is the fact that the current through an inductor lags the voltage across an inductor by $\pi/2$, or equivalently, the voltage across an inductor leads the current through the inductor by $\pi/2$. Inductive reactance increases with increasing frequency.

3.5 IMPEDANCE

The results of the previous sections relating the amplitudes and phases of the voltage across and the current through each of the three elemental ac components (resistor, capacitor, and inductor) must always be true. However, real ac circuits are made up of sometimes elaborate combinations of these primary building blocks. Suppose therefore that a box with a pair of input terminals contains a multicomponent network. We seek now to describe the relationship between the voltage across the terminals and the current flowing between the terminals.

In linear circuits with excitation at a single frequency, all ac voltages and currents oscillate at this same frequency and hence require two descriptive parameters: amplitude and phase. Kirchhoff's laws remain in effect, dictating the sums of voltages and currents in loops and nodes at every moment in time. But these are sums of quantities possessing the pair of attributes amplitude and phase. Thus, in going around a circuit, one would encounter expressions of the general type

$$A_1 \cos(\omega t + \theta_1) + A_2 \cos(\omega t + \theta_2),$$

which after some algebra involving trigonometric expansions can be expressed in the form $A \cos(\omega t + \theta)$. When sums and differences of many such terms are required, the algebra, although fundamentally straightforward, becomes very messy.

Fortunately, an algebraic equivalent exists for these procedures, and it yields tremendous dividends in terms of operational simplicity. It is the algebra of complex numbers.

Complex Algebra

A complex variable is written in the canonical form $z = a + jb$, where a and b are two real numbers, and the imaginary number j is defined by $j = \sqrt{-1}$. A second, often-used form for expressing a complex number is $z = Ae^{j\phi}$. DeMoivre's theorem gives the expansion $e^{j\phi} = \cos\phi + j\sin\phi$, and so for the same complex number z we must have $A\cos\phi = a$ and $A\sin\phi = b$. Thus, it is possible to switch back and forth, if needed, between these two forms of notation.

$$a = A\cos\phi$$
$$b = A\sin\phi$$

or

$$A = \sqrt{a^2 + b^2}$$
$$\phi = \tan^{-1}\frac{b}{a}.$$

These representations have geometric analogs, as indicated in Fig. 3.5.

The complex conjugate of $z = a + jb$ is $z^* = a - jb$, or in exponential notation, if $z = Ae^{j\phi}$, then $z^* = Ae^{-j\phi}$. The product of any complex number with its conjugate is always real. $zz^* = a^2 + b^2 = A^2$. Adding complex numbers is accomplished as follows:

$$z = z_1 + z_2 = (a_1 + a_2) + j(b_1 + b_2).$$

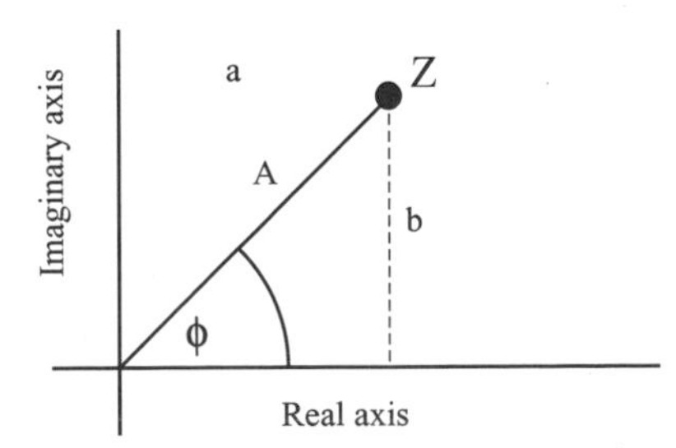

FIGURE 3.5. Complex impedance in terms of equivalent representations (a,b) and (A,ϕ).

In other words, the real parts add and the imaginary parts add. Subtraction is similar:

$$z = z_1 - z_2 = (a_1 - a_2) + j(b_1 - b_2).$$

Multiplication of

$$z = z_1 z_2 = a_1 a_2 + j^2 b_1 b_2 + j(a_1 b_2 + a_2 b_1)$$

becomes, since $j^2 = -1$,

$$z = z_1 z_2 = (a_1 a_2 - b_1 b_2) + j(a_1 b_2 + a_2 b_1).$$

Division is similar:

$$z = \frac{z_1}{z_2} = \frac{a_1 + jb_1}{a_2 + jb_2}.$$

This can be put in more standard form by multiplying the numerator and the denominator by the complex conjugate of the denominator. Thus,

$$z = \frac{a_1 + jb_1}{a_2 + jb_2} \frac{a_2 - jb_2}{a_2 - jb_2} = \frac{(a_1 a_2 + b_1 b_2)}{a_2^2 + b_2^2} + j \frac{(a_2 b_1 - a_1 b_2)}{a_2^2 + b_2^2}.$$

In exponential notation, multiplication and division are especially easy.

$$z = z_1 z_2 = A_1 e^{j\phi_1} A_2 e^{j\phi_2} = A_1 A_2 e^{j(\phi_1 + \phi_2)},$$

so the amplitudes multiply, whereas the phase angles add. Also,

$$z = \frac{z_1}{z_2} = \frac{A_1 e^{j\phi_1}}{A_2 e j \phi_2} = \left(\frac{A_1}{A_2}\right) e^{j(\phi_1 - \phi_2)},$$

so the amplitudes divide, whereas the angles subtract.

Complex Impedance

As already demonstrated, ac functions are characterized by a pair of parameters: amplitude and phase. A complex variable is also characterized by two attributes: amplitude and phase (or equivalently real and imaginary components). This correspondence gives rise to the following important result.

The expression $V = IR$, which was restricted to resistive dc circuits, may be replaced by the generalized form

$$\boxed{V = IZ}, \qquad (3.20)$$

in which the role of resistance has been superseded by the impedance Z. Equation (3.20) is correct provided the following associations are adhered to:

1. V and I are taken to be complex forms of the voltage and current. Functions originally expressed in a form such as $A_0 \cos(\omega t + \theta)$ are instead expressed as $A_0 e^{j(\omega t+\theta)}$.
2. When a pure resistor is involved, $\boxed{Z = R}$. This means that the impedance for a resistor is a real number whose magnitude equals the resistance.
3. When a pure capacitor is involved, $\boxed{Z = -j(1/\omega C)}$. This means that the impedance for a capacitor is an imaginary number whose magnitude equals the capacitive reactance.
4. When a pure inductor is involved, $\boxed{Z = +j(\omega L)}$. This means that the impedance for an inductor is an imaginary number whose magnitude equals the inductive reactance.
5. When several impedances are in series, the effective impedance of the combination is

$$\boxed{Z = Z_1 + Z_2 + Z_3 + \ldots}. \tag{3.21}$$

6. When several impedances are arranged in parallel, the effective impedance of the combination is

$$\boxed{\frac{1}{Z} = \frac{1}{Z_1} + \frac{1}{Z_2} + \frac{1}{Z_3} + \ldots}. \tag{3.22}$$

Example

These ideas can best be appreciated by using them in the analysis of a specific circuit. A series connection of a resistor, capacitor, and inductor is illustrated in Fig. 3.6. The applied voltage is $V = V_0 e^{j\omega t}$. Using Eq. (3.21), we have

$$Z = R + j\omega L - j(1/\omega C) = R + j(\omega L - 1/\omega C).$$

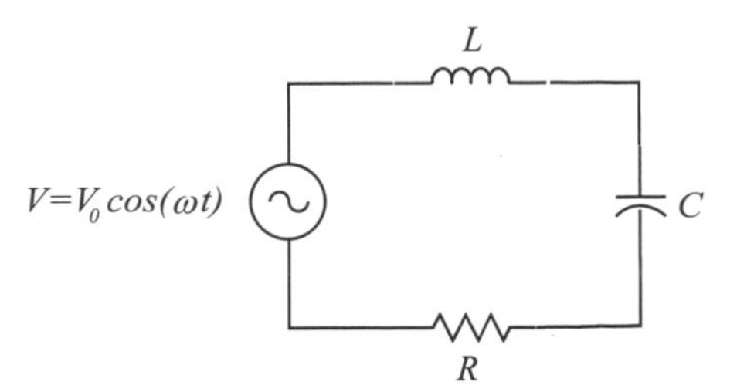

FIGURE 3.6. Series RLC circuit.

Then, from Eq. (3.20), the ac current in the circuit will be

$$I = \frac{V}{Z} = \frac{V_0 e^{j\omega t}}{R + j(\omega L - 1/\omega C)}.$$

Multiplying the numerator and the denominator by the complex conjugate of the denominator,

$$I = \frac{V_0 e^{j\omega t}\,[R - j(\omega l - 1/\omega C)]}{R^2 + (\omega L - 1/\omega C)^2}. \tag{3.23}$$

The square bracket contains the complex number $R - j(\omega L - 1/\omega C)$, which can be converted to the polar form $Z_0 e^{j\phi}$ using previously stated expressions:

$$Z_0 = \sqrt{R^2 + (\omega L - 1/\omega C)^2}, \tag{3.24}$$

$$\phi = \tan^{-1} \frac{-(\omega L - 1/\omega C)}{R}. \tag{3.25}$$

Returning to Eq. (3.23),

$$I = \frac{V_0 e^{j\omega t} Z_0 e^{j\phi}}{Z_0^2}$$

or

$$I = \frac{V_0}{Z_0} e^{j(\omega t + \phi)}. \tag{3.26}$$

In other words, the current $I = I_0 e^{j(\omega t + \phi)}$ has an amplitude

$$I_0 = \frac{V_0}{Z_0}, \tag{3.27}$$

which is simply the ratio of voltage amplitude divided by the magnitude of the impedance. Also, the current leads the applied voltage by $\phi = \tan^{-1}(1/\omega C - \omega L)$.

When $1/\omega C > \omega L$, ϕ will be positive and the current will lead the voltage (hence the behavior is capacitive at low frequencies). When $\omega L > 1/\omega C$, ϕ will be negative, and the current will lag the voltage (hence the behavior is inductive at high frequencies). The crossover from capacitive to inductive behavior occurs when $1/\omega C = \omega L$; that is, at the special frequency

$$\omega_0 = \frac{1}{\sqrt{LC}}. \tag{3.28}$$

The impedance at this particular frequency is seen to have an imaginary component of zero and hence Z is just the real number

$$Z = R. \tag{3.29}$$

This is a minimum value for the magnitude of the impedance. The circuit behaves as a pure resistance at this series resonant frequency $\omega = \omega_0$.

It is instructive now to consider the voltages across the individual components in this example. Returning to the fundamental relationship, Eq. (3.20), we have for the resistor,

$$V_R = I Z_R$$

or, since $I = I_0 e^{j(\omega t + \phi)}$ and $Z_R = R$,

$$V_R - I_0 R \, e^{j(\omega t + \phi)}. \tag{3.30}$$

This indicates that the voltage across the resistor has magnitude $I_0 R$ and that it is in phase with the ac current.

For the capacitor,

$$V_C = I Z_c,$$

with $Z_C = -j(1/\omega C)$. Hence,

$$V_C = (I_0/\omega C)(-j)\, e^{j(\omega t + \phi)}.$$

This can be rearranged with the aid of the useful identity $e^{j(-\pi/2)} = -j$ to become

$$V_C = \frac{I_0}{\omega C}\, e^{j(\omega t + \phi - \pi/2)}. \tag{3.31}$$

This indicates that the voltage across the capacitor has magnitude $I_0/\omega C$ and that it is lagging the current by $\pi/2$, as we would anticipate.

For the inductor,

$$V_L = I Z_L$$

with $Z_L = +j(\omega L)$. Hence,

$$V_L = (I_0 \omega L)(+j)\, e^{j(\omega t+\phi)}.$$

This can be rearranged with the aid of the useful identity $e^{j(+\pi/2)} = +j$ to become

$$V_L = I_0 \omega L\, e^{j(\omega t+\phi+\pi/2)}. \tag{3.32}$$

This indicates that the voltage across the inductor has magnitude $I_0 \omega L$ and that it is leading the current by $\pi/2$, as we would anticipate.

The total voltage across the series circuit should just be the sum of the component voltages.

$$V = V_R + V_C + V_L = I_0 \,[R - j/\omega C + j\omega L]\; e^{j(\omega t+\phi)}.$$

Simply substituting Eq. (3.27) for I_0 and Eq. (3.24) for Z_0 gives

$$V = V_0 \frac{R - j(1/\omega C - \omega L)}{\sqrt{R^2 + (\omega L - 1/\omega C)^2}}\; e^{j\phi}\, e^{j\omega t}.$$

But $e^{j\phi} = \cos\phi + j \sin\phi$, and using Eq. (3.25),

$$\cos\phi = \frac{R}{\sqrt{R^2 + (\omega L - 1/\omega C)^2}},$$

$$\sin\phi = \frac{1/\omega C - \omega L}{\sqrt{R^2 + (\omega L - 1/\omega C)^2}}.$$

Hence,

$$V = V_0 \frac{R - j(1/\omega C - \omega L)}{\sqrt{R^2 + (\omega L - 1/\omega C)^2}} \frac{R + j(1/\omega C - \omega L)}{\sqrt{R^2 + (\omega L - 1/\omega C)^2}}\; e^{j\omega t}.$$

Following some obvious cancellations, finally

$$V = V_0\, e^{j\omega t},$$

which is the correct expression for the applied voltage.

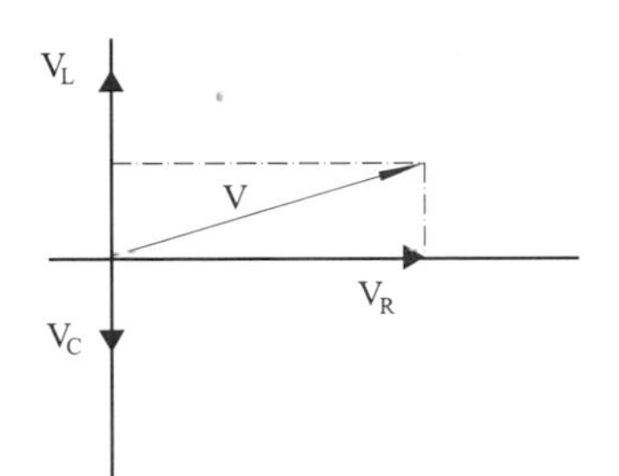

FIGURE 3.7. Vector diagram of the voltages across the resistor, capacitor, and inductor in the series RLC circuit.

The voltage summation just carried out was performed on complex quantities and was seen to yield a correct final result. Note, however, that the voltage *magnitudes* do *not* add: $V_0 \neq I_0 R + I_0/\omega C + I_0 \omega L$. Such a summation would be incorrect because the ac voltages are in essence vectors, and they must be added vectorially. Complex arithmetic carries out the equivalent of a vector addition. A graphical representation of the process is shown in Fig. 3.7.

Example

As a second example, consider a capacitor and a resistor in parallel, as shown in Fig. 3.8. The rule for adding parallel impedances, Eq. (3.22), gives

$$\frac{1}{Z} = \frac{1}{R} + \frac{1}{-j/\omega C}.$$

From this,

$$Z = \frac{-jR/\omega C}{R - j/\omega C} = \frac{-jR/\omega C}{R - j/\omega C} \frac{R + j/\omega C}{R + j/\omega C}$$

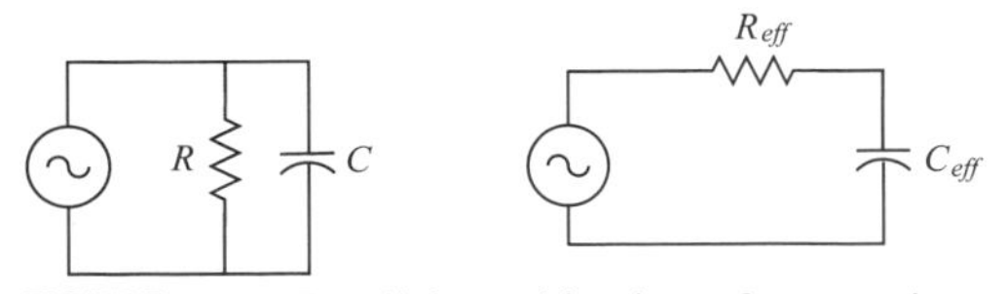

FIGURE 3.8. Parallel combination of a capacitor and resistor, and the equivalent representation as a series combination.

or

$$Z = \frac{R/(\omega C)^2}{R^2 + (1/\omega C)^2} - j\frac{R^2/\omega C}{R^2 + (1/\omega C)^2},$$

so

$$Z = \left[\frac{R}{1 + (\omega C R)^2}\right] - j\left[\frac{\omega C R^2}{1 + (\omega C R)^2}\right]. \tag{3.33}$$

This expression has the form

$$Z = R_{\text{eff}} - j/\omega C_{\text{eff}}. \tag{3.34}$$

In other words, the parallel combination of resistor R and capacitor C is equivalent to a series combination of an effective resistance R_{eff} and an effective capacitance C_{eff}, with

$$R_{\text{eff}} = \frac{R}{1 + (\omega C R)^2}, \tag{3.35}$$

$$C_{\text{eff}} = \frac{\omega C R^2}{1 + (\omega C R)^2}. \tag{3.36}$$

PROBLEMS

Problem 3.1. Consider the circuit in Fig. 3.9. The ac source frequency is 1 kHz. What is the magnitude of the impedance of the series combination L_1 R_1 C_1?

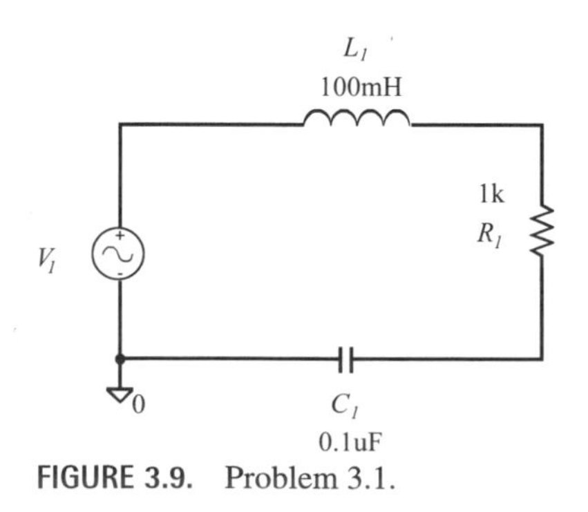

FIGURE 3.9. Problem 3.1.

[Ans. $|Z| = 1388\ \Omega$]. What is the phase shift between the source voltage and current? [Ans. $\phi = 43.9°$].

Problem 3.2. Consider the circuit in Fig. 3.10. Find the magnitude of the impedance seen by the source at a frequency of 100,000 Hz. Also find the phase shift between the source voltage and the source current. [Ans. $|Z| = 63{,}028\ \Omega$ and $\phi = 85.5°$].

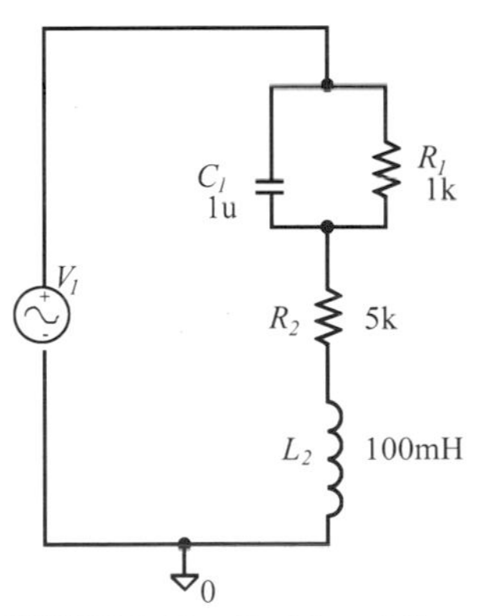

FIGURE 3.10. Problem 3.2.

CHAPTER 4

Bridge Circuits

4.1 DC BRIDGES

A number of sensors used in instrumentation applications are essentially variable resistors. That is, the observed resistance of such a sensing element would be determined by ambient physical conditions such as temperature, light level, or physical strain. For sensors of this type, dc bridge circuits are commonly employed either for explicit resistance measurement, or for monitoring deviations in resistance from some selected reference level. A dc bridge is a particular geometric configuration of resistors (known and unknown) driven by a voltage source, as illustrated in Fig. 4.1. The left-hand representation is the standard form of a bridge, whereas the right-hand diagram is functionally identical but is a slightly more conventional schematic style.

The so-called bridge voltage V_{out} can be easily derived by means of Kirchhoff's laws. Let I_C and I_D be two branch currents. Then,

$$V_{\text{bias}} = I_C(R_1 + R_3) = I_D(R_2 + R_4).$$

Also,

$$V_C = I_C R_3 = \frac{V_{\text{bias}} R_3}{R_1 + R_3},$$

$$V_D = I_D R_4 = \frac{V_{\text{bias}} R_4}{R_2 + R_4}.$$

These two voltages are measured with respect to a zero level assigned to the

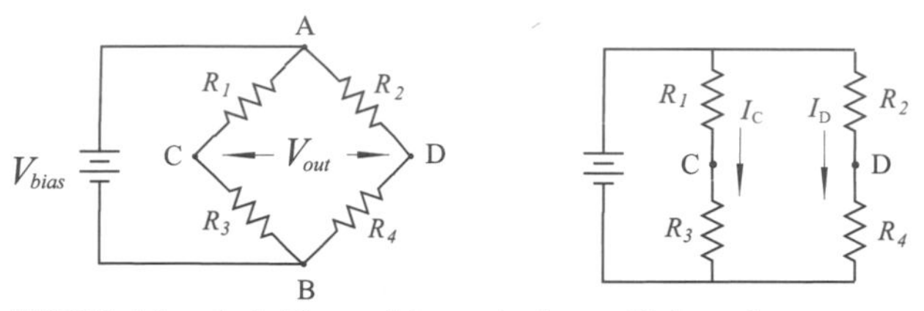

FIGURE 4.1. dc bridge and its equivalent with branch currents.

negative terminal of the battery. Hence,

$$V_{\text{out}} = V_C - V_D = V_{\text{bias}} \left[\frac{R_3}{R_1 + R_3} - \frac{R_4}{R_2 + R_4} \right]. \tag{4.1}$$

A balanced bridge is one for which the bridge voltage is zero: $V_{\text{out}} = 0$. From Eq. (4.1), this occurs when

$$\frac{(R_1 + R_3)}{R_3} = \frac{(R_2 + R_4)}{R_4}$$

or

$$\boxed{\frac{R_1}{R_3} = \frac{R_2}{R_4}}, \tag{4.2}$$

which is a standard result. The condition $V_{\text{out}} = 0$ is easily tested by placing a high input impedance voltmeter across the bridge output nodes.

Wheatstone Bridge

A Wheatstone bridge (Fig. 4.2) is specifically configured for the purpose of precisely determining the value of an "unknown" resistance.

The unknown, R_X, takes the place of one of the four resistors. Its companion in the same branch is a precision resistor R selected to be of approximately the same value as the unknown. Point C is a sliding contact to a length of resistive wire connected between A and B. This sliding contact is moved up and down until the meter reads zero voltage. The balance condition, Eq. (4.2), becomes

$$R_X = R \frac{R_a}{R_b}. \tag{4.3}$$

Notice that, in addition to the reference resistor R, it is only necessary to

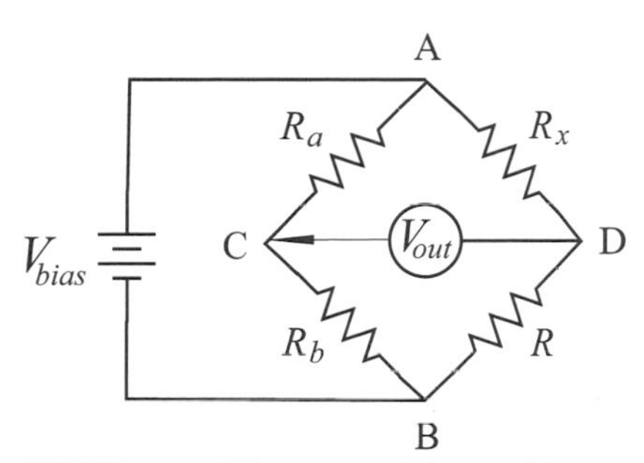

FIGURE 4.2. Wheatstone bridge with unknown R_x. R_a and R_b are an adjustable combination used for bridge balancing.

determine an accurate value for the ratio R_a/R_b in order to measure the unknown R_X. This resistance ratio is numerically equal to the ratio of the lengths of wire in the two arms CA and CB. An accurate reading of these lengths is usually a simple matter.

Unbalanced Bridge

A bridge can also be used to provide an output voltage in response to shifts in some physical parameter away from a stipulated operating or reference value. The three fixed resistors are usually selected to equal the sensor at this operating point. We denote the sensor resistance by R_X and the other resistors by R. Suppose the sensor is at position #3 in the bridge schematic given in Fig. 4.1 ($R_3 \Rightarrow R_X$)—the choice is arbitrary, but this position is as good as any other. Then, at the set condition of the system, $R_1 = R_2 = R_X = R_4 = R$ and $V_{\text{out}} = 0$. Now, if there is a change in pressure, strain, or whatever, then R_X will shift from its original value of R to, say, $R_X = R + \delta R$. From Eq. (4.1), the bridge voltage will be

$$V_{\text{out}} = V_{\text{bias}} \left[\frac{R + \delta R}{R + R + \delta R} - \frac{R}{R + R} \right]. \tag{4.4}$$

Define a relative change in sensor value

$$\epsilon \equiv \frac{\delta R}{R}. \tag{4.5}$$

Note that the sensor resistance may increase or decrease, depending on the nature of the change of its physical environment. Consequently, ϵ can be either a positive or a negative quantity.

$$V_{\mathrm{out}} = V_{\mathrm{bias}} \left[\frac{1+\epsilon}{2+\epsilon} - \frac{1}{2} \right] = V_{\mathrm{bias}} \left[\frac{\epsilon}{4+2\epsilon} \right]. \tag{4.6}$$

The factor in square brackets can be expressed as a power series in ϵ, resulting in

$$\boxed{V_{\mathrm{out}} = V_{\mathrm{bias}} \left[\frac{\epsilon}{4} - \frac{\epsilon^2}{8} + \frac{\epsilon^3}{16} \cdots \right]}. \tag{4.7}$$

Depending on the direction of the sensor resistance change (increase or decrease), ϵ can be positive or negative, as noted earlier. The polarity of the bridge voltage will thus be set by the sign of ϵ. If the relative shift in resistance is small ($|\epsilon| \ll 1$), then successive terms in this series rapidly become negligible, and a reasonable approximation might be

$$V_{\mathrm{out}} \approx V_{\mathrm{bias}} \frac{\epsilon}{4}. \tag{4.8}$$

Within the limits of this approximation, the bridge output voltage is nearly a linear function of the relative change in sensor resistance. Linearity is almost always a desirable property.

Sensitivity

In the previous section, a particular choice of resistors was made, namely $R_1 = R_2 = R_4 = R$ and $R_X \approx R = R_3$. The reason for such a choice will now be demonstrated.

Let us begin without any restriction on the resistor values. As noted earlier,

$$\frac{V_{\mathrm{out}}}{V_{\mathrm{bias}}} = \frac{R_3}{R_1 + R_3} - \frac{R_4}{R_2 + R_4}.$$

Suppose $R_3 = R_{30} + \delta R$, so that δR measures the shift in value of resistor R_3 from its nominal value R_{30}. Then,

$$\frac{V_{\mathrm{out}}}{V_{\mathrm{bias}}} = \frac{R_{30} + \delta R}{R_1 + R_{30} + \delta R} - \frac{R_4}{R_2 + R_4}.$$

But the bridge is presumed to have been originally balanced, so R_{30} must satisfy the condition

$$\frac{R_1}{R_{30}} = \frac{R_2}{R_4}.$$

Using this in the previous equation, the unbalanced output is

$$\frac{V_{\text{out}}}{V_{\text{bias}}} = \frac{R_{30} + \delta R}{R_{30}\left(\frac{R_2}{R_4}\right) + R_{30} + \delta R} - \frac{1}{1 + \frac{R_2}{R_4}}.$$

Consequently,

$$\frac{V_{\text{out}}}{V_{\text{bias}}} = \left[\frac{1+\epsilon}{1+\epsilon+q} - \frac{1}{1+q}\right] \tag{4.9}$$

with $q \equiv R_2/R_4$. To maximize the sensitivity S, the rate of change of bridge voltage with respect to the relative change in the unknown ϵ must be as large as possible.

$$S \equiv \frac{d(V_{\text{out}}/V_{\text{bias}})}{d\epsilon} = \frac{(1+\epsilon+q) - (1+\epsilon)}{(1+\epsilon+q)^2} = \frac{q}{(1+\epsilon+q)^2}.$$

Clearly, S depends on the parameter q. The optimal choice for q comes from

$$\frac{dS}{dq} = \frac{(1+\epsilon+q)^2 - 2q(1+\epsilon+q)}{(1+\epsilon+q)^4} = 0$$

or

$$(1+\epsilon)^2 + q^2 + 2q(1+\epsilon) - 2q(1+\epsilon) - 2q^2 = 0,$$

from which we obtain

$$q = (1+\epsilon) \tag{4.10}$$

so

$$\boxed{S_{\max} = \frac{1}{4(1+\epsilon)}}. \tag{4.11}$$

But the bridge is operated around an ambient condition for which $R_X \approx R$, and equivalently $\epsilon \approx 0$. Consequently, the optimal value for q from Eq. (4.10) is just 1, so the optimal choice in resistors R_2 and R_4 must be

$$R_2 = R_4, \tag{4.12}$$

and the corresponding sensitivity is

$$S_{\max} = \frac{1}{4}. \tag{4.13}$$

It will be noted that this value is consistent with our earlier expression for V_{out}, Eq. (4.8). These results explain the earlier choice in bridge components. The "best" arrangement is with all four resistors at near equal values.

4.2 AC BRIDGES

The four-element bridge can be adapted to impedance measurements simply by replacing resistors with true ac components, as illustrated in Fig. 4.3.

At balance, the voltage across the bridge (from C to D) is zero. Suppose the voltage at C is of the form $V_C = V_p \cos(\omega t + \phi_p)$ and the voltage at D is $V_D = V_q \cos(\omega t + \phi_q)$. Zero net bridge output voltage would then require $V_C = V_D$, which in turn requires $V_p = V_q$ and $\phi_p = \phi_q$. In other words, *both* amplitudes and phases must match for cancellation. Several types of ac null detector may be placed across the bridge output terminals. Simple earphones give a decreasing audible tone as balance is approached. Many modern oscilloscopes have sum and inversion functions so that $(V_C - V_D)$ can be displayed directly and balance achieved with visual feedback.

As was discussed earlier, amplitudes, and phases are correctly accounted for when currents, voltages, and impedances are expressed in complex form. Hence, $I_C Z_3 = I_D Z_4$. But $V_{\text{bias}} = I_C(Z_1 + Z_3) = I_D(Z_2 + Z_4)$. Combining

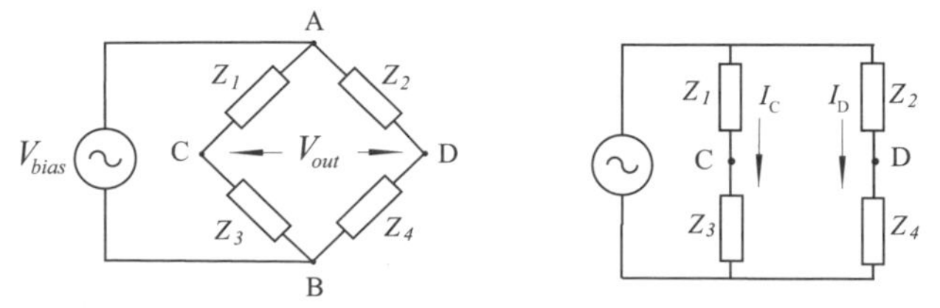

FIGURE 4.3. ac bridge with four impedances and ac excitation.

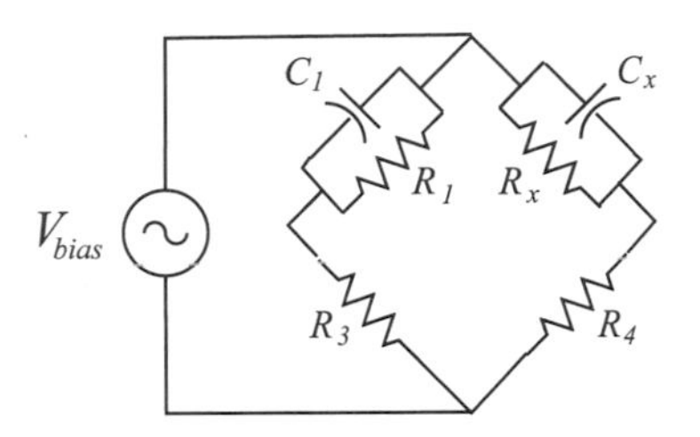

FIGURE 4.4. Capacitance bridge.

these expressions yields the condition for balance:

$$\boxed{\frac{Z_1}{Z_3} = \frac{Z_2}{Z_4}}. \tag{4.14}$$

The composition of possible "unknowns" is endless, but a few examples will illustrate the basic principles involved.

Nonideal Capacitance

Suppose the unknown component is a lossy capacitor, represented as an ideal capacitance C_X shunted by a resistance R_X. The bridge arrangement is shown in Fig. 4.4. In practice, the bridge is balanced by varying both the capacitor C_1 and the resistor R_1 until no ac signal is detected across the bridge output nodes.

The parallel combinations $C_1//R_1$ and $C_X//R_X$ have equivalent expressions

$$\frac{1}{Z_1} = \frac{1}{R_1} + \frac{1}{-j/\omega C_1},$$

$$\frac{1}{Z_X} = \frac{1}{R_X} + \frac{1}{-j/\omega C_X}.$$

These can be simplified [see Eqs. (3.35) and (3.36)] and combined with Eq. (4.14) to give

$$\frac{\left[\frac{R_1}{1+(\omega C_1 R_1)^2}\right] - j\left[\frac{\omega C_1 R_1^2}{1+(\omega C_1 R_1)^2}\right]}{R_3} = \frac{\left[\frac{R_X}{1+(\omega C_X R_X)^2}\right] - j\left[\frac{\omega C_X R_X^2}{1+(\omega C_X R_X)^2}\right]}{R_4}. \tag{4.15}$$

For this complex equation to be satisfied, the real parts of the two sides must be equal and the imaginary parts of the two sides must also be equal. Hence,

$$R_1 R_4[1 + (\omega C_X R_X)^2] = R_3 R_X[1 + (\omega C_1 R_1)^2], \tag{4.16}$$

$$\omega R_4 C_1 R_1^2[1 + (\omega C_X R_X)^2] = \omega R_3 C_X R_X^2[1 + (\omega C_1 R_1)^2]. \tag{4.17}$$

From this pair of equations is obtained

$$\boxed{R_X = \frac{R_4}{R_3} R_1}, \tag{4.18}$$

$$\boxed{C_X = \frac{R_3}{R_4} C_1}. \tag{4.19}$$

Therefore, the impedance of the lossy capacitor is completely determined by the balance condition. Note that these equations do not depend on either the amplitude or the frequency of the bridge excitation.

Nonideal Inductance

As a second example of an ac bridge, consider the problem of determining the impedance of a lossy inductor, modeled as an ideal inductor L_X in series with a resistance R_X. A commonly used arrangement is the Maxwell bridge shown in Fig. 4.5. The unknown impedance is mirrored in the opposite bridge arm by a parallel combination of a known capacitor and resistor. In this configuration, the bridge is balanced by adjusting variable resistor R_3 and variable capacitor C_3 until the output voltage reaches a null. Interestingly, here we have an unknown inductor being determined by a known capacitor. This strategy is attractive because precision capacitors are more readily available than precision inductors.

At balance,

$$\frac{R_1}{Z_3} = \frac{Z_X}{R_4},$$

so

$$R_1 R_4 = \left[\frac{R_3}{1+(\omega C_3 R_3)^2} - j\frac{\omega C_3 R_3^2}{1+(\omega C_3 R_3)^2}\right][R_X + j\omega L_X]. \tag{4.20}$$

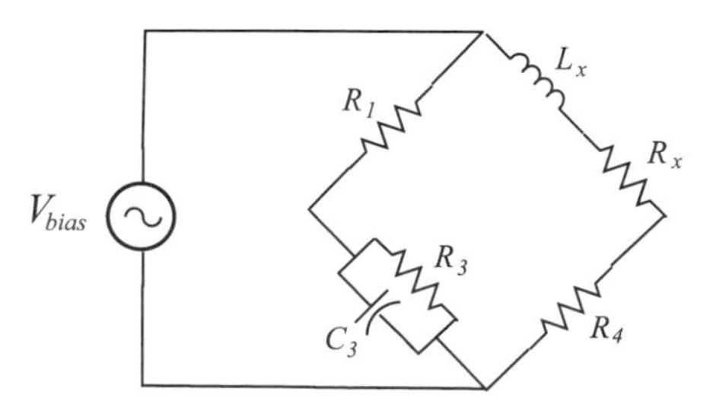

FIGURE 4.5. Maxwell bridge used to determine the ac properties of an unknown inductor.

As before, this is really two equations, one imposing equality of the real parts, the second applying to the imaginary parts.

$$R_1 R_4 = \frac{R_3 R_X}{1 + (\omega C_3 R_3)^2} + \frac{\omega^2 L_X C_3 R_3^2}{1 + (\omega C_3 R_3)^2}, \tag{4.21}$$

$$\frac{\omega R_X C_3 R_3^2}{1 + (\omega C_3 R_3)^2} = \frac{\omega L_X R_3}{1 + (\omega C_3 R_3)^2}. \tag{4.22}$$

From these, one obtains

$$\boxed{R_X = \frac{R_1 R_4}{R_3}}, \tag{4.23}$$

$$\boxed{L_X = R_1 R_4 C_3}. \tag{4.24}$$

Once again, the balance conditions do not depend on the bridge excitation frequency.

PROBLEMS

Problem 4.1. Figure 4.6 illustrates the arrangement for a Schering capacitance bridge.

Derive the following balance conditions:

$$R_x = R_4 \frac{C_3}{C_1},$$

$$C_x = C_1 \frac{R_3}{R_4}.$$

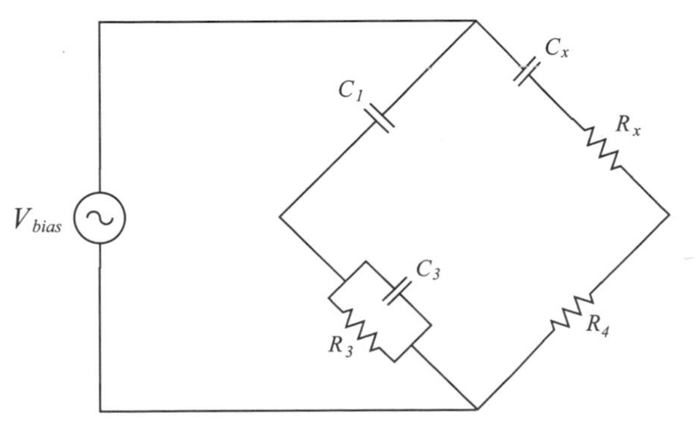

FIGURE 4.6. Problem 4.1.

Problem 4.2. Figure 4.7 illustrates the arrangement for a Hay inductance bridge.

Derive the following balance conditions:

$$R_x = \frac{\omega^2 C^2 R_1 R_2 R_3}{1 + \omega^2 C^2 R_1^2},$$

$$L_x = \frac{C R_2 R_3}{1 + \omega^2 C^2 R_1^2}.$$

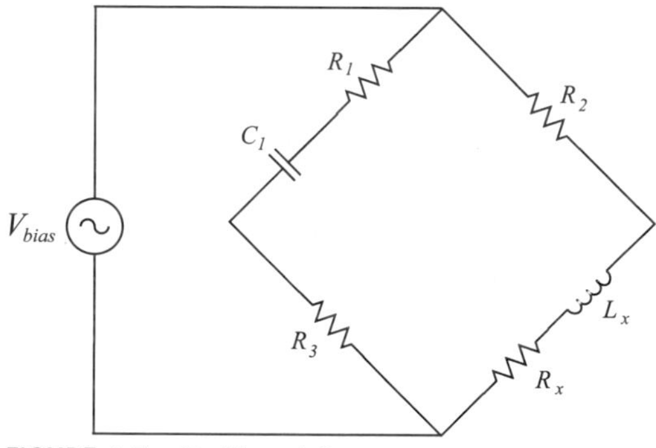

FIGURE 4.7. Problem 4.2.

C H A P T E R 5

Amplifiers

In the realm of instrumentation, signals coming from sensor outputs can be quite weak, and in such cases some signal enhancement clearly may be desirable. That is, signal amplification is needed. Indeed, amplification is one of the fundamental manipulations that can be performed on a signal. Amplifiers have played a key role throughout the history of electronics, appearing first in vacuum tube form, then as transistor-based designs, and most recently packaged into integrated circuits.

The advent of integrated circuit amplifiers has transformed the world of applied electronics (including instrumentation) because these amplifiers are complete units. The amplifier embedded within a package might contain dozens of transistors in a complex circuit, but the user need not be troubled by the details of the design. The "chip" only requires simple power supply connections and is promptly up and running. Consistency of performance and reliability are hallmarks of modern integrated circuits.

In its ideal form, an amplifier generates an output that is A times the input, regardless of the input signal strength or frequency. Real amplifiers have limitations in both respects, but for the moment we shall assume ideal behavior.

A differential amplifier (see Fig. 5.1) has two inputs, usually labelled + and −, or equivalently, *noninverting* and *inverting*.

The output voltage is given by the fundamental relationship

$$\boxed{V_{\text{out}} = A\,(V_+ - V_-)}, \tag{5.1}$$

where A is the amplifier gain. Of course, if the inverting input is grounded, then $V_{\text{out}} = A\,V_+$, so the behavior reverts to a conventional single-channel amplifier. In other words, the single-input amplifier can be thought of as simply a special case of the more general differential amplifier.

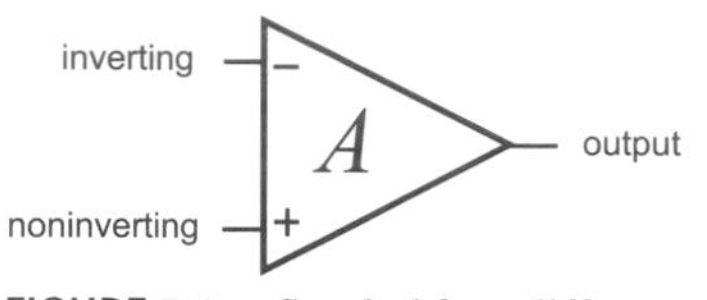

FIGURE 5.1. Symbol for a differential amplifier with intrinsic gain A.

An equivalent circuit model for a differential amplifier is shown within the dashed rectangle in Fig. 5.2. A source feeding the amplifier "sees" an input resistance r_{in}. Looking into the output terminal, the amplifier appears as a source equal to $A\ V_{\text{in}}$ feeding a series output resistance r_{o}. For an ideal amplifier, $r_{\text{in}} \to \infty$ and $r_{\text{out}} \to 0$.

The differential form of the amplifier has assumed preeminence because so many differential amplifiers are commercially available as integrated circuits. As a matter of fact, it is a rather special form of differential amplifier that is in widespread use: the so-called *operational amplifier*. An ideal op-amp is just an ideal differential amplifier with an exceedingly large gain ($A \to \infty$). Actual integrated circuit op-amps have typical gains of 10^4–10^6.

In use, a differential amplifier must be connected to the required power supplies. For many if not most IC devices, bipolar supplies are needed, say ±15 V, and so three pins on the IC package are assigned to $+V_{\text{bias}}$, $-V_{\text{bias}}$, and ground. Reality dictates that the amplifier output cannot exceed the positive supply level (sometimes termed the positive "rail"), nor can the output fall below the negative rail. This means, for example, that with ±15 V supplies, a differential input of 2 V coupled with a gain of $A = 100$ would only generate an output of approximately 15 V, not $100 \times 2 = 200$ V. For most applications, this saturating behavior must be avoided in order for the output to remain a linear function of the input. Nonlinearities generally lead to the undesirable effects of distortion.

Given the earlier remark about the enormous gain of operational amplifiers, and the observation concerning power supplies and saturation, it might seem that an op-amp would be almost useless for conditioning all but the tiniest signals. This would indeed be the case, except that op-amps are not normally employed

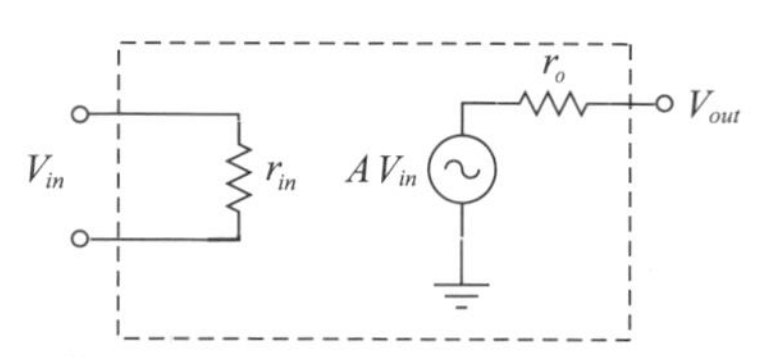

FIGURE 5.2. Equivalent internal circuit for the differential amplifier.

in bare form. With the addition of a few resistors, amplifiers possessing improved performance and reasonable overall gain can be achieved, as we shall now see.

5.1 NONINVERTING AMPLIFIER

Consider the schematic in Fig. 5.3. The combination of R_1 and R_F acts as a simple voltage divider on V_{out}, so

$$V_1 = \frac{R_1}{R_1 + R_F} V_{\text{out}}.$$

Also, from the fundamental relationship for a differential amplifier,

$$V_{\text{out}} = A\left[V_{\text{in}} - V_1\right].$$

Combining these two expressions,

$$V_{\text{out}} = A\left[V_{\text{in}} - \frac{R_1}{R_1 + R_F} V_{\text{out}}\right]$$

or

$$V_{\text{out}} = V_{\text{in}}\left[\frac{A}{1 + A\frac{R_1}{R_1+R_F}}\right],$$

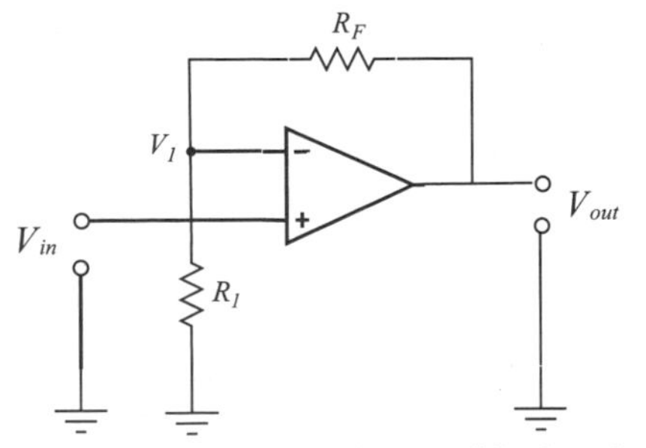

FIGURE 5.3. Noninverting amplifier based on a single op-amp.

and finally

$$V_{out} = V_{in}\left[\frac{1}{\frac{1}{A} + \frac{R_1}{R_1+R_F}}\right]. \tag{5.2}$$

The square bracket in this equation is a numerical factor relating output to input voltage for the particular configuration selected here. If the circuit in Fig. 5.3 is taken as a single entity with an input and an output, then this numerical factor is its gain. To distinguish this gain from the parameter A, which is an intrinsic property of the bare differential amplifier, A is termed the *open-loop gain*, and the amplification factor for the complete configuration is called the *closed-loop gain*, G. Hence,

$$G = \frac{1}{\frac{1}{A} + \frac{R_1}{R_1+R_F}}. \tag{5.3}$$

As noted earlier, an op-amp is a differential amplifier with enormous open-loop gain, in which case Eq. (5.3) has the limiting form

$$\boxed{\lim_{A\to\infty} G = 1 + \frac{R_F}{R_1}}, \tag{5.4}$$

so the closed-loop gain is set by the ratio of the two resistors. For a real op-amp, the open-loop gain might be something like 100,000 (an almost useless value, as was noted earlier), but the closed-loop gain can easily be set to more desirable levels such as 5, 10, or 50.

Because the square bracket in Eq. (5.2) is always a positive quantity, the polarity of the output voltage is the same as the polarity of the input. Hence, this configuration is referred to as a noninverting amplifier.

Input Resistance

Combining the schematic for the noninverting amplifier (Fig. 5.3) with the equivalent circuit for an op-amp (Fig. 5.2), we obtain as a composite representation Fig. 5.4. The driving source feeding V_{in} must also supply an appropriate input current I_{in}. The effective input resistance is thus

$$R_{in} = \frac{V_{in}}{I_{in}}. \tag{5.5}$$

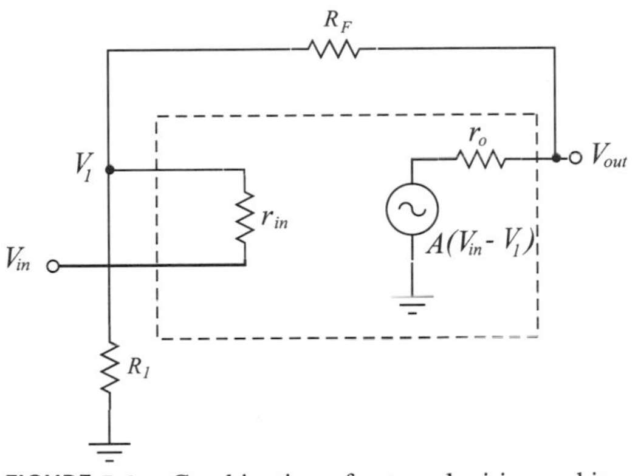

FIGURE 5.4. Combination of external wiring and internal equivalent circuit for a complete noninverting configuration.

The input current may be expressed

$$I_{\text{in}} = \frac{V_{\text{in}} - V_1}{r_{\text{in}}}. \tag{5.6}$$

Furthermore, if r_{in} is large compared to both R_1 and R_F,

$$V_1 = \frac{R_1}{R_F + R_1} V_{\text{out}} \tag{5.7}$$

and, using Eq. (5.2) for V_{out},

$$V_1 = \left[\frac{R_1}{R_F + R_1}\right] V_{\text{in}} \left[\frac{1}{\frac{1}{A} + \frac{R_1}{R_1 + R_F}}\right]. \tag{5.8}$$

Substituting this expression for V_1 into Eq. (5.6) and simplifying,

$$\boxed{R_{\text{in}} = r_{\text{in}} \left[1 + A \frac{R_1}{R_1 + R_F}\right]}. \tag{5.9}$$

As this equation makes clear, the effective input resistance (R_{in}) of the noninverting configuration is increased over the input resistance (r_{in}) of the bare op-amp, and if the open-loop gain A is large, the increase in this resistance will be substantial.

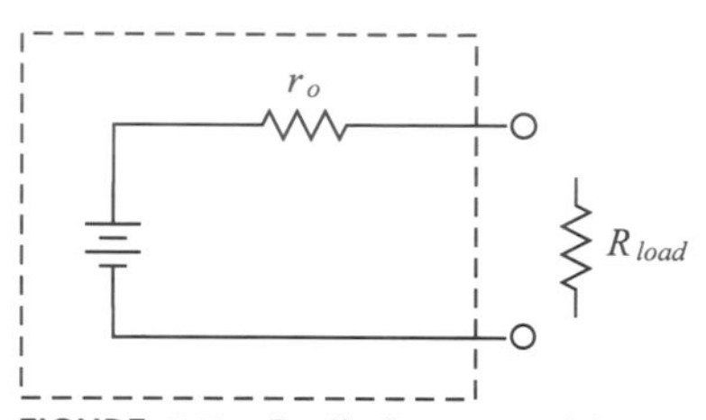

FIGURE 5.5. Preliminary consideration for determining amplifier output impedance.

Output Resistance

Suppose a battery and a resistor are connected in series, as indicated in Fig. 5.5. The output voltage will drop to exactly one-half if a load resistor equal in value to the internal resistance is added. This observation provides a method for determining output resistance:

- the output resistance is equal to that value of external load which will cause the ouput signal to drop by 50%.

Now apply this concept to the noninverting amplifier. As a first step, note that the equations leading to Eq. (5.2) must be modified when there is a load resistance.

$$V_1 = \frac{R_1}{R_1 + R_F} V_{\text{out}}$$

as before, but

$$\frac{A\left[V_{\text{in}} - V_1\right] - V_{\text{out}}}{r_{\text{o}}} = \frac{V_{\text{out}}}{R_L} + \frac{V_{\text{out}} - V_1}{R_F}. \tag{5.10}$$

Equation (5.10) expresses current conservation at the output node (see Fig. 5.6). From this pair of relations, the noninverting amplifier closed-loop gain can be derived:

$$G = \frac{1}{\left[\frac{R_1}{R_1+R_F} + \frac{1}{A}\right] + \frac{r_{\text{o}}}{AR_L} + \frac{r_{\text{o}}}{A(R_1+R_F)}}. \tag{5.11}$$

Comparing this to Eq. (5.3), it is apparent that the addition of a finite load resistance (R_L) on the amplifier output causes the closed-loop voltage gain to

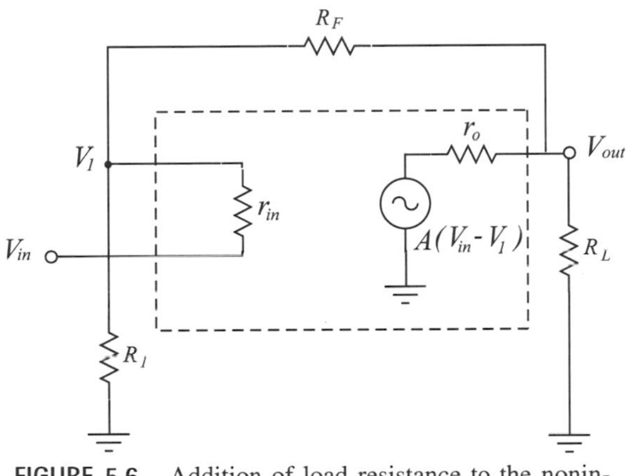

FIGURE 5.6. Addition of load resistance to the noninverting amplifier.

diminish. Incidentally, this expression differs slightly from Eq. (5.3), even when the limit $R_L \to \infty$ is reimposed. This is because the op-amp output resistance r_o is now accounted for, whereas it was not included in the slightly simplified derivation of Eq. (5.3).

Now let us determine the special value for a load resistance R_L such that the closed-loop gain from Eq. (5.11) becomes just half the value it has when $R_L \to \infty$—this will of course then equal the output resistance we are seeking. The algebra is a little tedious, but straightforward, and leads to

$$R_{\text{out}} = R_L = r_o \left[\frac{1}{1 + \frac{AR_1 + r_o}{R_1 + R_F}} \right]. \tag{5.12}$$

Neglecting r_o compared to AR_1, we arrive at

$$\boxed{R_{\text{out}} = r_o \left[\frac{1}{1 + A\frac{R_1}{R_1 + R_F}} \right]}, \tag{5.13}$$

which is the required expression for the output resistance of the noninverting amplifier. Clearly, a large open-loop gain A will cause the output resistance to become even smaller than the bare op-amp value of r_o.

Remarks

Equations (5.9) and (5.13) show that the input resistance is raised whereas the output resistance is lowered in the noninverting configuration. Both effects, if anything, are desirable in an amplifier. The origin of this behavior is the feedback provided by resistor R_F. Note that in the limit $R_F \to \infty$, these equations lead to $R_{\text{in}} = r_{\text{in}}$ and $R_{\text{out}} = r_{\text{o}}$, as expected.

Now, reconsider Eq. (5.8):

$$V_1 = \left[\frac{R_1}{R_F + R_1}\right]\left[\frac{1}{\frac{1}{A} + \frac{R_1}{R_1 + R_F}}\right] V_{\text{in}}.$$

It is immediately apparent from this expression that

$$\lim_{A \to \infty} V_1 = V_{\text{in}}. \tag{5.14}$$

Therefore, the potential at the inverting input to the op-amp is held continuously to almost exactly match the potential at the noninverting input.

$$\boxed{V_- \simeq V_+}. \tag{5.15}$$

This property is also a direct consequence of the negative feedback provided by R_F.

5.2 INVERTING AMPLIFIER

The second standard configuration for an op-amp is shown in Fig. 5.7. As always, $V_{\text{out}} = A\left[V_+ - V_-\right]$, and since $V_+ = 0$ here,

$$V_{\text{out}} = -AV_1.$$

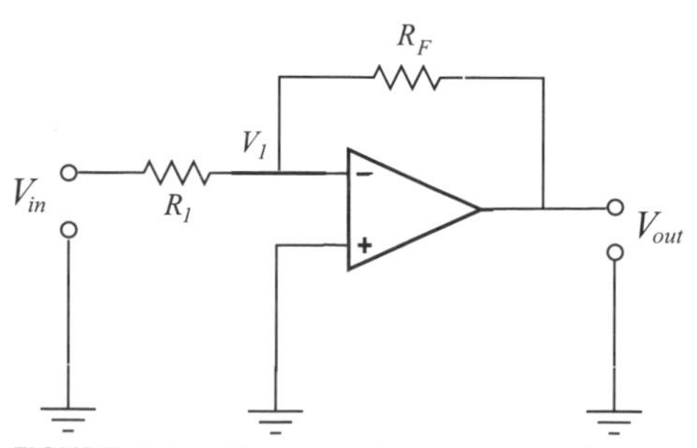

FIGURE 5.7. Component arrangement for an inverting amplifier based on a single op-amp.

Also, from current conservation at the inverting node,

$$\frac{V_{\text{in}} - V_1}{R_1} = \frac{V_1 - V_{\text{out}}}{R_F}.$$

From these two expressions is obtained

$$V_{\text{out}} = V_{\text{in}} \left[\frac{-\frac{R_F}{R_1}}{1 + \frac{1}{A}\left[\frac{R_1+R_F}{R_1}\right]} \right]. \tag{5.16}$$

Hence, the closed-loop gain is

$$G = -\left[\frac{R_F}{R_1}\right] \left[\frac{1}{1 + \frac{1}{A}\left[\frac{R_1+R_F}{R_1}\right]} \right]. \tag{5.17}$$

The negative sign here means that a positive input becomes a negative output, and conversely. Thus, this configuration is called an inverting amplifier. For op amps, the limiting gain is

$$\boxed{\lim_{A\to\infty} G = -\left[\frac{R_F}{R_1}\right]}. \tag{5.18}$$

Once again, the closed-loop gain is set by just the ratio of two resistors.

Input Resistance

The equivalent circuit for an inverting configuration is shown in Fig. 5.8. The input resistance is defined by

$$R_{\text{in}} = \frac{V_{\text{in}}}{I_{\text{in}}}.$$

But

$$I_{\text{in}} = \frac{V_{\text{in}} - V_1}{R_1}.$$

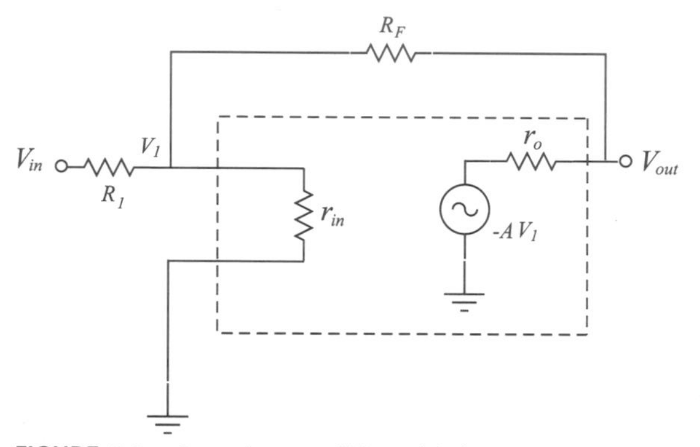

FIGURE 5.8. Inverting amplifier with internal equivalent circuit for the op-amp included.

Also, neglecting the voltage drop across r_0,

$$V_1 = -\frac{V_{\text{out}}}{A}$$

and V_{out} is given by Eq. (5.16). From these three relationships and assuming r_{in} to be very large compared to R_F, one obtains

$$R_{\text{in}} = R_1 \left[\frac{1 + \frac{1}{A}\left(\frac{R_1 + R_F}{R_1}\right)}{1 + \frac{1}{A}} \right], \tag{5.19}$$

which can also be written

$$\boxed{R_{\text{in}} = R_1 + \frac{R_F}{A + 1}}. \tag{5.20}$$

In the limit of large open-loop gain,

$$R_{\text{in}} = R_1. \tag{5.21}$$

Thus the internal resistance of the op-amp, r_{in}, is not present in the net input resistance of this configuration.

Output Resistance

The output resistance can be obtained by first determining the closed-loop voltage gain when an external resistance R_L is added to the output and then finding what value of this load will drop the output amplitude by half. The result of this process is that the output resistance of the inverting configuration is

$$\boxed{R_{\text{out}} = r_0 \left[\frac{1}{1 + A\frac{R_1}{R_1+R_F}} \right]}, \tag{5.22}$$

which is identical to the expression [Eq. (5.13)] for the output resistance in the noninverting amplifier.

Remark

In the previous sections on the noninverting and inverting amplifiers, the closed-loop gains [Eqs. (5.3) and (5.17)], input resistances [Eqs. (5.9) and (5.19)], and output resistances [Eqs. (5.13) and (5.22)] all contained a common term

$$\beta \equiv \frac{R_1}{R_1 + R_F} = \frac{1}{1 + \frac{R_F}{R_1}}, \tag{5.23}$$

which is often called the *feedback factor*, since it quantifies the proportional amount of negative feedback.

5.3 DIFFERENCE AMPLIFIER

A third important configuration is shown in Fig. 5.9. Current conservation at the inverting and noninverting nodes gives

$$\frac{V_a - V_-}{R_1} = \frac{V_- - V_{\text{out}}}{R_F} \tag{5.24}$$

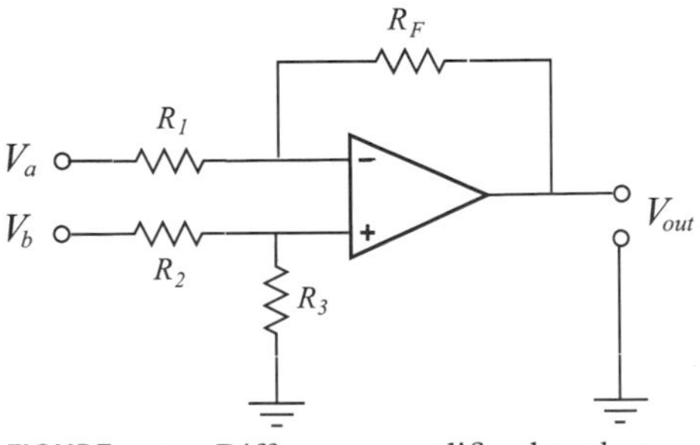

FIGURE 5.9. Difference amplifier based on a single op-amp.

or

$$V_- = \frac{R_1 R_F}{R_1 + R_F} \left[\frac{V_a}{R_1} + \frac{V_{\text{out}}}{R_F} \right], \tag{5.25}$$

and

$$\frac{V_b - V_+}{R_2} = \frac{V_+}{R_3} \tag{5.26}$$

or

$$V_+ = \frac{R_2 R_3}{R_2 + R_3} \frac{V_b}{R_2}. \tag{5.27}$$

Let us explicitly include at this point the assumption of a virtually ideal op-amp having extremely large open-loop gain and nearly infinite internal input resistance (r_{in}). Then, as we saw in the general result given in Eq. (5.15), negative feedback causes the potential at the inverting input to track the potential at the noninverting input: $V_+ = V_-$. Equating the right-hand sides of Eqs. (5.25) and (5.27),

$$\boxed{V_{\text{out}} = \left[\frac{1 + \frac{R_F}{R_1}}{1 + \frac{R_2}{R_3}} \right] V_b - \left[\frac{R_F}{R_1} \right] V_a}. \tag{5.28}$$

Hence, the output voltage is a weighted difference of the input voltages. If we make the particular choice

$$\frac{R_F}{R_1} = \frac{R_3}{R_2} = k, \tag{5.29}$$

then

$$V_{\text{out}} = k\,(V_b - V_a)\,. \tag{5.30}$$

In other words, this specific condition on the resistor ratios yields an output which is just a constant times the difference in the input voltages.

5.4 SUMMING AMPLIFIER

As a final important and useful configuration, consider the circuit shown in Fig. 5.10. Equating currents entering and leaving the connection node at the

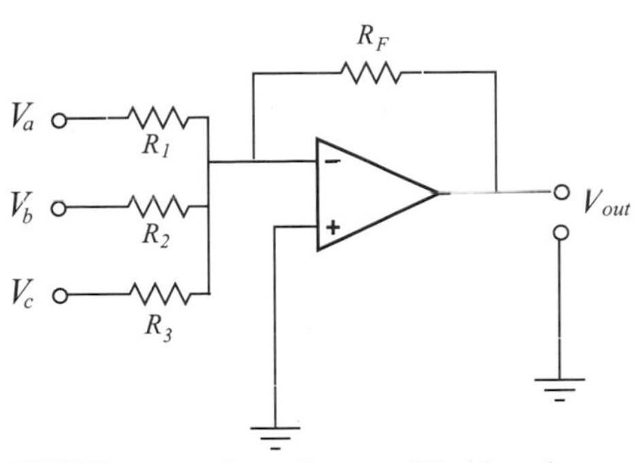

FIGURE 5.10. Summing amplifier based on a single op-amp.

inverting input,

$$\frac{V_a - V_-}{R_1} + \frac{V_b - V_-}{R_2} + \frac{V_c - V_-}{R_3} = \frac{V_- - V_{\text{out}}}{R_F}. \tag{5.31}$$

Again, the equality of potentials at the inverting and noninverting inputs is invoked: $V_+ = V_-$. But $V_+ = 0$ in this circuit, so

$$\frac{V_a}{R_1} + \frac{V_b}{R_2} + \frac{V_c}{R_3} = -\frac{V_{\text{out}}}{R_F}. \tag{5.32}$$

Finally,

$$\boxed{V_{\text{out}} = -\left[\left(\frac{R_F}{R_1}\right)V_a + \left(\frac{R_F}{R_2}\right)V_b + \left(\frac{R_F}{R_3}\right)V_c\right]}. \tag{5.33}$$

Therefore, the output voltage is a weighted sum of the input voltages (with a final inversion). By choosing appropriate resistor ratios, the weights can be set to any desired values, including unity.

This example included three input voltages, V_a, V_b, V_c, but the choice was arbitrary. In other words, any number of input voltages may be summed, and the extension of Eq. (5.33) is obvious.

5.5 FREQUENCY RESPONSE

In the previous sections of this chapter, the op-amp open-loop gain A has been treated as a constant. However, in reality this parameter varies with signal frequency. As might be expected of any ac function, a harmonic input signal of given frequency will experience both amplitude and phase changes in its passage

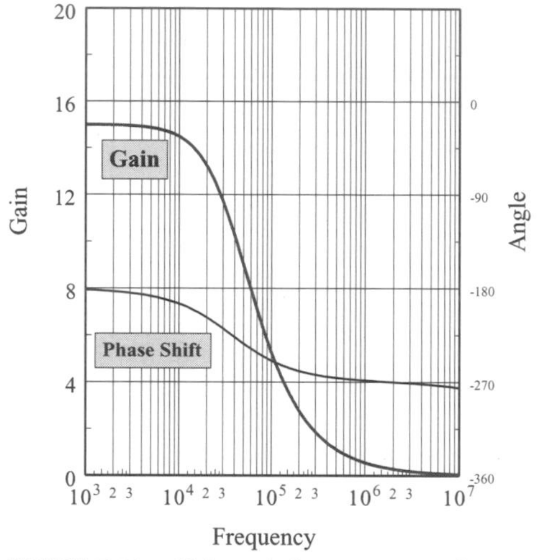

FIGURE 5.11. Gain and phase response for an inverting amplifier with $R_1 = 1$ kΩ and $R_F = 15$ kΩ.

through the op-amp. Figure 5.11 depicts these properties as they might be encountered for an inverting configuration based on a representative op-amp. The data were generated by a CAD circuit simulator package which accurately predicts real op-amp behavior.

In this sample design, R_1 was chosen to be 1 kΩ, and R_F was 15 kΩ. The expected closed-loop gain [Eq. (5.18)] would be -15. The gain magnitude is indeed equal to 15 for frequencies up to about 4000 Hz, after which there is a decided rolloff with increasing frequency. At $f \simeq 5 \times 10^5$ Hz, the gain equals unity. Above this frequency, $G \leq 1.0$ and the circuit actually attenuates the input signal rather than amplifying it.

Notice also that the phase shift begins at -180 degrees. This is due to the external conditions of the circuit, namely that it is in this case an inverting configuration (a noninverting circuit would initially exhibit 0 degrees). Figure 5.11 also reveals that additional amounts of shift develop as the frequency is changed. At high frequencies, the phase drops by an additional 90 degrees, resulting in a net value of about -270 degrees. This effect is caused by internal circuit characteristics of the op-amp itself.

It is common practice to express amplifier gain in units of decibels (dB), where the definition is contained in

$$G(\text{dB}) = 20 \log \left[\frac{V_{\text{out}}}{V_{\text{in}}} \right]. \tag{5.34}$$

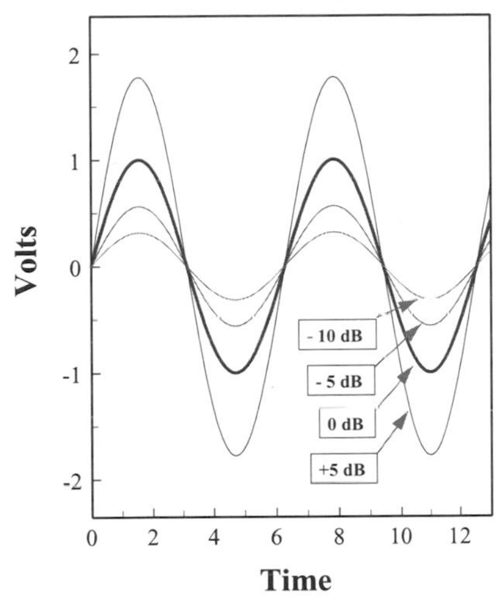

FIGURE 5.12. Illustration of relative magnitudes of a hypothetical ac signal increased and decreased in 5 dB increments.

As a visual aid to the decibel scale, Fig. 5.12 shows a hypothetical reference signal (in boldface), which serves as the 0 dB level, together with waveforms at +5 dB, −5 dB, and −10 dB. These units are quite nonlinear, as can be appreciated from the fact that 20 dB corresponds to a voltage ratio of 10, whereas 40 dB indicates a ratio of 100.

For the present example, the magnitude of the gain at low frequencies would be 20 log(15) = 23.5 dB.

As Fig. 5.13 illustrates, the high-frequency rolloff of the gain becomes almost linear when the vertical scale is in dB and the frequency axis is logarithmic.

The dashed line that approximates this linear limit has a slope of 20 dB/decade; that is, for each increase in frequency by a factor of 10 there is a corresponding drop in gain by a factor of 10. This line intercepts the low-frequency gain level (23.5 dB) at a value of f called the *corner frequency*, f_c; here $f_c \approx 4 \times 10^4$ Hz. Whereas the specific numbers that appear in Figs. 5.11 and 5.13 are particular to a given type of operational amplifier (324 in this example), the slope of the linear asymptote is the same for all op-amps. The reason for such a rate of decrease of gain will be made clear in the next section.

From the preceding discussion, it can be seen that to a first approximation amplifier gain is flat to a corner frequency, after which it rolls off at 20 dB/decade.

To push this illustrative example a little further, we now examine the effects of changing the feedback resistor in the inverting configuration, first to 5 kΩ

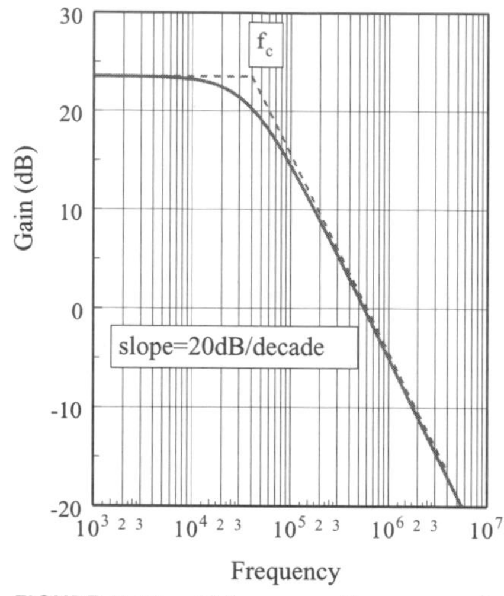

FIGURE 5.13. Gain versus frequency of the inverting amplifier using a vertical scale in decibels. The high-frequency rolloff is 20 dB per decade.

and then to 1 kΩ. In the first case, the magnitude of the low-frequency gain [Eq. (5.18)] should be 5 K/1 K = 5, which is 20 log(5) = 13.9 dB. In the second case, the gain is unity, which is equivalent to 20 log(1) = 0 dB. These two cases are combined with the previous example in Fig. 5.14.

Note that for designs possessing a smaller closed-loop gain, the corner frequency is higher. The corner frequency at unity gain (0 dB) can be estimated from the graph at about 2×10^5 Hz, an increase by a factor of nearly 5 over f_c when the gain was 15. Put another way, circuits with lower gain postpone rollover until higher frequencies. This can be viewed as a tradeoff of closed-loop gain for increased bandwidth.

The inverse of all of this is also true: higher closed-loop gains are matched by reduced bandwidth. The ultimate limit of this process is the bare op-amp itself, where the gain reverts to the open-loop value A. To quote a real-world example, the ubiquitous type 741 op-amp is listed as having $A \approx 2 \times 10^5$ and $f_c \approx 8$ Hz! The exceptionally small open-loop corner frequency coupled with the unmanageable gain (meaning its output will saturate with almost any input signal) makes an isolated op-amp fairly restricted in application.

Negative feedback therefore has at least two (there are actually more) beneficial effects: (1) reduction of net gain to more desirable levels; (2) increase in bandwidth to more usable levels.

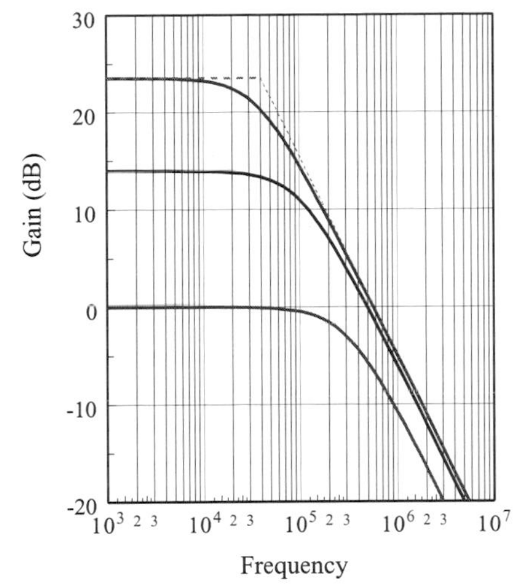

FIGURE 5.14. The previous example of an inverting amplifier, but with several different choices for the feedback resistor (and hence the gain). Notice that reduced gain is accompanied by increased bandwidth.

Complex Gain

To model the frequency dependence of an op-amp, the open-loop gain can be written as a complex function.

$$\boxed{A = A_0 \frac{1}{1 + j\left(\frac{f}{f_c}\right)}}. \tag{5.35}$$

Here, A_0 is the gain amplitude, f is the signal frequncy, and f_c is a second constant, which will turn out to be a corner frequency. In complex notation, a harmonic signal is of the general form $V_0 e^{j(\omega t + \theta)}$. Suppose then we take the product of an input signal $V_{\text{in}} = V_1 e^{j\omega t}$ with a gain as expressed in Eq. (5.35):

$$V_{\text{out}} = V_1 e^{j\omega t}\, A_0 \frac{1}{1 + j\left(\frac{f}{f_c}\right)}. \tag{5.36}$$

Observing that

$$\frac{1}{1 + j\left(\frac{f}{f_c}\right)} = a\, e^{j\phi} \tag{5.37}$$

with

$$a = \frac{1}{\sqrt{1 + \left(\frac{f}{f_c}\right)^2}}, \tag{5.38}$$

$$\phi = \tan^{-1}\left[-\frac{f}{f_c}\right]. \tag{5.39}$$

Then,

$$V_{\text{out}} = V_1 \left[\frac{A_0}{\sqrt{1 + \left(\frac{f}{f_c}\right)^2}}\right] e^{j(\omega t + \phi)}. \tag{5.40}$$

This demonstrates that the magnitude of the amplified signal at low frequency is equal to $V_1 A_0$ and that it rolls off with increasing frequency (as the square root in the denominator becomes increasingly large). Additionally, the output leads the input by the phase angle ϕ given in Eq. (5.39). The two functions a and ϕ are plotted in Fig. 5.15. The similarity to Fig. 5.11 is obvious. Note that the phase

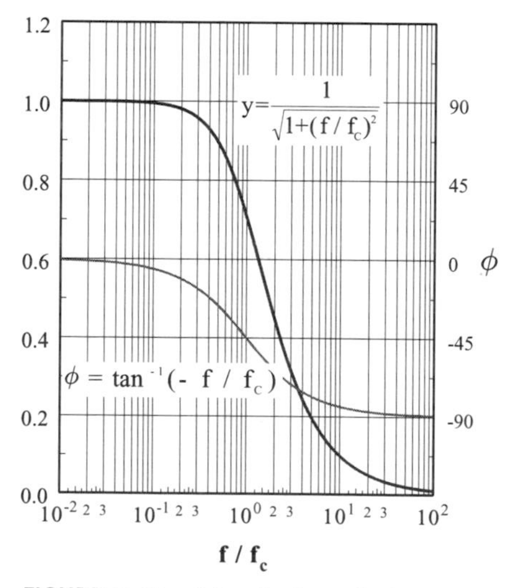

FIGURE 5.15. Magnitude and phase response expected from the complex representation of the op-amp open-loop gain function.

shift here, which reaches −90 degrees at high frequencies, is an intrinsic property of the amplifier model. The phase shift in Fig. 5.11 includes an additional −180 degrees because of the external inverting connection.

In the high-frequency limit, Eq. (5.38) becomes simply $a \to y = (x)^{-1}$, where $x = (f/f_c)$. In decibels, this is

$$y(\mathrm{dB}) = 20\log[x^{-1}] = -20\log[x],$$

so at $x = 1$ (the corner frequency) $y = 0$ dB; at $x = 10$, $y = -20$ dB; at $x = 100$, $y = -40$ dB, and so on. This function, plotted on a graph with a logarithmic x axis, is a straight line beginning at the corner frequency and dropping at a rate of 20 dB/decade. In other words, it is the same type of high-frequency rolloff that was illustrated in Fig. 5.13.

Clearly, the complex gain given in Eq. (5.35) successfully replicates both the amplitude and phase transformations inherent in real op-amps.

Gain Bandwidth

As a further insight, suppose the noninverting configuration is reexamined with the formerly fixed A replaced with expression (5.35). We begin with Eq. (5.3) and for convenience use the feedback factor β as defined in Eq. (5.23). Hence,

$$G = \frac{1}{\frac{1}{A} + \beta},$$

which becomes

$$G = \frac{1}{\frac{1+j\frac{f}{f_c}}{A_0} + \beta}$$

or

$$G = \frac{A_0}{(1 + A_0\beta) + j\left(\frac{f}{f_c}\right)}. \tag{5.41}$$

This equation may be rewritten in the standard form

$$G = G_0\left[\frac{1}{1 + j\left(\frac{f}{f_c^*}\right)}\right] \tag{5.42}$$

with

$$G_0 = \frac{A_0}{1 + A_0\beta}, \tag{5.43}$$

$$f_c^* = (1 + A_0\beta)\, f_c. \tag{5.44}$$

Comparing Eq. (5.35) for the open-loop condition and Eq. (5.42) for the closed-loop configuration, it is apparent that the closed-loop gain has been reduced from A_0 to $A_0/(1 + A_0\beta)$, while the corner frequency has been increased from f_c to $(1 + A_0\beta)f_c$. An important observation we can make from these results is that

$$G_0 f_c^* = \frac{A_0}{1 + A_0\beta}(1 + A_0\beta) f_c = A_0 f_c. \tag{5.45}$$

In other words, the product of gain and bandwidth is the same for the bare op-amp as it is for the complete noninverting configuration.

As an exercise, the preceding procedures could be repeated for the inverting configuration, beginning with Eq. (5.17).

PROBLEMS

Problem 5.1. A noninverting amplifier has a feedback resistor $R_F = 5$ K and has the inverting input coupled to ground through a 20 K resistor. The op-amp has an open-loop gain $A = 20{,}000$, an input resistance $r_{\text{in}} = 100$ K, and an output resistance $r_o = 5\ \Omega$.

1. What is the voltage gain? [Ans. 1.2499].
2. What is the overall input resistance? [Ans. 1600 Meg].
3. What is the net output resistance? [Ans. 0.000312 Ω].

Problem 5.2. An op-amp is specified as having an open-loop gain of 150,000 and an open-loop bandwidth (corner frequency) of 10 Hz.

1. What is the approximate open-loop gain at 10 kHz? [Ans. 150].
2. If this op-amp is used in a noninverting configuration with $R_F = 40$ K and $R_1 = 10$ K, what will be the closed-loop gain and bandwidth? [Ans. 5 and 300,000 Hz].

CHAPTER 6

Special-Purpose Circuits

In the previous chapter, the basic properties of operational amplifiers were summarized, and a number of standard configurations were discussed. It was shown, for example, that very simple arrangements of a few resistors in combination with an op-amp could provide noninverting or inverting amplification. These circuits are actually both usable and useful for boosting weak sensor signals and for performing analog addition and subtraction.

Many other special-purpose circuits exist that are capable of performing unique transformations on analog signals. Several of these are discussed in the following sections.

6.1 UNITY-GAIN BUFFER

The noninverting configuration was discussed in the previous chapter; the schematic is repeated here (Fig. 6.1). Assuming a very large open-loop gain, the closed-loop gain was found to be

$$G = 1 + \frac{R_F}{R_1}.$$

Suppose now this arrangement is forced to the limit $R_F \to 0$ and $R_1 \to \infty$. A schematic incorporating these limits is depicted in Fig. (6.2). The closed-loop gain obviously becomes

$$\boxed{G = 1.0}. \tag{6.1}$$

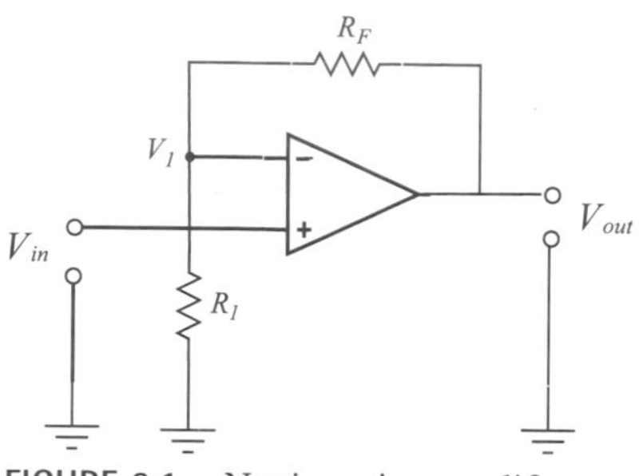

FIGURE 6.1. Noninverting amplifier.

Hence, output equals input. This may seem a rather pointless achievement until the input and output resistances are considered.

The input resistance is given by Eq. (5.9) in the appropriate limit, which is

$$\boxed{R_{\text{in}} = r_{\text{in}} [1 + A]}. \tag{6.2}$$

Typically, this will be extremely large, since r_{in} is large and A is very large. The output resistance is given by Eq. (5.13) in the appropriate limit. It is

$$\boxed{R_{\text{out}} = r_{\text{out}} \left[\frac{1}{1 + A} \right]}. \tag{6.3}$$

For r_{out} small and A very large, clearly R_{out} is extremely small.

This circuit thus has unity gain, extremely high input resistance, and extremely small output resistance. These properties make it ideal for use as a buffer to isolate one section of a circuit from another.

As an example, consider the schematic in Fig. 6.3. The circuitry (the purpose of which we are not concerned with here) contained in the box labeled B has some input resistance r_{in}. Now, if the value of r_{in} is **not** very large compared to the resistor R_x, then directly attaching box B across R_x (omitting the buffer) would produce a parallel combination, R_x and r_{in}, whose resistance would certainly be less

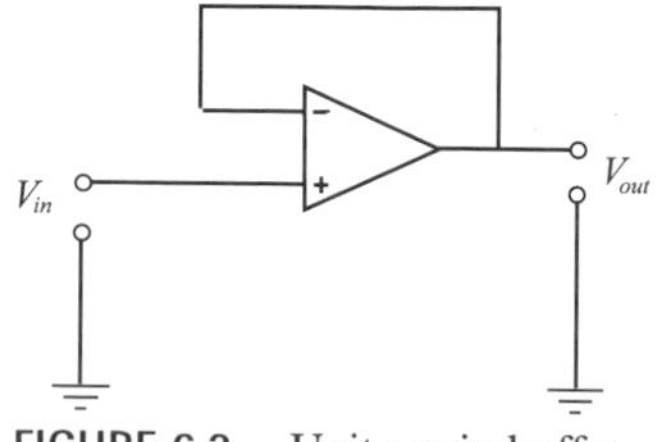

FIGURE 6.2. Unity-gain buffer.

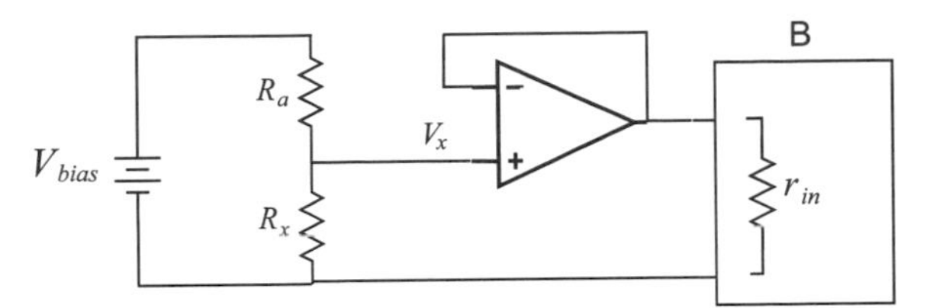

FIGURE 6.3. Use of a buffer to isolate one circuit from a second circuit whose input resistance is not large.

than the original R_x. This would decrease the voltage V_x so the subsequent signal transformations produced by B would not reflect the original signal across R_x.

On the other hand, if the buffer is inserted as shown, then R_x would be shunted only by a nearly infinite input resistance. Virtually no alteration in the signal V_x would occur.

6.2 INSTRUMENTATION AMPLIFIER

A special circuit with very useful properties is illustrated in Fig. 6.4.

Because the input pair of op-amps, A_1 and A_2, each have negative feedback resistors R, the potential at each noninverting input approximately equals the potential at the corresponding inverting input, as seen earlier in Eq. (5.15). This implies that the potential at the top of R_S is V_1, while the potential at the bottom of R_S is V_2.

Now, either $V_1 > V_2$ or $V_2 > V_1$. The operation of the circuit does not depend on which case applies, so without loss of generality we choose the situation $V_1 > V_2$. Consequently, $v_a > v_b$ and there will be a current flow I from v_a toward v_b down through the chain R, R_S, R.

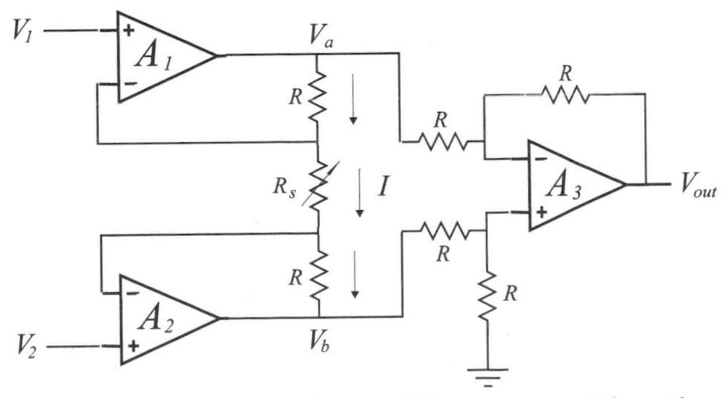

FIGURE 6.4. Instrumentation amplifier constructed from three op-amps. R_S is a single gain-setting resistor.

With these points in mind, the circuit operation may be derived as follows. Clearly,

$$IR_S = V_1 - V_2$$

or

$$I = \frac{V_1 - V_2}{R_S}.$$

Thus, since $v_a - V_1 = IR$,

$$v_a = V_1 \left[1 + \frac{R}{R_S} \right] - V_2 \left[\frac{R}{R_S} \right]. \tag{6.4}$$

Likewise, $V_2 - v_b = IR$, so

$$-v_b = V_1 \left[\frac{R}{R_S} \right] - V_2 \left[1 + \frac{R}{R_S} \right]. \tag{6.5}$$

Amplifier A_3 is configured in a standard difference arrangement and has an output given by Eq. (5.30),

$$V_{\text{out}} = k\,(v_b - v_a),$$

where $k = R/R = 1$ in this case. Using Eqs. (6.4) and (6.5),

$$\boxed{V_{\text{out}} = \left[1 + 2\frac{R}{R_S} \right] (V_2 - V_1)}. \tag{6.6}$$

Thus, the overall circuit action is to provide differential amplification of the two inputs and to do so with a gain that is set by the resistor values R and R_S.

In an instrumentation amplifier, op-amps A_1, A_2, A_3 together with all resistors R are considered as internal to the device, whereas R_S may be viewed as the single external component: a gain-setting resistor. (Instrumentation amps are available commercially packaged as single-chip integrated circuits; in this form, R_S literally is an external component.) With this technique, gains of several hundred are easily achieved. Because the signal voltages V_1 and V_2 are fed directly to op-amp noninverting terminals, the input impedance of an instrumentation amplifier is very large ($\sim 10^9\ \Omega$).

Finally, it should be noted that any common voltage that might be present in both inputs V_1 and V_2 will be canceled in the output because of the differencing action in Eq. (6.6). For this reason, difference amplifiers, and instrumentation amplifiers in particular, are very useful when the signals of interest are superimposed on large, shared baselines.

Because of the imperfections of real circuits and amplifiers, the amplified output is better expressed by something like

$$V_{\text{out}} = G_D\,(V_2 - V_1) + G_C\left(\frac{V_2 + V_1}{2}\right). \tag{6.7}$$

That is, there is both a differential gain G_D, as in Eq. (6.6), and a common gain G_C, which operates on the average input. Ideally, G_C should be near zero, but a finite value implies that the output voltage will be contaminated by a buried component whose origin is not the signal of interest.

The ability to cancel shared baselines is quantified by a parameter known as the *common-mode rejection ratio* (CMRR). It is defined as

$$\text{CMRR} = \frac{G_D}{G_C}. \tag{6.8}$$

Obviously, in the ideal case with $G_C = 0$, CMRR would be infinite. Actual instrumentation amps can achieve CMRRs of about 10^5 or larger.

6.3 LOG AND ANTILOG AMPLIFIERS

Suppose an instrumentation application generates a signal which at times is quite small, whereas at other times it is comparatively large. This property is known as wide dynamic range. Audio signals often fall in this category, ranging from very soft to very loud. Wide dynamic range poses difficulties when further electronic processing of the signal is planned because either the low levels may drop out or the high levels may overload the electronics. A workaround is found in the technique of *compression*, where the signal is in effect rescaled, say with a *logarithmic converter*. After processing, an inverse scaling, or *decompression*, is applied. Clearly, an *antilog converter* would be needed for this task.

Alternatively, suppose that a transducer signal (transducers will be treated in Part III) is an exponential function of the physical parameter being monitored. This could, for example, apply to a thermistor temperature sensor. In such a case, the transducer output can be *linearized* by passing it through a circuit with inverse scaling properties—a logarithmic converter.

The key to obtaining the desired nonlinear output from an op-amp circuit is the use of an appropriate nonlinear feedback element. A semiconductor diode is described by an expression of the form

$$I = I_0 \left[e^{qV/kT} - 1 \right], \tag{6.9}$$

where I is the current flowing through the diode, I_0 is the so-called reverse bias saturation current (a constant for any particular device), q is the electronic charge, V is the voltage across the diode, k is Boltzmann's constant (1.38×10^{-23} joule per kelvin), and T is the diode temperature expressed in absolute degrees (kelvin). At or near room temperature ($T \approx 300$ K), the factor kT/q is approximately 0.025 V. Hence, for even modest diode voltages

$$I \simeq I_0 \, e^{qV/kT}. \tag{6.10}$$

Log

When a diode is placed as illustrated in Fig. 6.5, the voltage output may be derived as follows. Clearly, $V_- = 0$ and so $I = V_{\text{in}}/R$. But this current also flows through the diode, and furthermore the diode voltage is expressed by $V = -V_{\text{out}}$. Hence,

$$\frac{V_{\text{in}}}{R} = I_0 \, e^{-qV_{\text{out}}/kT}$$

or

$$V_{\text{out}} = -\frac{kT}{q} \ln\left(\frac{V_{\text{in}}}{I_0 R}\right), \tag{6.11}$$

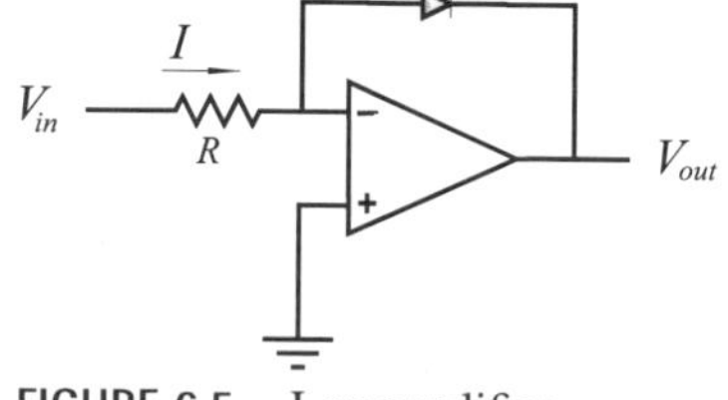

FIGURE 6.5. Log amplifier.

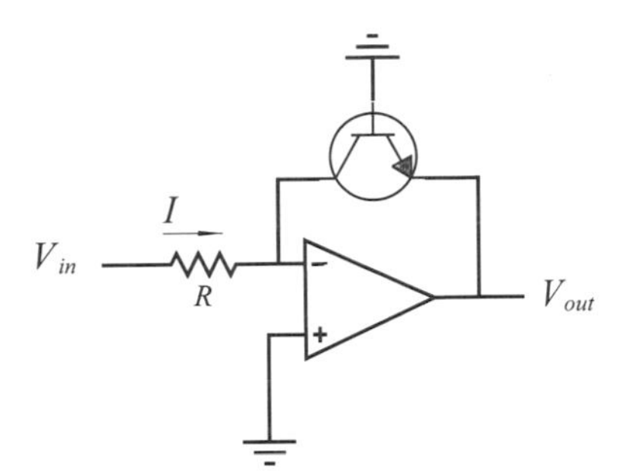

FIGURE 6.6. Log amplifier using an NPN transistor in place of a diode.

and finally

$$\boxed{V_{\text{out}} = -\left[\frac{kT}{q}\right]\ln(V_{\text{in}}) + \left[\frac{kT}{q}\ln(I_0 R)\right]}. \tag{6.12}$$

This expression is of the form $V_{\text{out}} = -a \ln V_{\text{in}} + b$, so the output voltage is a logarithmic function of the input voltage.

It is also possible to use an NPN transistor connected as shown in Fig. 6.6 to achieve essentially the same behavior. Note that the collector and base are both at ground potential, and that with positive V_{in} and consequently negative V_{out}, the base-emitter junction is forward-biased.

Antilog

For antilog operation, the resistor and diode are simply interchanged as indicated in Fig. 6.7. Now the equation for V_{out} is derived as follows. The forward diode drop is V_{in}. Hence,

$$I \simeq I_0\, e^{\frac{qV_{\text{in}}}{kT}}. \tag{6.13}$$

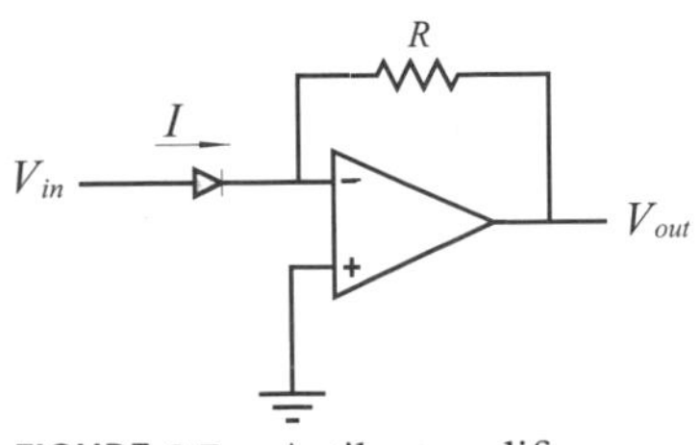

FIGURE 6.7. Antilog amplifier.

Also,

$$IR = -V_{\text{out}}. \tag{6.14}$$

Thus,

$$\boxed{V_{\text{out}} = -\,[I_0 R]\; e^{\left(\frac{q}{kT}\right)V_{\text{in}}}}. \tag{6.15}$$

In this case, the output voltage is an exponential (antilog) function of the input.

6.4 CONSTANT CURRENT SOURCE

Figure 6.8 shows an op-amp circuit that provides a current I to a load R_L. Because $V_- \approx V_+$ is assured by the negative feedback and $V_+ = 0$, the current through the input resistor is just

$$I = \frac{V_{dc}}{R}. \tag{6.16}$$

Virtually all of this current then flows through the load on account of the extremely high inpedance at the input terminals of the op-amp. In other words, the load current I is set only by V_{dc} and R and is independent of the particular value of the load resistance. Hence, the circuit depicted in Fig. 6.8 acts as a constant current source.

6.5 VOLTAGE AND CURRENT CONVERSION

Voltage-to-Current Converter

If the dc voltage input to Fig. 6.8 is replaced by a variable source as in Fig. 6.9, then the output current becomes

$$\boxed{I_{\text{out}} = \frac{V_{\text{in}}}{R}}. \tag{6.17}$$

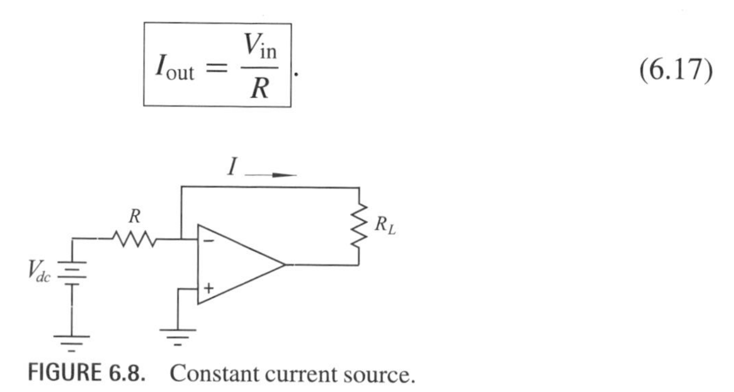

FIGURE 6.8. Constant current source.

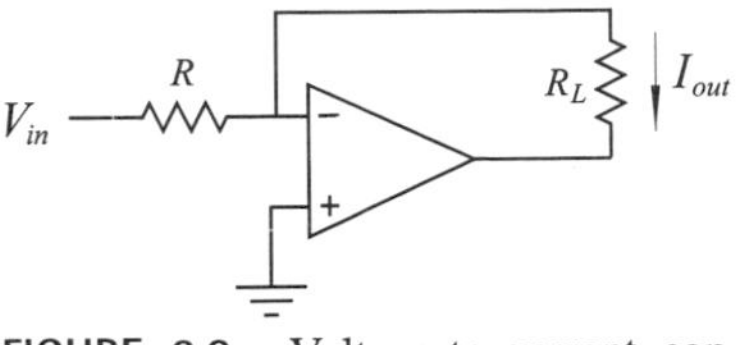

FIGURE 6.9. Voltage-to-current conversion circuit.

Thus, the input voltage has been transformed into an output current flowing through the load resistor.

Another possibility for voltage-to-current conversion is the arrangement in Fig. 6.10. For this circuit, the current I_{out} flows through R_L and R, so

$$I_{\text{out}} = \frac{V_-}{R}.$$

But $V_- = V_+ = V_{\text{in}}$,

$$I_{\text{out}} = \frac{V_{\text{in}}}{R}. \tag{6.18}$$

This is exactly the same as expression (6.17), so the two variants provide identical voltage-to-current conversion.

Differences exist in other aspects of the circuits. The input impedance of Fig. 6.10 is very high—a desirable condition for a voltage sensor in the same way that an ideal voltmeter has high input impedance, whereas the design of Fig. 6.9 is basically an inverting amplifier with an input impedance [see Eq. (5.21)] of only R. On the other hand, the dynamic range of the inverting configuration is larger.

Current-to-Voltage Converter

The inverse of the operation discussed in the previous section is current-to-voltage conversion. Analyzing the circuit shown in Fig. 6.11, and observing that

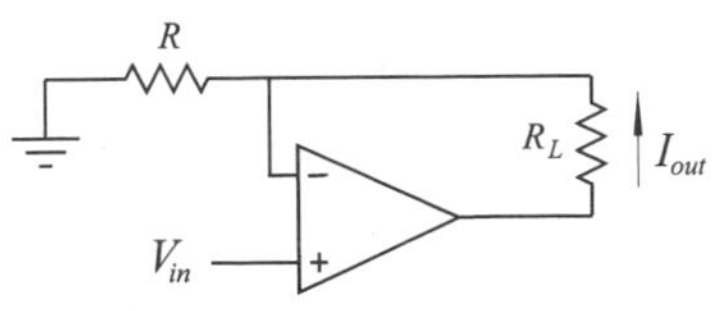

FIGURE 6.10. A second arrangement for voltage-to-current conversion.

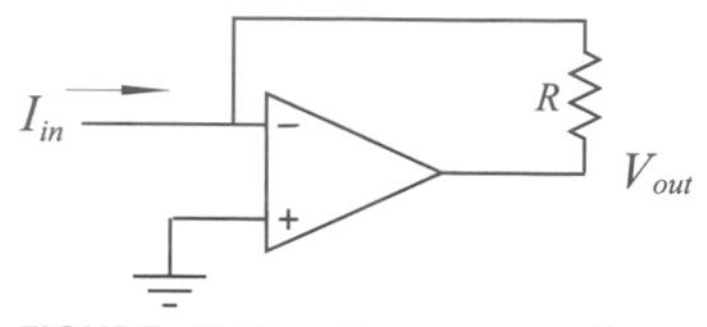

FIGURE 6.11. Current-to-voltage conversion.

the input current flows through the output resistor R,

$$\boxed{V_{\text{out}} = -I_{\text{in}} R}, \tag{6.19}$$

where $V_- = V_+ = 0$ has been applied. This configuration can be regarded as a limiting form of the standard inverting amplifier (Chapter 5) for which the input impedance was [Eq. (5.21)] R_1, which is zero in this case. Thus, the impedance that this converter presents to the source of current I_{in} is extremely small—a desirable property in the same sense that an ideal ammeter has nearly zero impedance.

6.6 ANALOG INTEGRATION AND DIFFERENTIATION

In the earlier section on log and antilog circuits, it was seen how the equivalent of either of two particular mathematical operations could be performed on given input voltages. Summing and differencing amplifiers, discussed in the previous chapter, are of course the analogs of addition and subtraction. Electronic counterparts of other mathematical operators also exist, including the functions of multiplication, division, and square root. In this section, we consider integration and differentiation.

Integrators

The quantity of charge on a capacitor and the potential difference across the capacitor are related: $Q = CV_C$. Therefore,

$$\frac{dQ}{dt} = C\frac{dV_C}{dt}.$$

But the rate of change of the charge residing on the capacitor plates is simply a current flowing onto or off of those plates, so

$$I_C = C\frac{dV_C}{dt} \tag{6.20}$$

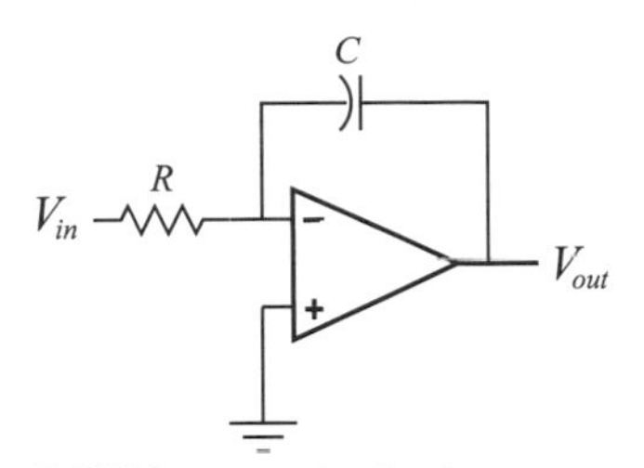

FIGURE 6.12. Analog integrator.

is the capacitor current. Notice that in the steady state (i.e., dc conditions) $I_C = 0$, which means simply that a capacitor can "carry" only ac currents.

Considering Fig. 6.12, this capacitor current is seen to be just V_{in}/R because, as usual, $V_- = V_+ = 0$. Hence,

$$\frac{V_{\text{in}}}{R} = -C\frac{dV_{\text{out}}}{dt}, \tag{6.21}$$

where we have used the fact that $V_{\text{out}} = -V_C$. From Eq. (6.21), it is evident that

$$\boxed{V_{\text{out}} = -\frac{1}{RC}\int V_{\text{in}}\, dt}, \tag{6.22}$$

so the output voltage is a time integration of the input voltage.

To illustrate the operation of an integrator, the circuit shown in Fig. 6.13 was created in the software simulation package PSpice (a product of MicroSim Corporation). Notice that a second op-amp has been added—it has a gain of 1 and is included simply to invert the integrator output. In this particular case, the overall response should be

$$V_{\text{out}} = \frac{1}{RC}\int V_{\text{in}}\, dt.$$

The voltage input was chosen to be a pulse sequence of amplitude 1.0 V, pulse width of 1.5 sec, and period of 2.5 sec. Figure 6.14 is a plot of the PSpice circuit simulation results. The output voltage is clearly seen to be just the integral of the input signal. Two further comments are in order. First, the PSpice schematic includes a timed switch which is placed across the integrating capacitor. This switch opens 0.01 sec after the simulation begins and guarantees that C_1 is initially discharged. Second, it should be remembered that op-amp output voltages cannot exceed power supply levels. With certain input waveforms it is quite easy to saturate the integrator output.

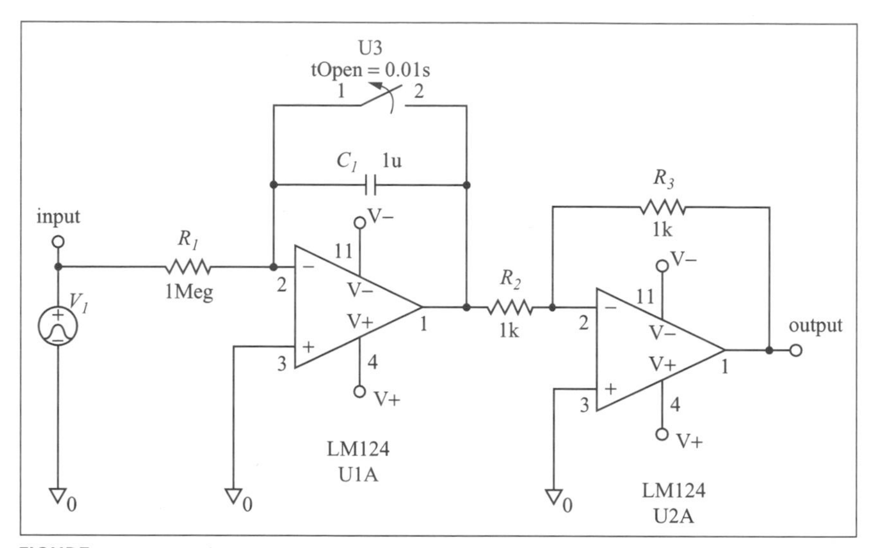

FIGURE 6.13. PSpice schematic of an analog integration circuit. The second op-amp is simply acting as an inverter.

Differentiators

By simply interchanging the resistor and capacitor in the integrator, a differentiator is formed. To see this, consider Fig. 6.15. As before, the capacitor current is the rate of change of charge buildup on the capacitor, so

$$I_{\text{in}} = C\frac{dV_{\text{in}}}{dt}. \tag{6.23}$$

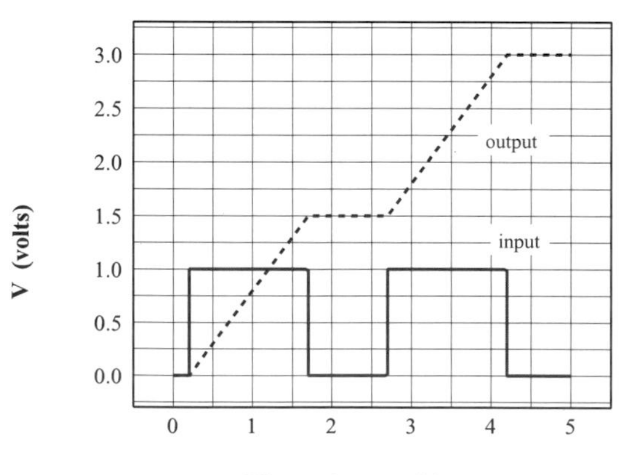

FIGURE 6.14. PSpice simulation results for the integrator.

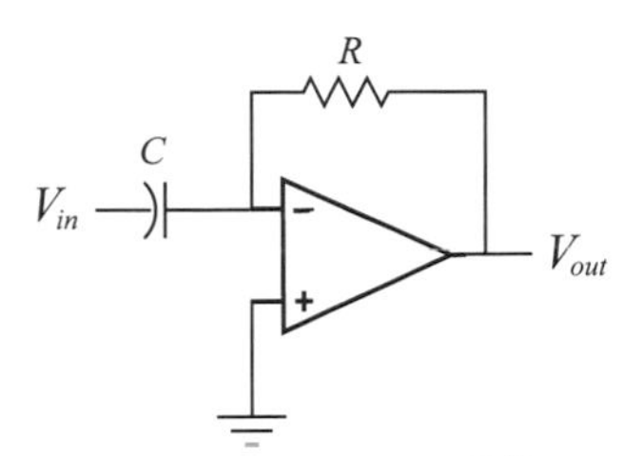

FIGURE 6.15. Analog differentiator.

But

$$V_{\text{out}} = -I_{\text{in}} R,$$

and thus

$$\boxed{V_{\text{out}} = -RC\frac{dV_{\text{in}}}{dt}}. \tag{6.24}$$

Therefore, this circuit "differentiates" the input signal.

As with the integrator, the operation can be demonstrated with the aid of a PSpice simulation. The schematic for this example is shown in Fig. 6.16. The 1 nF capacitor across R_1 is included to prevent ringing oscillations at the output of the first op-amp which otherwise occur when the input voltage pulse reaches its corner points. Here, too, an inverting stage has been added for convenience in

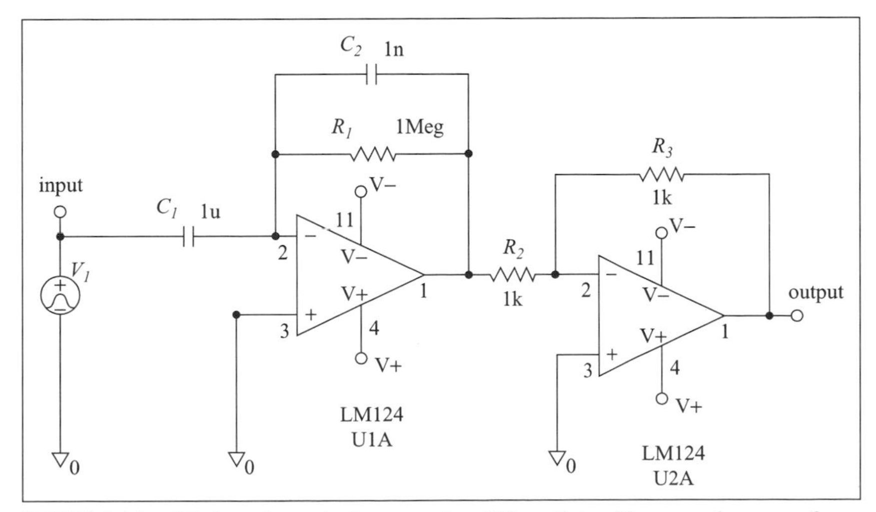

FIGURE 6.16. PSpice schematic for an analog differentiator. The second op-amp functions simply as an inverter.

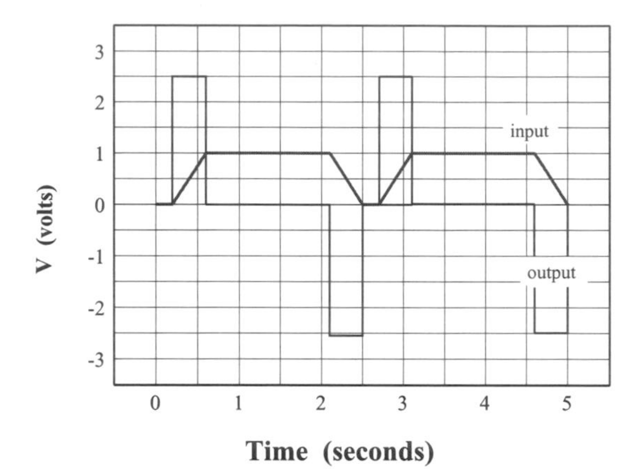

FIGURE 6.17. PSpice simulation results for the analog differentiator.

representing the output, which is plotted in Fig. 6.17. Where the input signal is flat (constant), the derivative is zero and the output is zero. During the linear ramp phases of the input, the slope is constant, and the output rises or falls abruptly to the value 2.5 V.

PROBLEMS

Problem 6.1. Consider the sawtooth waveform shown in Fig. 6.18. The vertical axis is in volts and the horizontal axis is in seconds.

1. If this signal is fed to an integrator circuit with $R = 1$ Meg and $C = 5\ \mu$F, describe the resulting output waveform. Caution: the integral of a linear function is not itself a linear function.
2. If this signal is fed to a differentiator circuit with $R = 1$ Meg and $C = 5\ \mu$F, describe the resulting output waveform.

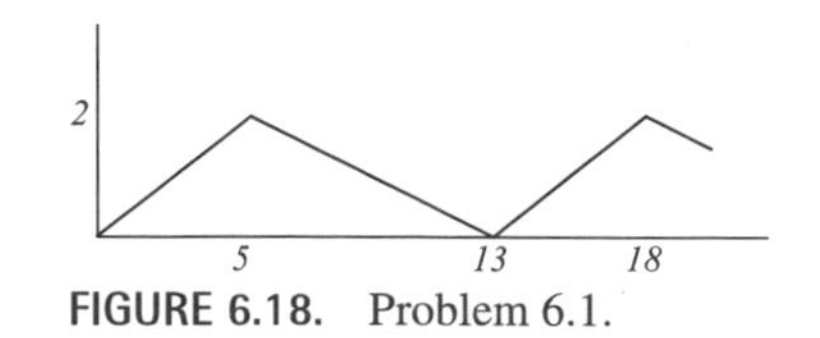

FIGURE 6.18. Problem 6.1.

Problem 6.2. The PSpice circuit shown in Fig. 6.19 is made up of three stages: a log amplifier, an inverter, and an antilog amplifier. Suppose the input waveform is the same as shown in the previous problem. Calculate and plot the expected waveforms at Out1 and Out2.

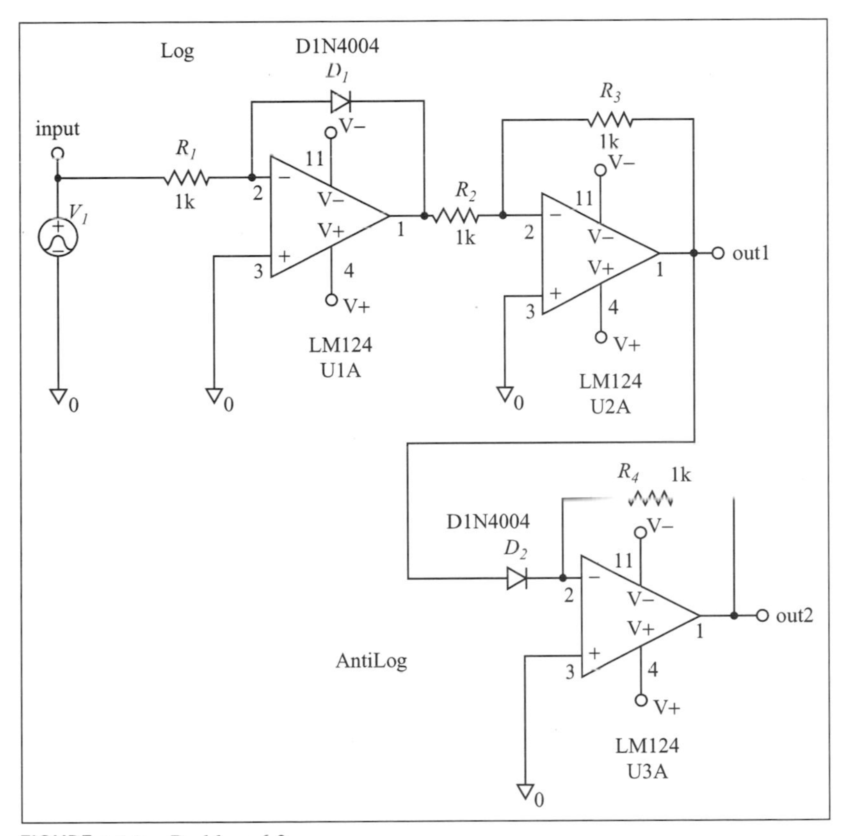

FIGURE 6.19. Problem 6.2.

CHAPTER 7

Waveform Generators

Instrumentation systems frequently require a source of repetitive voltage waveforms. The repetition rate, stated in cycles per second, or Hz, can range from one waveform in many seconds, minutes, or hours, to hundreds, thousands, or millions of waveforms per second. At the extreme ends of the spectrum—that is, exceptionally slow or fast signals—specialized generating circuits and techniques may be required. This chapter deals primarily with the more common intermediate range of approximately a few cycles per second to nearly one megahertz.

Any periodic waveform satisfies the condition $v(t+T) = v(t)$, where T is the repeat interval, or *period*. There are a number of standard waveform shapes that are encountered in applications. The most common is certainly a sinusoidal voltage signal of the general form $v(t) = A\sin(2\pi ft + \phi)$, where A is the amplitude, f is the frequency, and ϕ is a phase shift indicating a displacement in time between the zero point of the "clock" and the sine wave itself. The period and frequency are of course related through $\omega = 2\pi f$. In addition to the sine wave, there are rectangular pulses and square waves, as well as triangular and sawtooth shapes, and finally so-called arbitrary waveforms.

The next section deals with the generation of sinusoidal voltage signals. Other shapes are then discussed.

7.1 OSCILLATORS

Feedback is the process of routing a portion of the output of a circuit back to the input. In Fig. 7.1, the block represents a network, which introduces an attenuation α and a phase shift ϕ in the feedback loop.

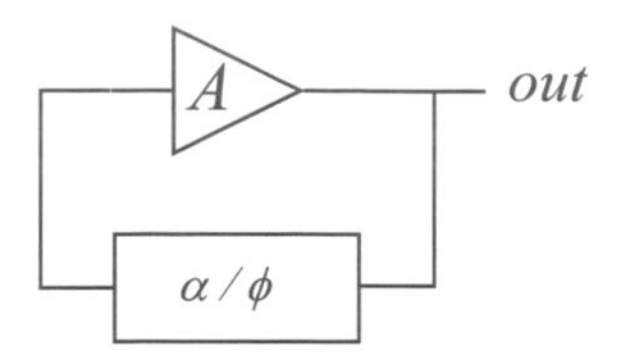

FIGURE 7.1. Amplifier with feedback loop. The feedback network introduces an attenuation α and phase shift ϕ.

In the two preceding chapters, the feedback path in the various amplifiers was purely resistive. Because of this, the associated external phase shift was zero. The feedback path terminated at the op-amp inverting input, so the net effect was a form of partial signal cancellation. This arrangement constitutes negative feedback and was an essential aspect of the inverting and noninverting configurations discussed previously. Negative feedback reduces the overall gain, improves stability, and increases bandwidth.

The opposite situation, in which some portion of the output is returned in-phase to the amplifier, would be expected to have correspondingly opposite results: decreased stability and increased gain. The most common situation where instability is actually a target of the design process is that of electronic oscillators. Positive feedback is thus the key to insuring that oscillations occur in circuits. An oscillator is specified by a number of attributes, such as frequency range, output, stability, and so forth. In many instances, enhancements in one attribute can only be achieved at the expense of others, or in overall circuit complexity. Because of this, no single oscillator design emerges as "best." On the other hand, a few designs have achieved the status of standards because they represent good compromises among the competing requirements. Several of these classics will now be presented as a means of illustrating oscillator fundamentals.

Wien-Bridge Oscillator

The schematic depicted in Fig. 7.2 represents a Wien-bridge oscillator.

The network in the box labeled Z will pass an attenuated and phase-shifted portion of the output back to the noninverting op-amp input. For sustained oscillations to occur, the losses induced in Z must be just compensated by the closed-loop gain of the amplifier. Too little gain and the oscillations will damp away; too much gain and the oscillations will grow uncontrollably. In addition, the phase shift brought about by passage through Z must equal 0 or destructive interference will occur. This process is analogous to the synchronized pumping required to maintain the periodic motion of a swing.

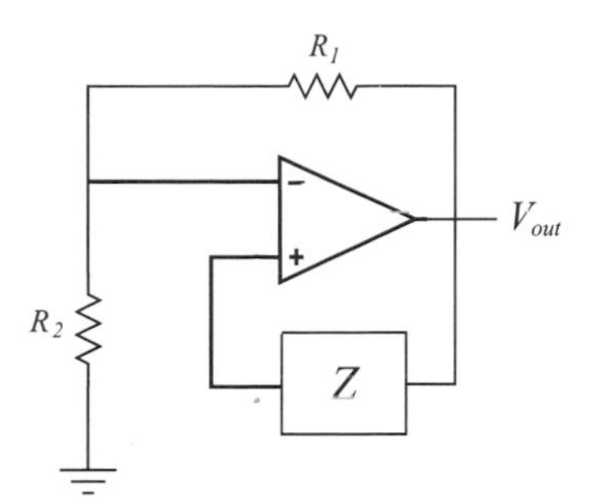

FIGURE 7.2. Oscillator using positive feedback to the noninverting input.

The signal returned through Z may be viewed simply as the voltage input to the noninverting configuration comprised of the op-amp, R_1 and R_2. Thus [see Eq. (5.4)],

$$V_{\text{out}} = \left[1 + \frac{R_1}{R_2}\right] V_+.$$

In a Wien-bridge oscillator, Z is a so-called lead–lag network of the type shown in Fig. 7.3, where V_{out} and V_+ match the symbols in the previous diagram.

In complex notation,

$$\frac{V_+}{V_{\text{out}}} = \frac{Z_b}{Z_a + Z_b} \tag{7.1}$$

with

$$Z_a = R - j\left(\frac{1}{\omega C}\right) \tag{7.2}$$

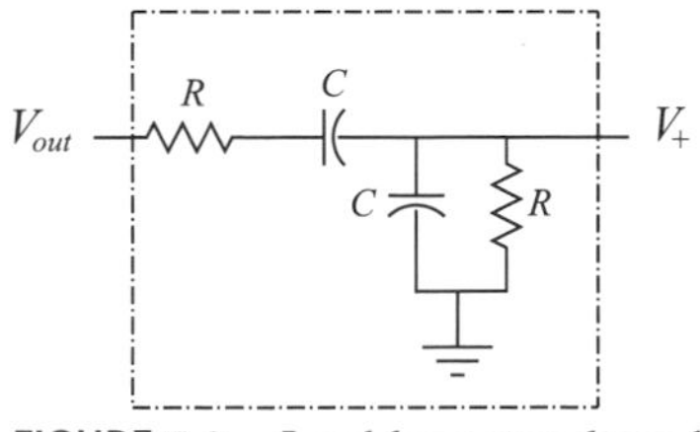

FIGURE 7.3. Lead-lag network used in the feedback path of the Wien-bridge oscillator.

and

$$Z_b = [R^{-1} + j\omega C]^{-1}. \tag{7.3}$$

After a little algebra, one finds for the magnitudes

$$\boxed{\left|\frac{V_+}{V_{\text{out}}}\right| = \frac{1}{\sqrt{9 + \left(R\omega C - \frac{1}{R\omega C}\right)^2}}} \tag{7.4}$$

and for the phase between V_+ and V_{out}

$$\boxed{\phi = \arctan\left[\frac{\frac{1}{R\omega C} - R\omega C}{3}\right]}. \tag{7.5}$$

The right-hand sides of Eqs. (7.4) and (7.5) are obviously frequency-dependent. At very low or high frequencies, the phase shift has limiting values of $+\frac{\pi}{2}$ (V_+ leads V_{out}) and $-\frac{\pi}{2}$ (V_+ lags V_{out}), respectively. The relative amplitude drops away at both low and high frequencies and reaches a maximum at the special value

$$\boxed{\omega_0 = \frac{1}{RC}} \tag{7.6}$$

for which $|V_+| = \frac{1}{3}\,|V_{\text{out}}|$. This can easily be seen by rewriting Eq. (7.4) as

$$\left|\frac{V_+}{V_{\text{out}}}\right| = \frac{1}{\sqrt{9 + \left(\frac{\omega}{\omega_0} - \frac{\omega_0}{\omega}\right)^2}}. \tag{7.7}$$

Similarly,

$$\phi = \arctan\left[\frac{\frac{\omega_0}{\omega} - \frac{\omega}{\omega_0}}{3}\right], \tag{7.8}$$

and it is evident that the phase shift is zero at $\omega = \omega_0$.

The lead–lag circuit thus displays resonance at the frequency ω_0. The Wien bridge of Fig. 7.2 will exhibit sinusoidal oscillations at this same frequency if, as discussed earlier, the amplifier closed-loop gain equals 3. This requirement is easily met by choosing $R_1 = 2R_2$.

The properties of the lead–lag network can be illustrated by a specific example. The schematic in Fig. 7.4 was drawn in PSpice. The resulting simulation data

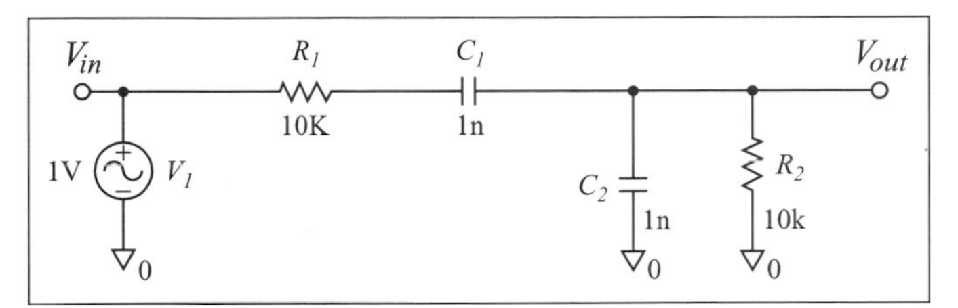

FIGURE 7.4. PSpice schematic of a lead–lag network.

are plotted in Fig. 7.5, and it is clear that a resonance occurs at $f = 15.9$ kHz, in agreement with Eq. (7.6). Also, as expected, at this frequency the phase shift of the network is 0.

Improved Wien-Bridge Oscillator

The preceding discussion assumed ideal behavior from the components, including the op-amp itself. In reality, deviation of resistor values from nominal, op-amp imperfections, and thermal drift, all imply that the Wien bridge might not actually generate stable oscillations at the resonant frequency. Revising basic oscillator designs so that stability ensues is a subject in itself. The following discussion is meant only to illustrate the process.

The filament in an incandescent bulb is a resistive element chosen for its ability to operate at elevated temperatures. An ideal resistor is described by constant R. In contrast, self-heating causes the filament resistance to increase in a manner

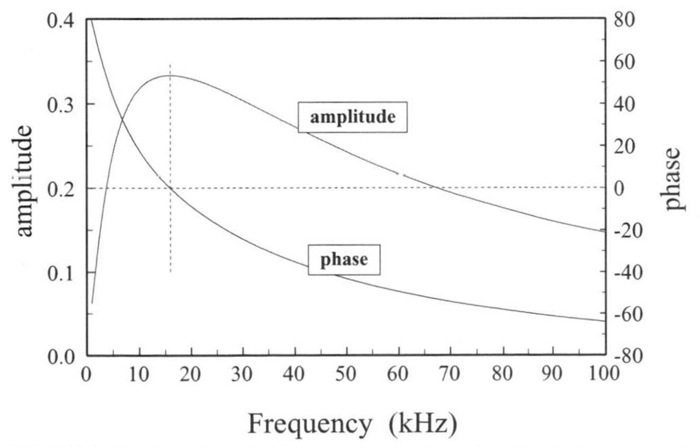

FIGURE 7.5. Results of a PSpice simulation of the lead–lag network. The vertical dotted line marks the resonant frequency at which the network is most transparent and the phase shift is zero.

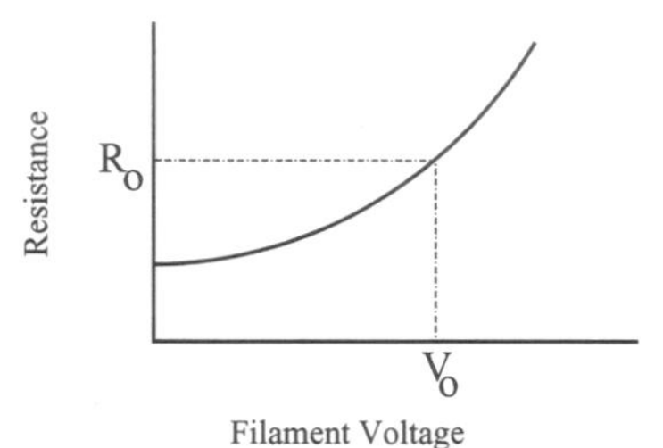

FIGURE 7.6. Hypothetical characteristic of the filament in an incandescent lamp. Points along the curve correspond to higher filament temperatures, and hence higher resistances.

depicted in Fig. 7.6. Suppose the operating point R_0, V_0 lies on the characteristic of a particular bulb, as indicated.

The bulb can now be substituted for one of the gain-setting resistors of the Wien bridge oscillator, as shown in Fig. 7.7.

With the nominal resistance of R_0, the closed-loop gain is 3, as required.

When the circuit is first turned on, the bulb is cold and its resistance R_{filament} is at a minimum well below R_0. The closed-loop gain $(1 + 2R_0/R_{\text{filament}})$ is thus greater than 3, which overcompensates for the lead–lag attenuation of $\frac{1}{3}$. This means that inevitably some small fluctuation will be amplified into larger and larger oscillations, so the circuit is self-starting.

Now, consider what happens if the system begins to drift away from the conditions for sustained oscillation. Suppose the net gain climbs above 3. Then, the oscillator output will increase, and so will the voltage appearing across the bulb. This in turn will result in an increase in filament resistance Fig. 7.6, and

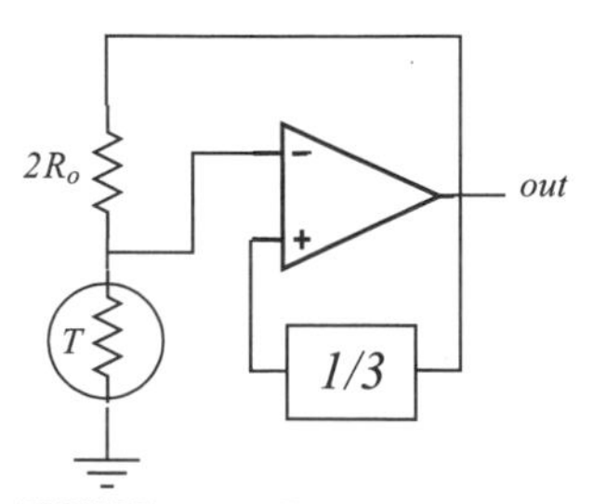

FIGURE 7.7. Improved Wien-bridge oscillator employing an incandescent lamp as a form of automatic gain control for self-starting and stability.

hence a decrease in closed-loop gain, so the output will drop and the system will move back to its intended bias point.

If, instead, the gain drifted below 3, then the oscillator output would temporarily drop, R_{filament} would become less than its target value R_0, and the closed-loop gain would increase, bringing the system back to its intended bias point.

Hence, this improved design for the Wien-bridge oscillator is thus both self-starting and stable.

Phase-Shift Oscillator

In the Wien-bridge oscillator, a portion of the output was fed back to the noninverting amplifier input. This then required that the network phase shift be zero for positive signal reinforcement. It is also possible to direct the feedback signal to the inverting amplifier input, provided the phase shift is, in such a case, 180 degrees. This is necessary because the inverting input itself introdces a further 180 degrees, bringing the combined total up to the positive feedback target of 360 degrees. Such an approach is taken in the phase-shift oscillator (Fig. 7.8).

The oscillator output is fed back to the inverting amplifier consisting of the op-amp together with R_1 and R_f. The closed-loop gain is $-\frac{R_f}{R_1}$ [see Eq. (5.18)].

The feedback network is the chain of resistors and capacitors. An analysis of the $R - C$ ladder yields

$$\frac{V_{\text{out}}}{V_{\text{in}}} = \frac{1}{\left(1 - \frac{5}{(2\pi f RC)^2}\right) - j\left(\frac{6}{2\pi f RC} - \frac{1}{(2\pi f RC)^3}\right)}, \tag{7.9}$$

where V_{in} is the voltage at the input to the ladder (i.e., the op-amp output). The factor in brackets following j is zero for

$$\boxed{f_0 = \frac{1}{2\pi\sqrt{6}RC}}. \tag{7.10}$$

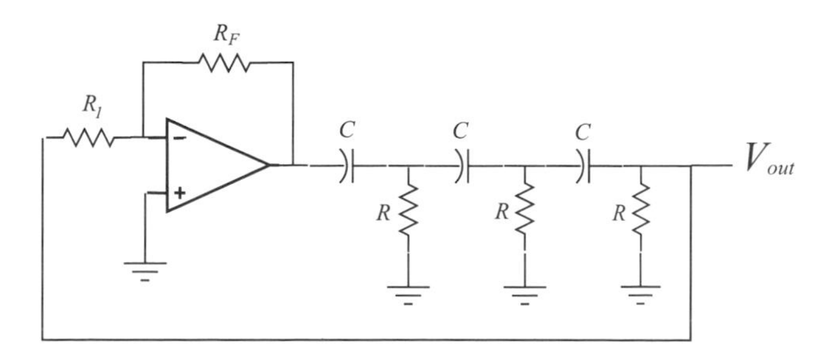

FIGURE 7.8. Phase-shift oscillator.

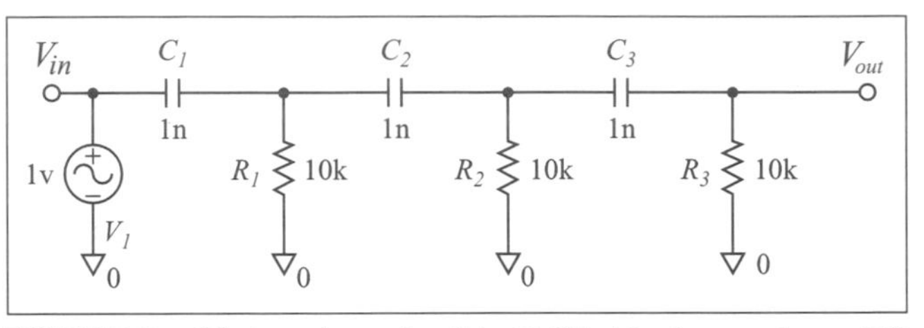

FIGURE 7.9. PSpice schematic of the R-C ladder from a phase-shift oscillator.

At this frequency, the relative output is

$$\frac{V_{\text{out}}}{V_{\text{in}}} = -\frac{1}{29}. \tag{7.11}$$

Therefore, the phase-shift oscillator will run at the frequency given by Eq. (7.10) when the compensating amplifier gain is set by

$$\frac{R_f}{R_1} = \frac{1}{29}. \tag{7.12}$$

As an example, consider the PSpice schematic in Fig. 7.9. The simulation results (Fig. 7.10) indicate a phase shift of 180 degrees at a frequency of 6.5 kHz, in agreement with Eq. (7.10). The relative output at this frequency is $\frac{1}{29}$, again as expected.

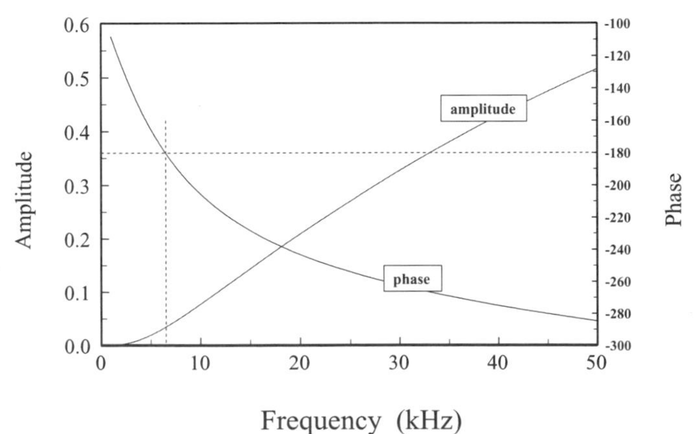

FIGURE 7.10. PSpice simulation results for the R-C ladder. The dotted vertical line marks the oscillator frequency f_0 at which the amplitude is $\frac{1}{29}$ and the phase shift is 180 degrees.

7.2 PULSE GENERATORS

As already noted, the Wien-bridge and phase-shift circuits produce sinusoidal voltage oscillations at the design frequencies. Continuous streams of square pulses are another commonly needed format. Such pulses may be *unipolar*, ranging up and down between a baseline of zero and some peak value, or *bipolar*, where the excursions are bounded between $\pm V_{\text{peak}}$. The repeat time for each pulse cycle is the *period*, and the ratio of time in the high-voltage state to the total period is the *duty cycle* of the pulse stream.

To illustrate the process of pulse generation, we now examine two well-known circuits.

Relaxation Oscillator

In the circuit shown in Fig. 7.11, the op-amp output is applied as positive feedback through the voltage divider composed of R_2 and R_3. Because of the positive feedback, the op-amp output will be driven rapidly to positive saturation $V_{\text{out}} = +V_{\text{max}}$ when $V_+ > V_-$ and will jump to negative saturation $V_{\text{out}} = -V_{\text{max}}$ when $V_+ < V_-$.

The voltage at the noninverting input is

$$V_+ = V_{\text{out}} \frac{R_3}{R_2 + R_3}. \tag{7.13}$$

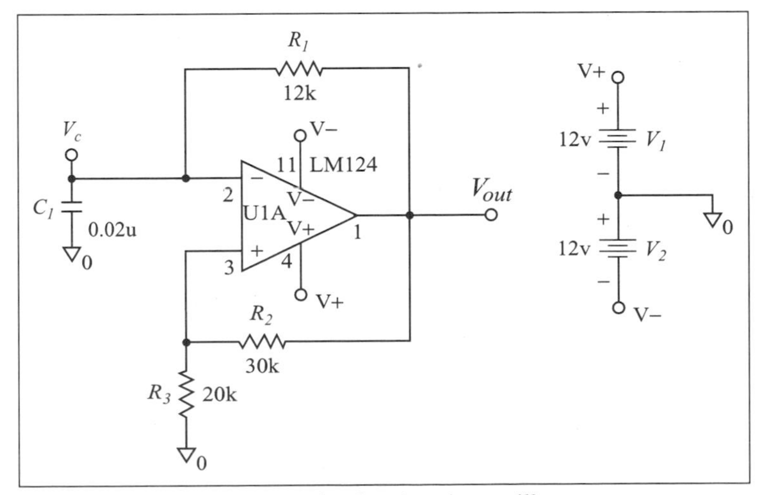

FIGURE 7.11. PSpice schematic of a relaxation oscillator.

To begin, suppose that V_{out} is at $+V_{max}$. Then V_+ will be the fraction of V_{max} specified by Eq. (7.13). The output voltage is also applied to the combination R_1, C_1, with the result that the capacitor charges through R_1. The time constant for this process is $\tau = R_1C_1$. The capacitor voltage V_{cap} (which is just V_-) will increase until it reaches $V_{max}\frac{R_3}{R_2+R_3}$, at which time the op-amp output will almost instantaneously switch down to $-V_{max}$. The voltage at the noninverting input to the op-amp is still given by Eq. (7.13) but is now a negative quantity, and the capacitor will begin to discharge through R_1. This process will continue until the capacitor voltage has dropped from $+[V_{max}\frac{R_3}{R_2+R_3}]$ to $-[V_{max}\frac{R_3}{R_2+R_3}]$. Then, the op-amp will again be driven into positive saturation and its output will switch to $+V_{max}$, and so on.

Thus, the capacitor voltage appears as a sequence of alternating charge and discharge intervals, and the op-amp output is a matching series of pulses with amplitude $\pm V_{max}$.

Consider a charging segment. The equation for the capacitor voltage as a function of time is

$$V_{cap}(t) = V_{max} - \left[V_{max}\frac{R_3}{R_2 + R_3} + V_{max}\right] e^{-\frac{t}{\tau}}. \tag{7.14}$$

As required, the initial voltage is $V_{cap}(0) = -[V_{max}\frac{R_3}{R_2+R_3}]$, while the limiting value is $V_{cap}(t \to \infty) = V_{max}$. The time ($T_1$) required for V_{cap} to rise to the upper switching value is obtained by setting the left-hand side of Eq. (7.14) to $+[V_{max}\frac{R_3}{R_2+R_3}]$. The result is

$$e^{-\frac{T_1}{\tau}} = \frac{V_{max}\left[1 - \frac{R_3}{R_2+R_3}\right]}{V_{max}\left[1 + \frac{R_3}{R_2+R_3}\right]} \tag{7.15}$$

or

$$e^{\frac{T_1}{\tau}} = \frac{R_2 + 2R_3}{R_2}. \tag{7.16}$$

From this, the period of the square waves, $T = 2T_1$, is obtained:

$$\boxed{T = 2(R_1C_1)\ln\left(1 + \frac{2R_3}{R_2}\right)}. \tag{7.17}$$

The results from a PSpice simulation are plotted in Fig. 7.12. The capacitor charging and discharging cycles are clearly seen. The expected period of the

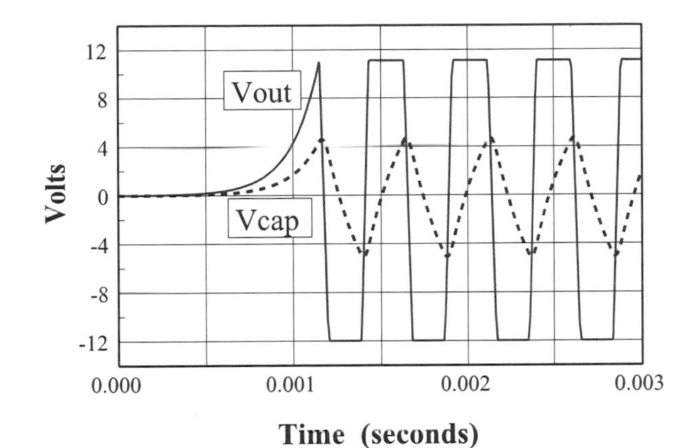

FIGURE 7.12. PSpice simulation results for the relaxation oscillator. An initial start-up phase lasts about a millisecond.

oscillations would be

$$T = 2\,(12\text{ K} \times 0.02\ \mu\text{F})\ \ln\left(1 + \frac{2 \times 20\text{ K}}{30\text{ K}}\right) = 0.407\text{ msec},$$

which agrees reasonably with the data in the figure.

555 Timer

For straightforward applications requiring square waves, the so-called 555 chip has become the integrated circuit of choice. This device is available from a number of manufacturers, one or two to a DIP package, and in both bipolar and CMOS versions. It is very economical and easy to use.

As depicted in Fig. 7.13, the 555 contains a pair of op-amp comparators, a flip-flop, an output amplifier, and a discharge transistor. There are in addition three matched resistors (5 KΩ) running from the positive power supply V_{cc} to ground. The external connections and components shown in this figure are particular to operating the 555 in its *astable* mode—that is, in a free-running state.

Basically, the oscillations are the result of repeated charge and discharge cycles of capacitor C. Charging toward V_{cc} through $R_1$and R_2 takes place when the transistor is "off." Discharging toward ground through just R_2 occurs when the transistor is "on."

Suppose a charging interval is in progress. The transistor is "off" and V_{cap} is rising. Due to the internal resistor chain, the comparators are set to trip at $\frac{1}{3}V_{cc}$ and $\frac{2}{3}V_{cc}$. When V_{cap} finally reaches $\frac{2}{3}V_{cc}$, the output of the upper comparator

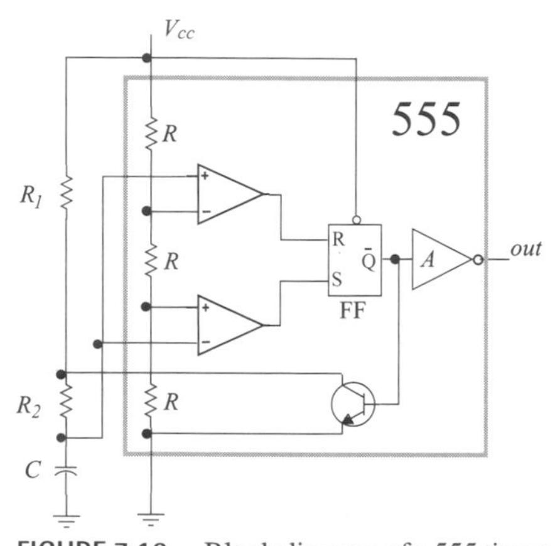

FIGURE 7.13. Block diagram of a 555 timer. The shaded boundary frames the chip contents; the remaining components are external to the IC.

will switch positive, thus resetting the flip-flop and causing $\bar{Q} = 1$. The resulting positive voltage on the base of the transistor switches it "on," initiating a discharge sequence.

The capacitor voltage will decay until it reaches $\frac{1}{3}V_{CC}$, at which point the lower comparator output will abruptly go positive, setting the flip-flop to $\bar{Q} = 0$. This low voltage on the base of the transistor will turn it "off," thus initiating a new charge cycle.

The equation for the capacitor voltage as it rises from $\frac{1}{3}V_{CC}$ towards a limiting value of V_{CC} is

$$V_{\text{cap}}(t) = V_{CC} - \left[-\frac{1}{3}V_{CC} + V_{CC} \right] e^{-\frac{t}{\tau_1}}, \tag{7.18}$$

where the charging time constant is $\tau_1 = (R_1 + R_2)C$. But of course the process abruptly terminates at the upper trip point of $\frac{2}{3}V_{CC}$. Hence, the time, T_1, for a charging segment can be obtained from

$$\frac{2}{3}V_{CC} = V_{CC} - \left[-\frac{1}{3}V_{CC} + V_{CC} \right] e^{-\frac{T_1}{\tau_1}}.$$

Thus,

$$T_1 = [(R_1 + R_2)C] \ln(2). \tag{7.19}$$

Similarly, the discharge through R_2 follows

$$V_{\text{cap}}(t) = \left[\frac{2}{3}V_{cc}\right] e^{-\frac{t}{\tau_2}}. \tag{7.20}$$

This terminates when V_{cap} reaches $\frac{1}{3}V_{cc}$ and the time required is

$$T_2 = [R_2 C] \ln(2). \tag{7.21}$$

The time T for a complete charge/discharge cycle for the 555 is then

$$\boxed{T = [(R_1 + 2R_2)\,C] \ln(2)}. \tag{7.22}$$

The 555 output will be high while $\bar{Q} = 0$; that is, during charging intervals. The ratio of the time during which the output is high to the total period of the repetitive square waves is the duty cycle.

$$\text{duty cycle} = \frac{T_1}{T_1 + T_2} = \frac{R_1 + R_2}{R_1 + 2R_2}. \tag{7.23}$$

Since $T_1 > T_2$, as is evident from Eqs. (7.19) and (7.21), the duty cycle can never fall below 50%, whatever the values of the two resistors may be.

A PSpice simulation with components $R_1 = 2.2$ K, $R_2 = 4.7$ K, and $C = 22$ nF produced the waveforms shown in Fig. 7.14. For these values, we expect $T_1 = 0.105$ msec, $T_2 = 0.072$ msec, and $T = 0.177$ msec, in good agreement with the observed data.

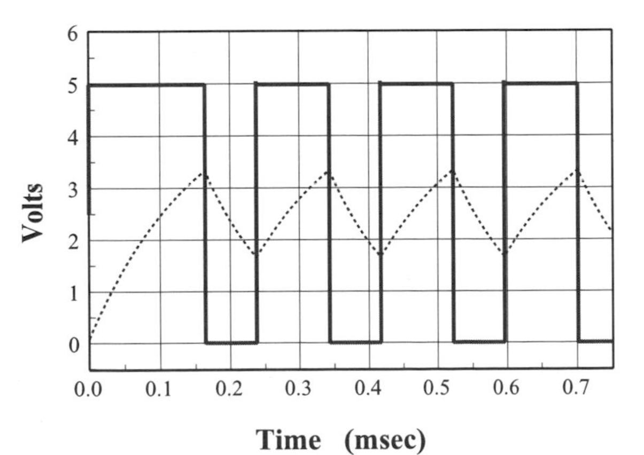

FIGURE 7.14. PSpice simulation results for a free-running 555 oscillator. The square wave is the final chip ouput; the other waveform is taken from the capacitor and shows charging and discharging cycles.

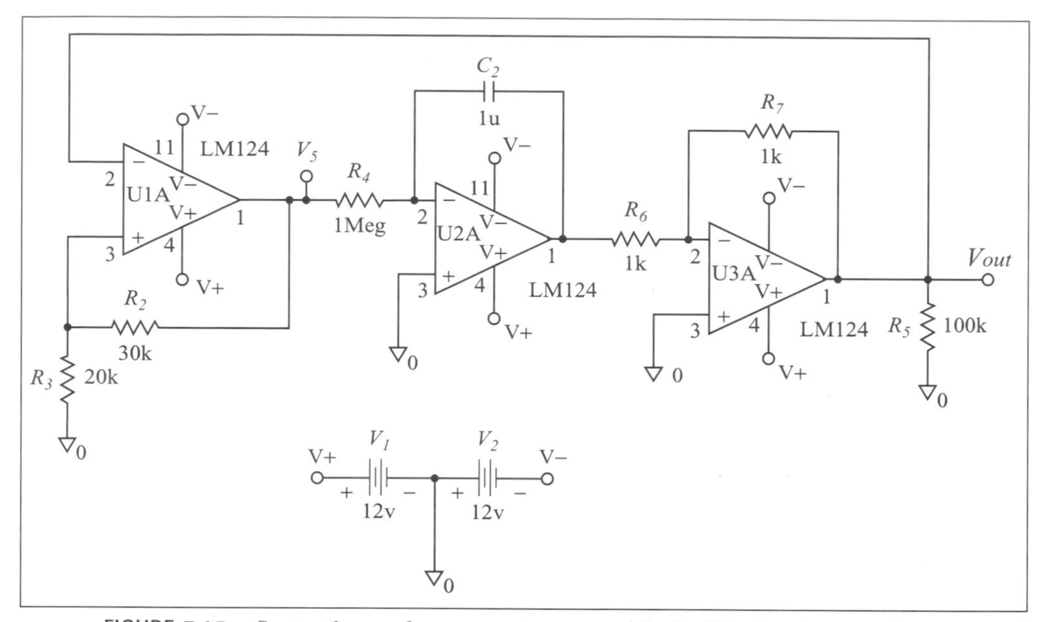

FIGURE 7.15. Sawtooth waveform generator composed of a Schmitt trigger, an integrator, and an inverter.

Sawtooth Waveform

When a dc voltage is applied to an integrator (see previous chapter), the output will be a linear ramp. This property can serve as the basis of a sawtooth waveform generator. Consider the PSpice schematic shown in Fig. 7.15.

As in the relaxation oscillator, the first op-amp acts as what is called a Schmitt trigger, its output rapidly switching to $\pm V_{\max}$ whenever the inverting input either just drops below or just rises above the fraction $\frac{R_3}{R_2+R_3}$ of the present output. Thus, there are two trip points

$$\text{UTP} = +\frac{R_3}{R_2 + R_3} V_{\max},$$

$$\text{LTP} = -\frac{R_3}{R_2 + R_3} V_{\max}.$$

The net result is that the output will remain at $+V_{\max}$ as long as the inverting input remains below the upper trip point (UTP). Once UTP is exceeded at the inverting input, the output will switch to $-V_{\max}$ and remain at that value until the inverting input drops below the lower trip point (LTP). Again, the output is constant as long as the inverting input remains below UTP, when switching once more will take place. The voltage interval between UTP and LTP is known as the *hysteresis* of the Schmitt trigger.

Let us suppose that the first op-amp is delivering $+V_{\text{max}}$ to the second op-amp, which is wired as an integrator. The output of this second op-amp satisfies Eq. (6.21), which after inversion by the third unit becomes

$$\frac{dV_{\text{out}}}{dt} = \frac{V_{\text{max}}}{R_4 C_2}, \tag{7.24}$$

so V_{out} ramps up linearly with slope $\frac{V_{\text{max}}}{R_4 C_2}$. This rising output will ultimately reach the upper trip point (UTP), and when it does the first op-amp will switch to $-V_{\text{max}}$ and the final output will ramp down according to

$$\frac{dV_{\text{out}}}{dt} = -\frac{V_{\text{max}}}{R_4 C_2}. \tag{7.25}$$

When the lower trip point (LTP) is reached, the first op-amp will return to $+V_{\text{max}}$ and ramping up will occur. Thus, the V_{out} will consist of alternating up and down ramps, which range from UTP to LTP. The time T_{up} or T_{down} required for either up or down ramping can be determined from the slopes.

$$\boxed{T_{\text{up}} = T_{\text{down}} = \left[\frac{2R_3}{R_2 + R_3}\right] R_4 C_2}. \tag{7.26}$$

For the example shown in the schematic, this gives

$$T_{\text{up}} = T_{\text{down}} = 0.8 \text{ sec}.$$

The results of a PSpice simulation of the circuit are plotted in Fig. 7.16.

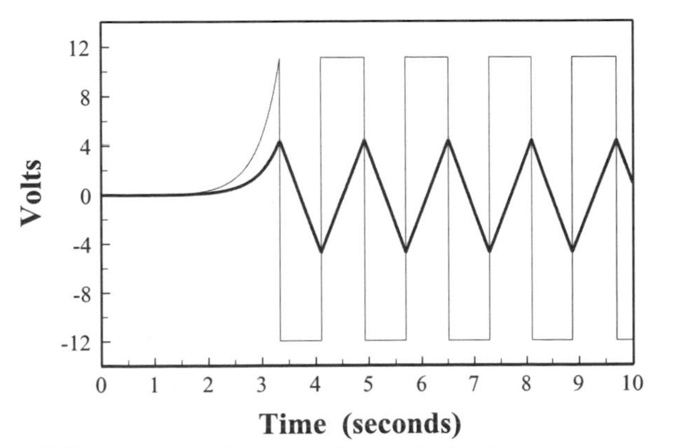

FIGURE 7.16. PSpice simulation results for the sawtooth generator. Both the Schmitt output (square) and final output (sawtooth) are shown. A start-up transient is present.

After an initial start-up, the expected triangular output waveform is observed. This sawtooth ranges between the upper and lower trip points, which for this example are at ±40% of the op-amp saturation voltage. The circuit also provides a square wave at the output terminal of the first op-amp, as illustrated in Fig. 7.16.

7.3 CRYSTAL OSCILLATORS

The oscillators discussed earlier in this chapter shared at least one important feature—a frequency-selective element placed in a feedback loop. For the Wien-bridge circuit, this element was a lead–lag network. Generally speaking, oscillators perform best when the frequency selection is "sharp." This usually results in precise tuning and stability.

Quartz crystals are a common choice for the required frequency-selective devices in oscillators. As a material, quartz is piezoelectric. This means that its particular atomic structure results in the generation of an electric potential whenever the crystal is mechanically deformed. The reverse is also true: the application of an electric field across the crystal results in a mechanical deformation. A consequence of these two properties is that a quartz crystal may be shaped by cleaving, cutting, grinding, and polishing until its precise dimensions support an electromechanical resonance. That is, for some specific frequency of electrical excitation, the wavelength of the mechanical oscillations induced by the piezoelectric effect matches a physical dimension of the crystal. Like the acoustic waves in an organ pipe, the system is "tuned" by its shape.

A quartz crystal for use in electronics is a two terminal device—a pair of leads emerge from the package. An equivalent circuit that captures most of the essential electrical properties of a crystal is shown in Fig. 7.17.

This consists of a series branch with resistance, capacitance, and inductance, together with a capacitive parallel branch. The component values for this equivalent circuit are of course dependent on the specific crystal.

As an example, the device model QZS32768 from the PSpice simulation library was examined; its impedance properties are shown in Fig. 7.18. There is a series resonance at 32,768 Hz and a parallel resonance slightly above that, as

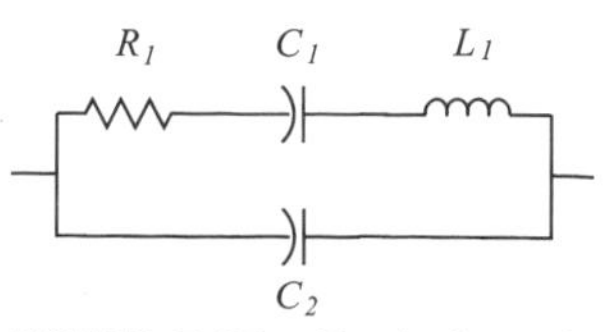

FIGURE 7.17. Equivalent circuit of a quartz crystal.

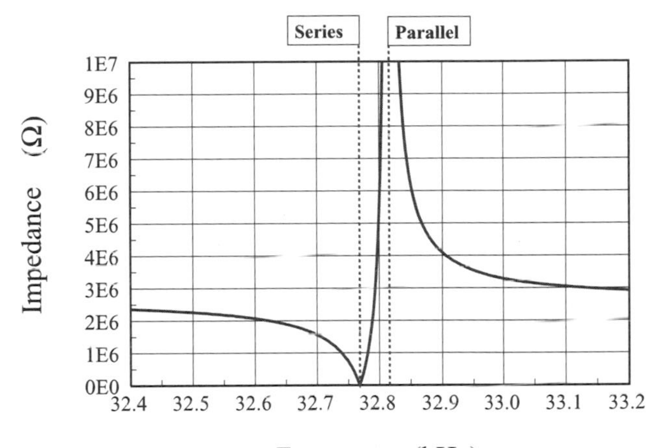

FIGURE 7.18. Impedance of PSpice library model QZS32768 crystal.

indicated in the figure. Note the sharpness of the resonances. (Figure 7.18 has a considerably expanded horizontal scale.)

In Fig. 7.19, this crystal is embedded in the positive feedback loop of an op-amp. For this PSpice simulation, the pulse generator V_3 was found necessary to kick-start the oscillator. Only a very brief single pulse was required for this purpose.

The results of the simulation are shown in Fig. 7.20. Both the signal V_{xtal} (light trace) and V_{out} (bold trace) are shown. As expected, the oscillations occur at the resonance of 32,768 Hz (period of 30.52 μsec).

7.4 REMARKS

The preceding sections have described just a few typical sine wave and sawtooth waveform generators. Many other circuits serve as the basis of both custom and commercial designs, and there are a number of books that provide more depth on this topic. These references can be sources of valuable technical detail, including coverage of specialized areas such as high-frequency and low-frequency oscillators, as well as techniques for improving stability, particularly with respect to thermal drift.

The advent of microprocessors and custom integrated circuits has drastically altered the face of instrumentation, especially in terms of the range of commercial products now available. These are in reality complex systems (although they are not necessarily costly), and not the sort of simple circuit that could easily be replicated on a workbench.

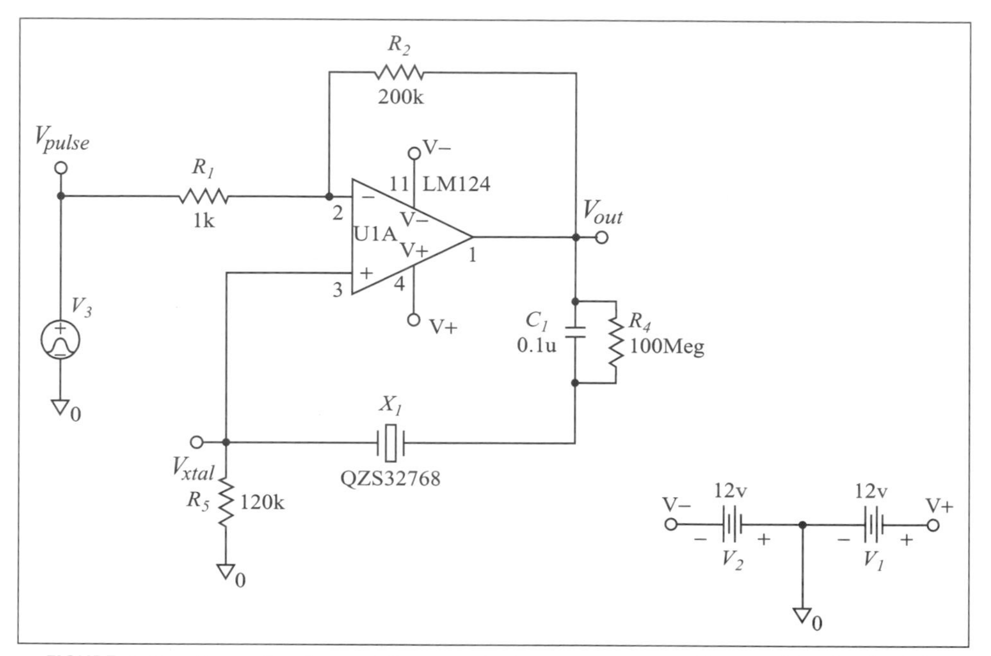

FIGURE 7.19. Example of an oscillator employing a quartz crystal in a positive feedback loop. The pulse generator was needed to initiate oscillations.

Arbitrary waveform generators fall within this category. As the name implies, an arbitrary waveform generator is able to create a repetitive signal of any desired shape. The user must deliver to the instrument a prescription of the waveform, usually as a string of perhaps thousands of numbers representing a discrete sampling of the waveshape. This data array then resides in system memory within the

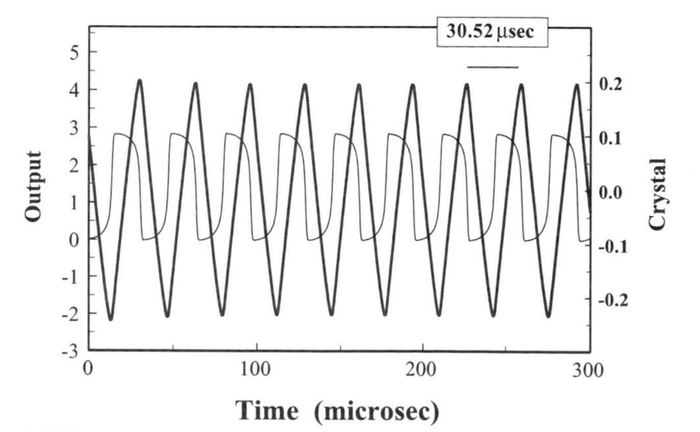

FIGURE 7.20. PSpice simulation of a crystal oscillator. The circuit output is the bold waveform. Also indicated is the time-dependent voltage at the top of R_5.

instrument and is used in combination with digital-to-analog converters (DAC) and controllers to synthesize the final signal.

Performance is dependent upon the speed of the DACs, the number of bits allocated to each memory location, and the size of memory. For example, suppose an instrument has a memory array consisting of 1024 8-bit words. The vertical resolution of any synthesized waveform will then be 1 part in $2^8 = 256$, or better than $\frac{1}{2}$%. Further, suppose for this example that the maximum conversion speed of the DACs is 100 nsec. One complete sweep of the stored waveform would then take at least $1024 \times 10^{-7} = 0.1024$ msec. This corresponds to a maximum output frequency from the generator of 9765 Hz.

PROBLEMS

Problem 7.1. Choose component values for the lead–lag network shown in Fig. 7.4 so that it can be used in a Wien-bridge oscillator with a target frequency of 50 kHz.

Problem 7.2. Choose component values for the R-C ladder shown in Fig. 7.9 so that it can be used in a phase-shift oscillator intended to run at 25 kHz.

Problem 7.3. Select new component values in the relaxation oscillator shown in Fig. 7.11 so that the output has a period of 1.00 msec.

Problem 7.4. Select external components for the 555 timer so that it oscillates with a period of 0.50 msec .

Problem 7.5. Using the schematic of Fig. 7.15, design a sawtooth waveform generator that will run at 5 Hz.

CHAPTER 8

Filters

For our purposes, a filter is any electrical network through which passes a complex time-varying signal $F(t)$ and acts on this signal to selectively alter its harmonic makeup.

Fourier's theorem was briefly mentioned at the beginning of Chapter 3. The theorem shows how any periodic waveform $[F(t+T) = F(t)]$ can be synthesized from a sum of appropriately chosen harmonics of the fundamental periodicity $[\omega_n = n\omega_0 = n\frac{2\pi}{T}]$. Because of this, any signal may be thought of in terms of its equivalent Fourier spectrum—the frequency distribution of amplitudes C_n and phase shifts θ_n as implied by Eq. (3.4).

A filter, then, effectively modifies the C's and θ's in some prescribed manner. The standard filter categories are: low-pass, high-pass, band-pass, and band-stop.

As the names suggest (see Fig. 8.1), certain frequencies are either blocked or passed through.

From an instrumentation point of view, there are many reasons for filtering a raw signal. A common motivation is the need to remove noise. Because noise is typically broadband and particularly noticeable at high frequencies, carefully chosen low-pass filtering can clean up the signal in situations where the information content is primarily at the low end of the spectrum. Low-pass filtering can also reveal slow trends that may be masked by clutter at higher frequencies. Band-pass filters can be useful in isolating a specific frequency interval within which some phenomenon is expected to occur, such as mechanical resonances in a vibrating structure monitored by strain gages.

Ideal filters usually have abrupt transitions ("brick walls") at the edges of pass and stop bands. That is, frequencies are either completely passed through or totally blocked. Such perfection can never be achieved in practice. Designs that manage to produce relatively sharp transitions in the frequency domain are accompanied by ripple just below and just above the edge. The sharper the

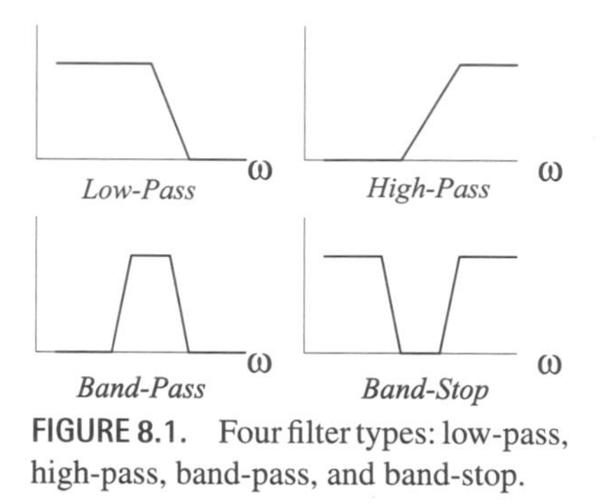

FIGURE 8.1. Four filter types: low-pass, high-pass, band-pass, and band-stop.

transition, the more the ripple. Improved sharpness without increased ripple can only be achieved with increased complexity in the filter design and with the addition of more reactive components to the filter (higher "order"). The essence of filter design is in balancing the tradeoffs between these competing factors and in assessing which attributes are most important for a given instrumentation application.

Electrical filters constructed entirely from resistors, capacitors, and inductors are termed *passive*, in contrast to filters employing transistors or op-amps, which are *active*.

8.1 PASSIVE FILTERS

Passive filters can be assembled from as few as two components, although increasingly complex designs permit more precise control of the frequency response.

First-Order Filters

Low-pass RC filter

Consider a series RC circuit as shown in Fig. 8.2. The voltage across the capacitor is given by the complex expression

$$\frac{V_{\text{cap}}}{V_{\text{in}}} = \frac{I\left(\frac{-j}{\omega C}\right)}{I\left(R + \frac{-j}{\omega C}\right)}, \tag{8.1}$$

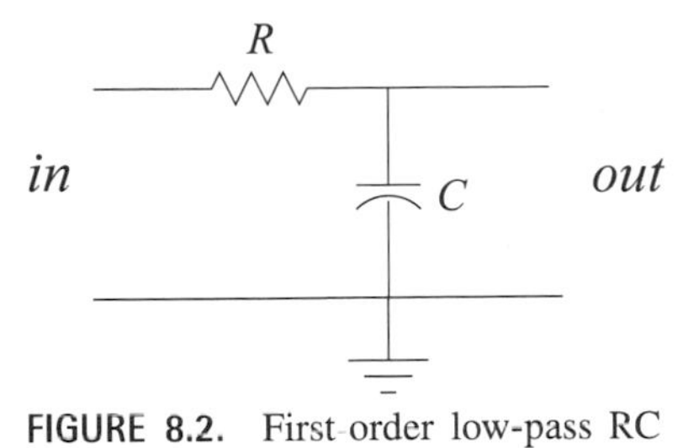

FIGURE 8.2. First-order low-pass RC filter.

where I is the ac current flowing around the loop and V_{in} represents the input source voltage. This equation may be arranged to

$$\frac{V_{\text{cap}}}{V_{\text{in}}} = \frac{1}{1 + j\omega CR} \tag{8.2}$$

or, in complex form with amplitude and phase,

$$\frac{V_{\text{cap}}}{V_{\text{in}}} = A\, e^{j\phi}. \tag{8.3}$$

In this equation,

$$\boxed{A = \frac{1}{\sqrt{1 + \omega^2 C^2 R^2}}} \tag{8.4}$$

and

$$\boxed{\phi = -\tan^{-1}(\omega CR)}. \tag{8.5}$$

Two limits are of interest. At low frequency, $A \simeq 1$, whereas at high frequencies $A \to (\omega CR)^{-1}$.

In the section on frequency response in Chapter 5, the concept of decibels was introduced. In these units, the limits are $A \simeq 0$ dB, and $A \to -20 \log(\frac{\omega}{\omega_c})$ dB. The corner frequency is defined as

$$\boxed{\omega_c = \frac{1}{RC}}. \tag{8.6}$$

As an example, let $R = 1$ kΩ and $C = 0.1$ μF. Then, $\omega_c = 10^4$ radians/sec, which corresponds to a corner frequency of $f_c = 1592$ Hz. PSpice results are plotted in Fig. 8.3, where this corner point is marked with a solid square. Note

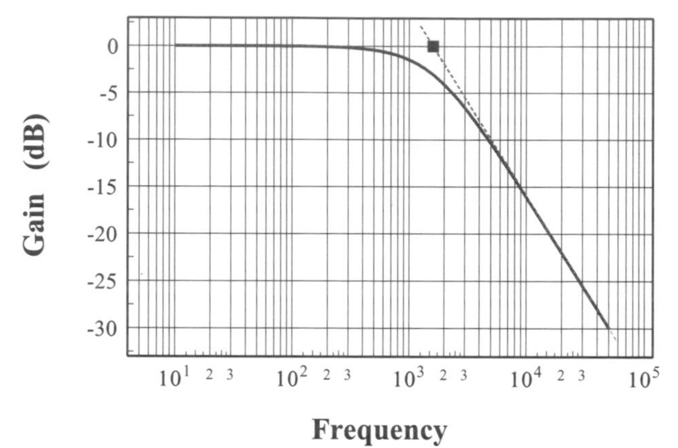

FIGURE 8.3. Frequency response for the RC low-pass filter. The vertical scale is logarithmic (decibels).

that at f_c the first-order response (gain) has dropped by 3 dB. The nearly linear rolloff well above f_c is quite apparent.

Clearly, this simple RC network functions as a low-pass filter. Well below the corner frequency, there is almost no attenuation of the input. Note, however, that there is no abrupt cutoff, but rather high frequencies are increasingly attenuated at the rate of 20 dB per decade. This means that any jump in frequency by a factor of 10 is accompanied by a matching drop of amplitude also by a factor of 10. This rate of 20 dB per decade can also be stated in equivalent terms as a rolloff of 6 dB per octave, an octave being any frequency jump by a factor of 2. These attenuation features are characteristic of a so-called *first-order* response.

Equation (8.5) describes the frequency-dependent phase shift that accompanies the filter attenuation. This behavior for the same PSpice simulation is illustrated in Fig. 8.4. In the limit $\omega \to 0$, the phase shift approaches zero. At high frequencies, this equation yields $\phi \to -90$ degrees. The negative sign in the phase angle indicates that the output from the low-pass filter lags the input. At the corner frequency of 1592 Hz, the phase shift is -45 degrees.

Low-pass RL filter

A simple two-component first-order low-pass filter can also be constructed from a resistor and an inductor. In this case, the positions of resistor and inductor are as illustrated in Fig. 8.5.

An analysis similar to that of the preceding section leads to the expression

$$\frac{V_R}{V_{\text{in}}} = \frac{1}{1 + j\,\frac{\omega L}{R}} \tag{8.7}$$

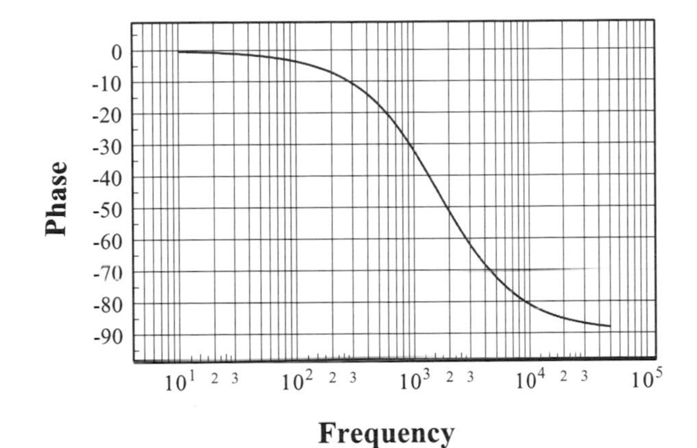

FIGURE 8.4. Phase response of first-order low-pass RC filter.

and thus

$$\frac{V_R}{V_{\text{in}}} = A\, e^{j\phi}, \tag{8.8}$$

with

$$\boxed{A = \frac{1}{\sqrt{1 + \frac{\omega^2 L^2}{R^2}}}} \tag{8.9}$$

and

$$\boxed{\phi = -\tan^{-1}\left(\frac{\omega L}{R}\right)}. \tag{8.10}$$

The corner frequency will be

$$\boxed{\omega_c = \frac{R}{L}}. \tag{8.11}$$

As an example, let $R = 1\ \text{k}\Omega$ and $L = 0.1$ H. This choice will set the corner frequency at the same value, $f_c = 1592$ Hz, as for the previous RC filter. It is

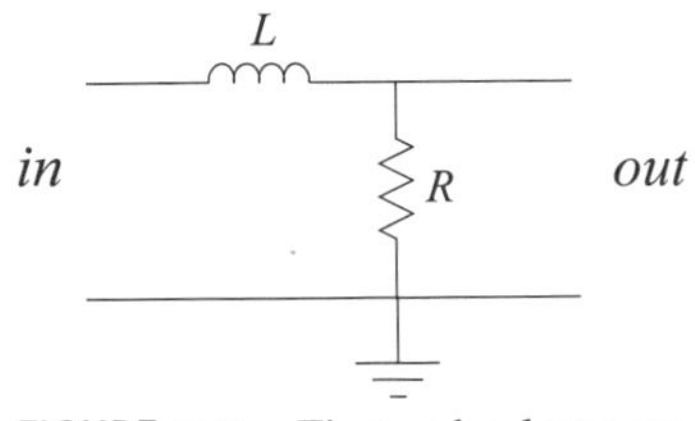

FIGURE 8.5. First-order low-pass LR filter.

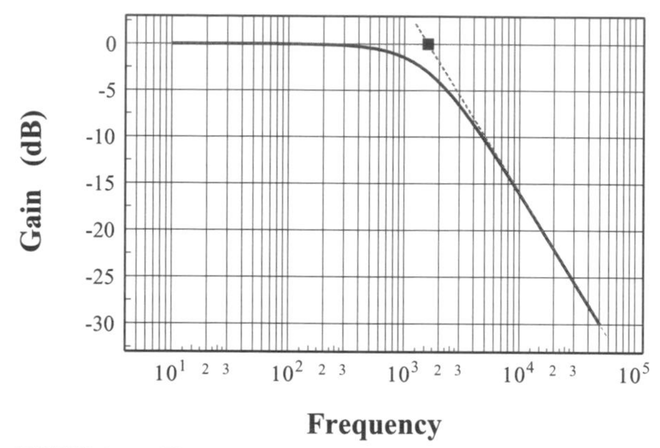

FIGURE 8.6. Frequency response of the RL filter.

immediately evident in the PSpice simulation results of Fig. 8.6 that the response characteristics of a low-pass *RL* filter are identical to those of a low-pass *RC* circuit (Fig. 8.3).

High-pass RC filter

Suppose the two components in the circuit of Fig. 8.2 are interchanged as in Fig. 8.7. The output voltage taken across the resistor is

$$\frac{V_R}{V_{\text{in}}} = \frac{1}{1 - \frac{j}{\omega CR}}. \tag{8.12}$$

As in the previous cases, let

$$\frac{V_R}{V_{\text{in}}} = A\, e^{j\phi}. \tag{8.13}$$

Then,

$$\boxed{A = \frac{1}{\sqrt{1 + \frac{1}{\omega^2 C^2 R^2}}}} \tag{8.14}$$

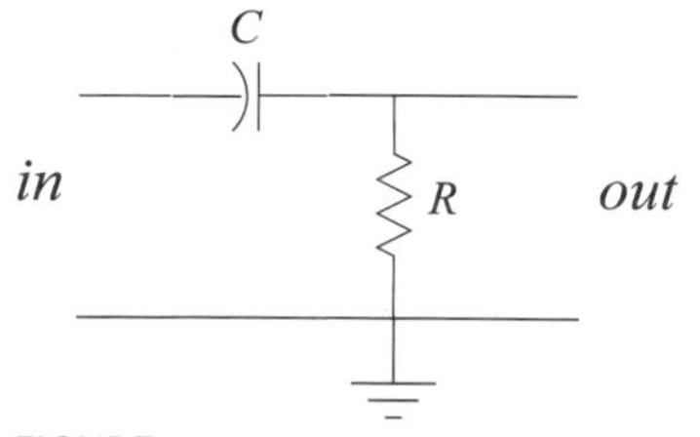

FIGURE 8.7. First-order high-pass RC filter.

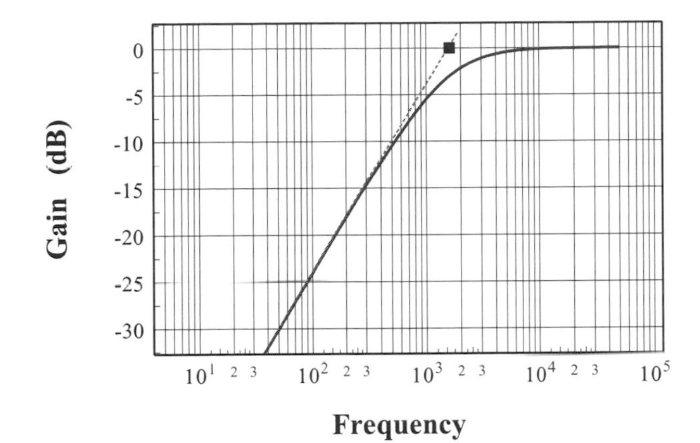

FIGURE 8.8. Frequency response of the first-order high-pass filter. As with the low-pass characteristic, the response is down 3 dB at the corner frequency.

and

$$\boxed{\phi - \tan^{-1}\left(\frac{1}{\omega CR}\right)}. \tag{8.15}$$

These two expressions can be compared to Eqs. (8.4) and (8.5), which applied to the low-pass filter. Note that in the present situation, the limits are the reverse of those previously found—that is, now at high frequencies $A \simeq 1$, whereas in the limit of low frequencies $A \to \omega CR$, which drops towards zero. Data from a PSpice simulation with $R = 1\ \text{k}\Omega$ and $C = 0.1\ \mu\text{F}$ are plotted in Fig. 8.8, where it is clear that this is a high-pass filter with the same corner frequency $\omega_c = \frac{1}{RC}$ as for the earlier low-pass circuit (see Fig. 8.3).

The phase shift as expressed by Eq. (8.15) approaches 90 degrees as $\omega \to 0$ and drops towards zero at very high frequencies, as if the curve in Fig. 8.4 were displaced vertically by 90 degrees. Because the phase factor ϕ is positive, the output from the high-pass filter leads the input.

Band-pass filter

It is readily apparent that a bandpass filter can be simply constructed from the product of suitable low-pass and high-pass filters, as indicated in Fig. 8.9. To illustrate this procedure, consider the circuit shown in Fig. 8.10.

Here, a high-pass filter made up of R_1 and C_1 is followed by a low-pass filter made up of R_2 and C_2. The order of the two does not matter. To choose a specific example, let $R_1 = 1$ K, $C_1 = 1.0\ \mu$F, $R_2 = 1$ K, and $C_2 = 0.1\ \mu$F. In this case,

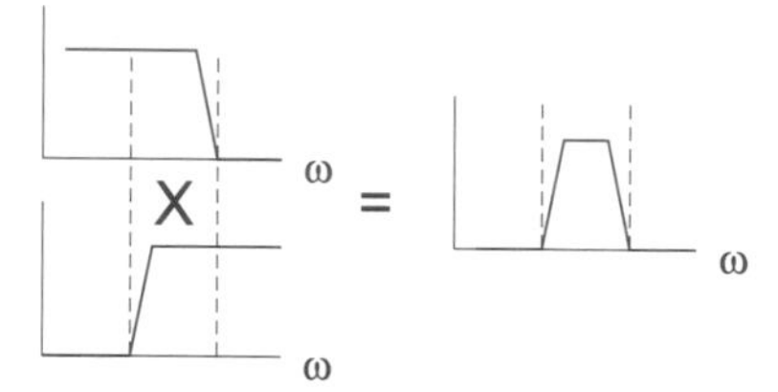

FIGURE 8.9. Synthesis of a band-pass filter from the product of a low-pass filter and a high-pass filter.

the lower corner frequency is set by the high-pass relationship

$$f_{cL} = \frac{1}{2\pi R_1 C_1} = 159.2 \text{ Hz},$$

and the upper corner point determined from the low-pass components is at

$$f_{cU} = \frac{1}{2\pi R_2 C_2} = 1592 \text{ Hz}.$$

A PSpice simulation produced the data plotted in Fig. 8.11. The expected band-pass behavior is clearly evident with dropoffs of 20 dB per decade at low and high frequencies.

The resulting phase shift from this composite filter is seen in Fig. 8.12. Because of the series connection, the phase shifts from each of the two filter blocks will simply add. Thus, it is expected that the overall phase shift of the band-pass design will be the sum of the phases expressed by Eqs. (8.5) and (8.15). At low frequencies, the phase-shifting properties of the high-pass filter dominate, while at high frequencies the phase shift is controlled by the low-pass section. As the figure demonstrates, below the passband midpoint the filter output signal leads the input signal; above the passband midpoint, the opposite is true. The

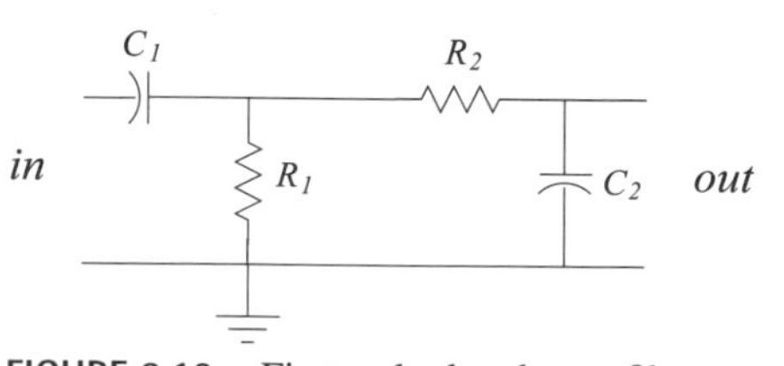

FIGURE 8.10. First-order band-pass filter using a combination of low-pass and high-pass sections.

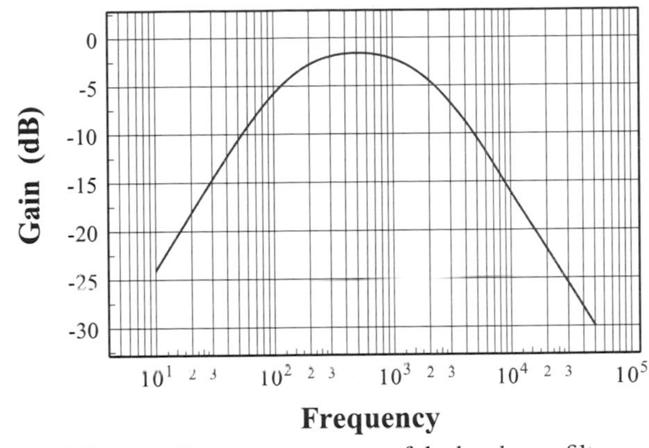

FIGURE 8.11. Frequency response of the band-pass filter.

neighborhood near the center of the passband is characterized by small phase shifts—clearly a desirable attribute.

Band-stop filter

The idea underlying the band-pass design was illustrated in Fig. 8.9. The product (series connection) of a low-pass and a high-pass filter generates an overall band-pass characteristic. A similar argument shows that a sum (parallel connection) of a low-pass module and a high-pass module can create a band-stop design, as suggested in Fig. 8.13.

In principle, this is straightforward, but in practice there is a complication. This arises from the need for an electronic summing operation to link the two filters.

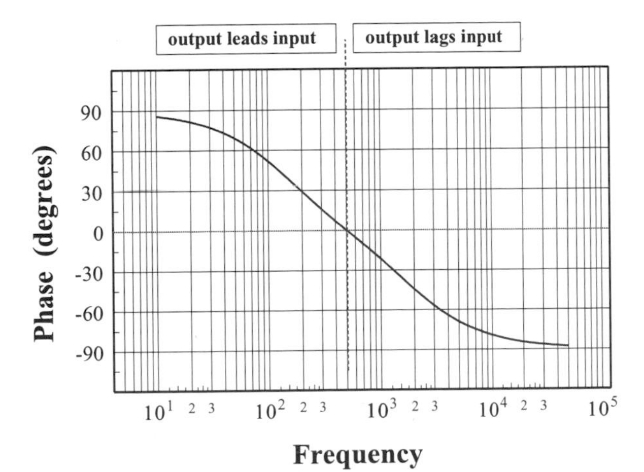

FIGURE 8.12. Phase response of the band-pass filter.

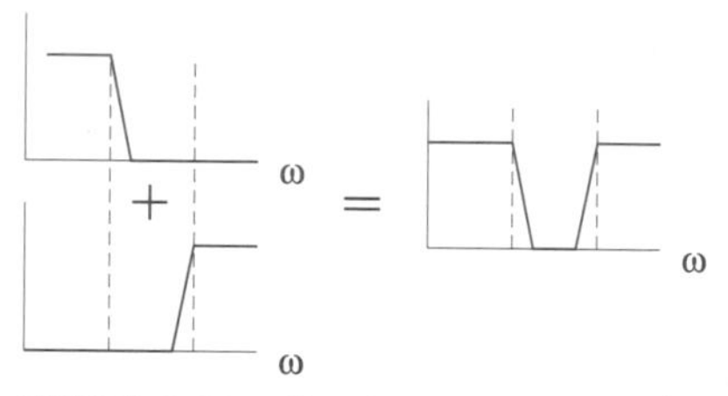

FIGURE 8.13. Synthesis of a band-stop filter from the sum of a low-pass filter and a high-pass filter.

But recall the summing amplifier (see Fig. 5.10) that was discussed in Chapter 5. As illustrated by the example in Fig. 8.14, this circuit allows the realization of the band-stop function. The combination R_1, C_1 is a low-pass first-order filter, whereas R_2 and C_2 comprise a high-pass first-order filter. The filter outputs are buffered by $U1A$ and $U2A$ and then summed by $U3A$. The corner points $f_c = (2\pi RC)^{-1}$ for these two filters are, for this example, 15.92 Hz and 3183 Hz, respectively. PSpice simulation results are shown in Fig. 8.15.

Note that for this plot a linear rather than logarithmic (decibel) amplitude scale has been chosen. This more clearly indicates the connection between the low-pass, high-pass, and band-stop (sum) characteristics. The center frequency of this filter is 226 Hz.

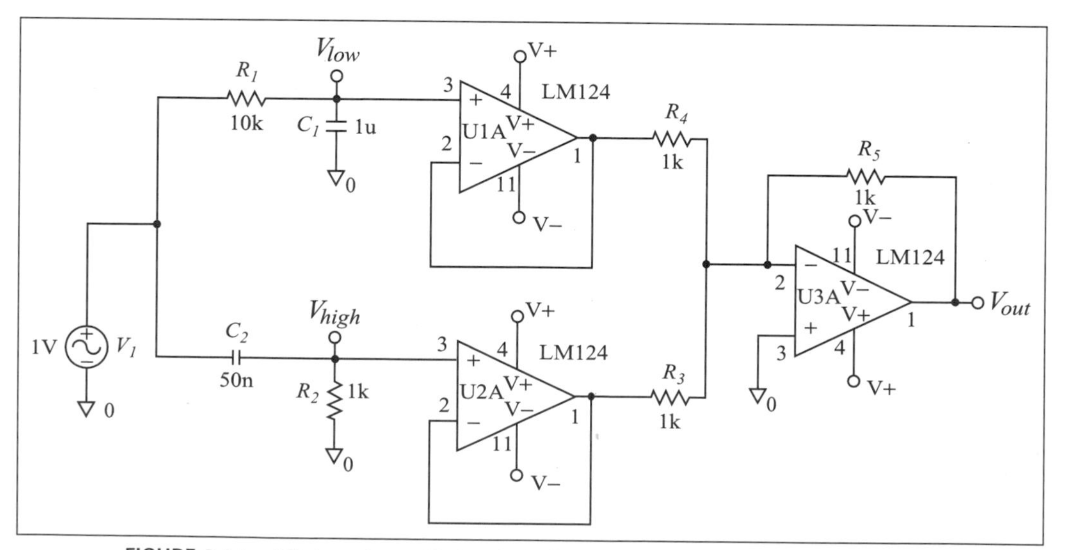

FIGURE 8.14. PSpice schematic showing a low-pass filter and a high-pass filter being combined in a summing amplifier to form a band-stop filter.

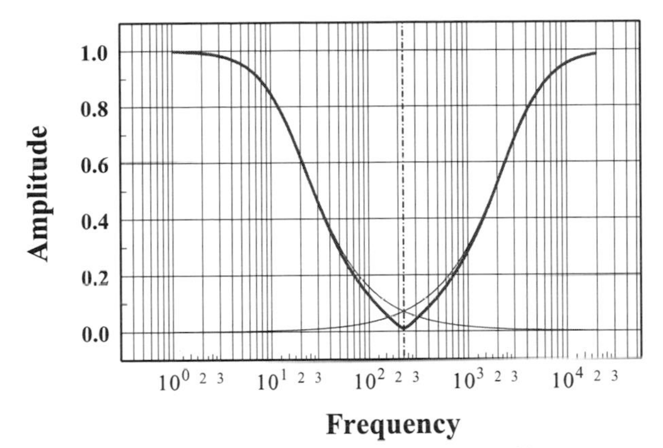

FIGURE 8.15. PSpice results for the band-stop filter.

Second-Order Filters

Cascaded filters

The band-pass filter discussed earlier was composed of a pair of *RC* units connected in series to form a composite network. It is, of course, possible to chain together any number of filter modules, just as one may chain together amplifier modules to increase gain. Suppose we have a series array of elements as depicted in Fig. 8.16.

Each unit operates on its input in the usual manner so that

$$
\begin{aligned}
V_1 &= V_{\text{in}} \left(A_1\, e^{j\phi_1}\right) \\
V_2 &= V_1 \left(A_2\, e^{j\phi_2}\right) \\
V_3 &= V_2 \left(A_3\, e^{j\phi_3}\right) \\
&\text{etc.}
\end{aligned}
$$

so

$$\frac{V_{\text{out}}}{V_{\text{in}}} = (A_1 A_2 A_3 \ldots\ldots)\; e^{j(\phi_1+\phi_2+\phi_3+\ldots..)}. \tag{8.16}$$

This expression shows that the final magnitude is determined by the product of the individual gains, while the net phase shift is the sum of the individual phase

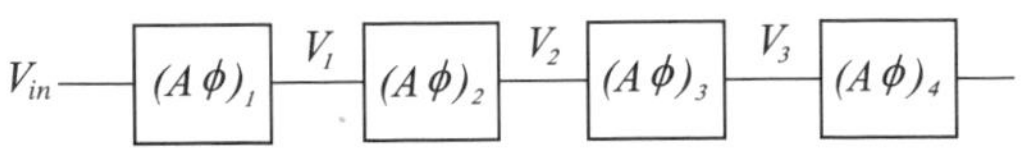

FIGURE 8.16. Cascaded modules, each with gain A and phase shift ϕ.

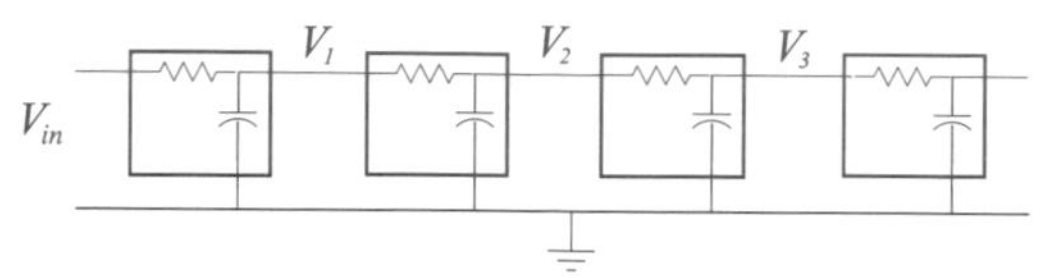

FIGURE 8.17. Cascaded first-order low-pass RC filters.

shifts. Noting that

$$20 \log (A_1 A_2 A_3.....) = 20 \log A_1 + 20 \log A_2 + 20 \log A_3 + ..., \tag{8.17}$$

it is apparent that if the voltage ratio is measured in decibels, as defined by Eq. (5.34), then the effective gain for the chain is

$$G\,(\text{dB}) = G_1 + G_2 + G_3 + ... \,. \tag{8.18}$$

Therefore, when measured in dB, the amplification (or attenuation) factors simply add.

Let us now apply these ideas using simple low-pass *RC* filter modules. To select a specific example, choose $R = 1\ \text{k}\Omega$ and $C = 0.1\ \mu\text{F}$ as in Fig. 8.2, and link the units together as in Fig. 8.17.

Actually, there is a slight problem here that must be addressed. As drawn in Fig. 8.17, any given segment will be loaded by the effect of the succeeding stages. In other words, the filters are not acting as a purely multiplicative chain in the spirit of Eq. (8.16) because each A is modified by the fact that additional circuitry is placed across its output. This can be corrected by inserting buffers between filter segments, as illustrated in Fig. 8.18.

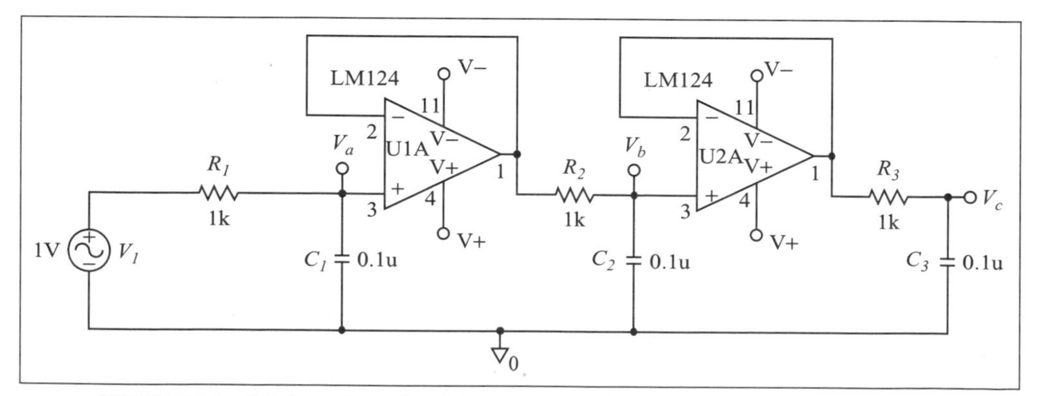

FIGURE 8.18. PSpice schematic of three cascaded first-order low-pass RC filters. The sections are separated by unity-gain buffers.

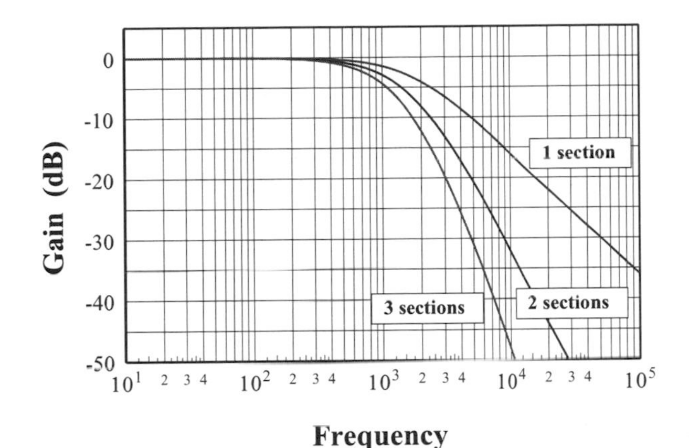

FIGURE 8.19. PSpice simulation results for the cascaded low-pass filters.

The PSpice outputs from each of the three stages V_a, V_b, and V_c are plotted in Fig. 8.19. The rolloffs are as anticipated, 20, 40, and 60 dB per decade (or 6, 12, and 18 dB per octave), indicating first-order, second-order, and third-order filter characteristics. Thus, the edges of the characteristic can be sharpened by adding filter stages.

Low-pass RLC filter

A second-order filter is created when two independent reactive components are present. In the last example, this was caused by the pair of capacitive blocks. A slightly different filter can be formed from series R and L, together with a shunting capacitance, as in Fig. 8.20. Then,

$$\frac{V_{\text{out}}}{V_{\text{in}}} = \frac{-\frac{j}{\omega C}}{R + j\omega L - \frac{j}{\omega C}} \tag{8.19}$$

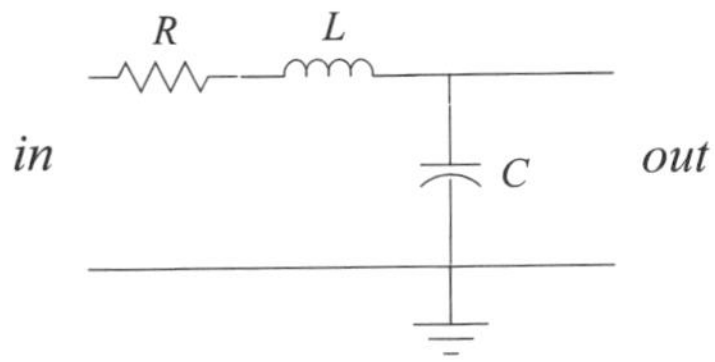

FIGURE 8.20. Second-order RLC filter.

or

$$\boxed{\frac{V_{\text{out}}}{V_{\text{in}}} = \frac{1}{(1-\omega^2 LC) + j(\omega RC)}}. \tag{8.20}$$

Suppose we define

$$\boxed{Q = \frac{1}{R}\sqrt{\frac{L}{C}}} \tag{8.21}$$

and

$$\boxed{\omega_0 = \frac{1}{\sqrt{LC}}}. \tag{8.22}$$

Then, Eq. (8.20) becomes

$$\frac{V_{\text{out}}}{V_{\text{in}}} = \frac{1}{\left(1-\left[\frac{\omega}{\omega_0}\right]^2\right) + j\left(\frac{\omega}{\omega_0}\frac{1}{Q}\right)}. \tag{8.23}$$

In this form, it is clear that although the circuit has three components, in fact there are only two independent variables: ω_0 and Q. The magnitude of the complex voltage ratio in Eq. (8.23) is

$$\left|\frac{V_{\text{out}}}{V_{\text{in}}}\right| = \frac{1}{\sqrt{\left(1-\left[\frac{\omega}{\omega_0}\right]^2\right)^2 + \left(\frac{\omega}{\omega_0}\frac{1}{Q}\right)^2}}. \tag{8.24}$$

The behavior of this function at high frequencies is dominated by the factor $[\frac{\omega}{\omega_0}]^4$, which occurs in the square root, and hence

$$\lim_{\omega\to\infty}\left|\frac{V_{\text{out}}}{V_{\text{in}}}\right| = \left(\frac{\omega_0}{\omega}\right)^2. \tag{8.25}$$

Noting then that $20\log|\frac{V_{\text{out}}}{V_{\text{in}}}|$ tends to $40\log(\frac{\omega_0}{\omega})$, the high-frequency rolloff is seen to be 40 dB per decade. Hence, this is a second-order low-pass filter.

For the components used in this example ($R = 1$ K, $L = 0.1$ H, $C = 0.1\ \mu$F), $Q = 1$ and $\omega_0 = 10^4$ ($f_0 = 1592$ Hz). The results from a PSpice simulation are shown in Fig. 8.21. Both decibel (left scale; solid curve) and linear (right scale;

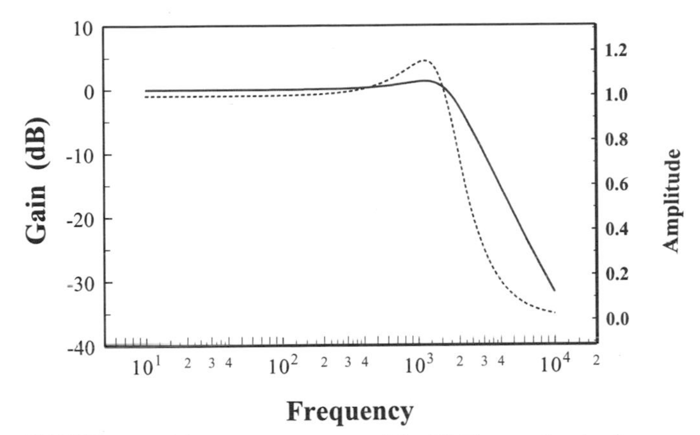

FIGURE 8.21. Frequency response of the RLC second-order low-pass filter showing the peak for $Q = 1$. The dashed curve is the same function but plotted with a linear scale (right) to bring out detail around the peak.

dashed curve) representations are given. The linear plot emphasizes the resonant peak in the filter function, whereas the logarithmic scaling reveals the expected rolloff at high frequencies.

8.2 ACTIVE FILTERS

Most of the passive filters just discussed were first-order—they utilized a single reactive element and possessed rolloffs of 20 dB per decade. More sharply defined pass- and stop-bands require higher-order circuits, which typically are of the active type. This is an extensive and complex topic, and we proceed by narrowing our focus to a single class of active filter named after R.P. Sallen and E.L. Key [1] but also commonly referred to as VCVS (voltage-controlled-voltage-source) circuits. First, the operating principles of a second-order active filter of this type will be developed, and then higher-order configurations will be presented.

Second-Order Filters

Low-pass

The basic arrangement of a second-order low-pass Sallen and Key active filter is shown in Fig. 8.22. From the schematic,

$$V_a - V_+ = I_a R_2, \tag{8.26}$$

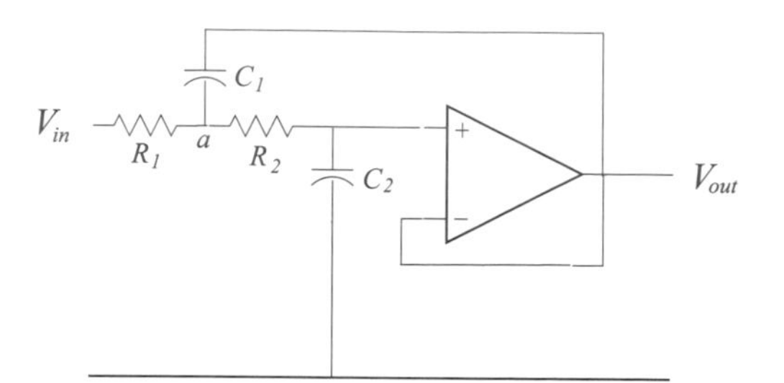

FIGURE 8.22. Second-order Sallen and Key low-pass active filter.

where I_a denotes the current flowing through resistor R_2 from node a towards the noninverting op-amp input. Because of the nearly infinite input impedance at this + node, I_a is constrained to flow through C_2, so $I_a = I_2$. Furthermore, the op-amp is wired as a unity-gain buffer, so $V_{\text{out}} = V_+$. Then, Eq. (8.26) becomes

$$V_a - V_{\text{out}} = I_2 R_2. \tag{8.27}$$

The current I_{in} in resistor R_1 satisfies

$$V_{\text{in}} - V_a = I_{\text{in}} R_1. \tag{8.28}$$

But also

$$I_2 = I_{\text{in}} + I_1. \tag{8.29}$$

Combining Eqs. (8.27)–(8.29),

$$V_{\text{in}} = V_{\text{out}} - I_1 R_1 + I_2 (R_1 + R_2). \tag{8.30}$$

To complete the derivation of the filter response, suitable expressions are needed for I_1 and I_2. These are easily obtained. The current in C_1 obeys

$$V_{\text{out}} - V_a = I_1 (-jX_1) \tag{8.31}$$

with $X_1 = \frac{1}{\omega C_1}$. Comparing Eq. (8.27) with Eq. (8.31),

$$I_1 = I_2 \left(-j \frac{R_2}{X_1} \right). \tag{8.32}$$

Likewise, the current in C_2 obeys

$$V_{out} = V_+ = I_2(-jX_2), \tag{8.33}$$

so

$$I_2 = j\frac{V_{out}}{X_2}, \tag{8.34}$$

where $X_2 = \frac{1}{\omega C_2}$ is the capacitive reactance.

Finally, substituting Eqs. (8.32) and (8.34) into Eq. (8.30),

$$V_{out} = V_{in}\left[\left(1 - \frac{R_1 R_2}{X_1 X_2}\right) + j\left(\frac{R_1 + R_2}{X_2}\right)\right]^{-1}. \tag{8.35}$$

Thus,

$$\boxed{\frac{V_{out}}{V_{in}} = \frac{1}{[1 - \omega^2 R_1 R_2 C_1 C_2] + j[\omega C_2(R_1 + R_2)]}}. \tag{8.36}$$

Notice the exact correspondence in form between this equation and Eq. (8.20), which was developed from a second-order passive *RLC* filter. In other words, this active low-pass filter, which contains only resistors and capacitors, effectively simulates an equivalent inductance. Generally speaking, capacitors are preferred over inductors as circuit components because of their wide availability, low cost, and small size.

Equations (8.20) and (8.36) point to the correspondences

$$LC \Leftrightarrow R_1 R_2 C_1 C_2$$

$$RC \Leftrightarrow C_2(R_1 + R_2)$$

from which, and employing Eqs. (8.21) and (8.22),

$$\boxed{Q = \frac{\sqrt{R_1 R_2 C_1}}{\sqrt{C_2}\,(R_1 + R_2)}}, \tag{8.37}$$

$$\boxed{\omega_0 = \frac{1}{\sqrt{R_1 R_2 C_1 C_2}}}. \tag{8.38}$$

The circuit of Fig. 8.22 is commonly generalized by the addition of gain-setting resistors R_3 and R_4, as shown in Fig. 8.23. Now, the op-amp is wired as

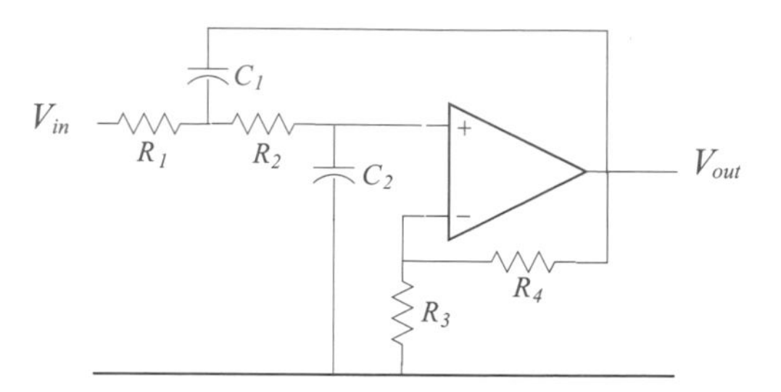

FIGURE 8.23. Second-order active low-pass filter with gain-setting resistors added.

a noninverting amplifier with closed-loop gain $G = 1 + \frac{R_4}{R_3}$. This changes the result of Eq. (8.36) to [2]

$$\frac{V_{\text{out}}}{V_{\text{in}}} = \frac{G}{\left[1 - \omega^2 R_1 R_2 C_1 C_2\right] + j\left[\omega R_1 C_1 (1 - G) + \omega C_2 (R_1 + R_2)\right]}, \tag{8.39}$$

which of course returns to the earlier result when $G = 1$. The filter ω_0 is still given by Eq. (8.38), but the Q changes somewhat to

$$Q = \frac{\sqrt{R_1 R_2 C_1 C_2}}{R_1 C_1 (1 - G) + C_2 (R_1 + R_2)}. \tag{8.40}$$

In a practical sense, there are too many adjustable parameters in these equations. That is, there are seemingly infinitely many ways of achieving any given filter response. A usable design process must narrow the options in some reasonable fashion.

Normalized filters

The following is one of several standardized schemes for defining a canonical Sallen and Key low-pass filter [3]. Let $R_1 = R_2 = 1\ \Omega$. Specify the corner frequency of the normalized filter as $\omega_c = 1$ radian/sec (ω_c will equal ω_0 in some cases, but not generally). Let $R_3 = \infty$ and $R_4 = 0$, so that $G = 1$. Even with all these restrictions, there remain infinitely many acceptable pairs C_1, C_2, but certain precise combinations lead to especially desirable filter characteristics. The significance of some choices is illustrated in Fig. 8.24. The curve that results from $C_1 = 1.414$ F and $C_2 = 0.707$ F rolls off smoothly, dropping to $\frac{1}{\sqrt{2}}$ (i.e., -3 dB) at the corner frequency $f_c = 1/2\pi = 0.159$ Hz. In fact, these particular component values happen to satisfy the criterion of maximal flatness within the passband; this is known as a *Butterworth filter*. The detailed procedures for deducing capacitance and/or resistance values from conditions on

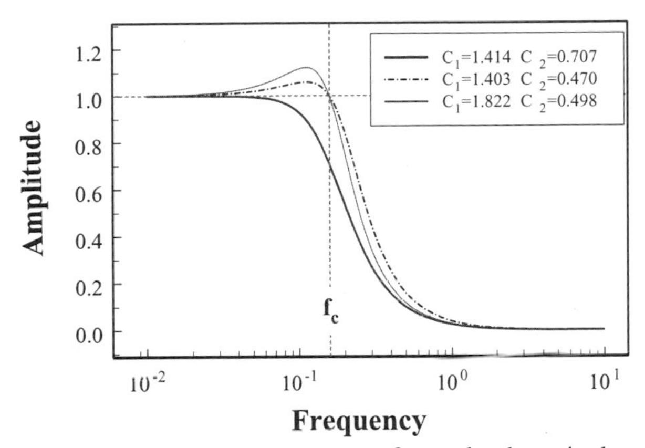

FIGURE 8.24. Frequency response of second-order active low-pass filter illustrating Butterworth (smooth lower curve) and Chebyshev (peaked) characteristics.

flatness, or from polynomial properties (next paragraph), constitute a lengthy and specialized topic for which the reader is directed to other sources [3, 4, 5].

The two other curves in the figure exhibit a peaked response. For the capacitance values selected, these are second-order *Chebyshev filter* characteristics (also known as equal ripple), and they achieve an increased abruptness in falloff by paying the price of ripple in the passband. By permitting larger ripple, the transition region can be narrowed, as the figure illustrates.

For these two cases, as shown in the logarithmically scaled plot of Fig. 8.25, the peak amplitude is 0.5 dB when $C_1 = 1.403$ F and $C_2 = 0.470$ F, and is 1.0 dB when $C_1 = 1.822$ F and $C_2 = 0.498$ F. As expected for second-order filters, the

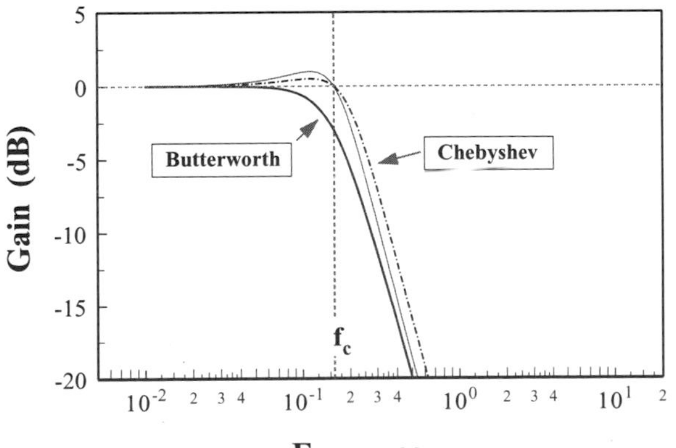

FIGURE 8.25. Decibel plot of second-order Butterworth and Chebyshev filter characteristics. Well beyond the corner frequency, the rolloff is 40 dB per decade.

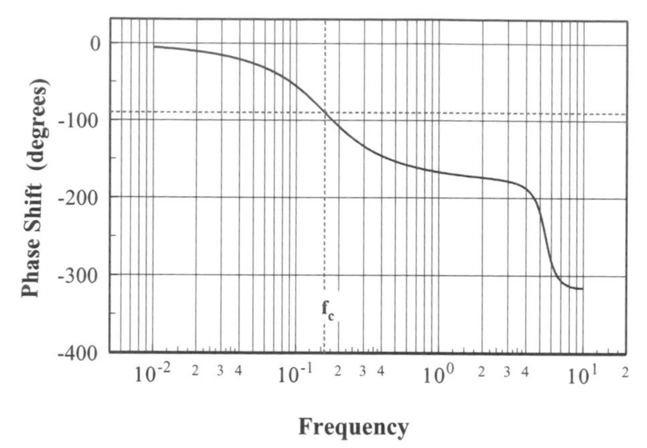

FIGURE 8.26. Phase shift in a second-order Butterworth low-pass filter.

high-frequency rolloff is 40 dB per decade. Note that for the Chebyshev filter, the frequency f_c is not the place at which the rolloff has dropped the amplitude by 3 dB, as is the case for second-order Butterworth and first-order passive filters, but rather it marks the end of the ripple zone.

To be more precise, f_c is defined for the Chebyshev characteristic as the frequency at which the response drops below the ripple band. For even-order filters (2nd order, 4th order, etc.), the ripple consists of one or more peaks which lie entirely above the 0 dB level. In odd-order Chebyshev filters (3rd order, 5th order, etc.), the ripple lies entirely below the 0 dB level. An example illustrating this property is given in the section on third-order filters.

The phase response of a second-order Butterworth filter is shown in Fig. 8.26.

The following table summarizes results from this section (remember that $R_1 = R_2 = 1.0\ \Omega$).

	C_1(F)	C_2(F)	$\omega_0(\text{sec}^{-1})$	Q
Butterworth	1.41421	0.70711	1.0	0.7071
Chebyshev 0.5 dB	1.40259	0.47013	1.2315	0.8636
Chebyshev 1.0 dB	1.82192	0.49783	1.0500	0.9565

Design example

All of the preceding capacitor combinations C_1, C_2 in Farads are for normalized filters with a corner frequency of 1 radian/sec and $R_1 = R_2 = 1\ \Omega$. Suppose a second-order low-pass Butterworth filter is desired, which has a corner frequency of 1500 Hz. The filter characteristic can be scaled appropriately upward in frequency if the normalized Butterworth capacitors $C_1 = 1.41421$ F and

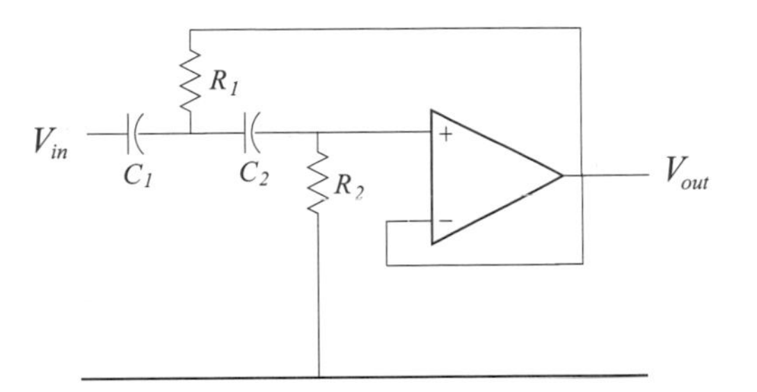

FIGURE 8.27. Second-order high-pass active filter.

$C_2 = 0.70711$ F are each made smaller by the desired factor of 2π (1500). This gives new values $C_1 = 1.501 \times 10^{-4}$ F and $C_2 = 7.503 \times 10^{-5}$ F. At this point, it is possible to rescale the two resistors from their original 1 Ω values to something more reasonable, such as perhaps 10 kΩ. The new corner frequency can be left unaffected after magnifying the resistors if the capacitors are divided by the same numerical factor (in this case 10^4). Hence, we arrive at the final filter values: $R_1 = 10$ K, $R_2 = 10$ K, $C_1 = 15.01$ nF, $C_2 = 7.503$ nF.

High-pass

High-pass active filters can be created simply by interchanging the roles of resistors and capacitors in a manner reminiscent of the passive *RC* circuit. Thus, Fig. 8.27 represents a second-order, high-pass Sallen and Key design. The following table gives the component values for several filter types. In these cases, $C_1 = C_2 = 1.0$ F and the corner frequency is again normalized to $\omega_c = 1.0$ radian/sec.

	$R_1(\Omega)$	$R_2(\Omega)$	$\omega_0(\mathrm{sec}^{-1})$	Q
Butterworth	0.70711	1.41421	1.0	0.4714
Chebyshev 0.5 dB	0.71281	2.12707	0.8121	0.4336
Chebyshev 1.0 dB	0.54586	2.00872	0.9550	0.4099

As an example of an active high-pass circuit, Fig. 8.28 shows the frequency response of the normalized 1.0 dB Chebyshev filter as specified by the resistor values in the third line of the table.

Band-pass

The general layout for a second-order Sallen and Key band-pass filter is shown in Fig. 8.29. One normalizing scheme [3] that is suitable is to set $C_1 = C_2 = 1.0$ F. Further, set $R_1 = 1.0\ \Omega$. The gain of the amplifier is of course $G = 1 + \frac{R_4}{R_5}$, so only the ratio $\frac{R_4}{R_5}$ together with R_2 and R_3 remain as undetermined quantities.

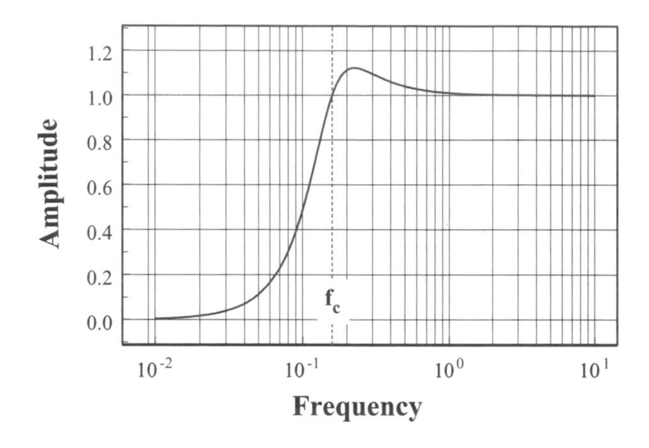

FIGURE 8.28. Second-order 1.0 dB Chebyshev high-pass filter frequency response.

Choose $R_4 = R_5$. It is desired to have the center frequency of the passband equal to unity: $\omega_c = 1.0$ radian/sec. The following table shows some possible choices for the remaining resistors and the resulting values of Q.

$R_2(\Omega)$	$R_3(\Omega)$	Q
0.74031	2.35078	2
0.63439	2.57630	5
0.60471	2.63567	10

The parameter Q measures the sharpness of the peak and is defined as the ratio $\frac{\omega_c}{\Delta\omega}$, where ω_c, as above, is the center frequency and $\Delta\omega$ is the peak width.

Design example

Using the first line in the table, we can design a filter with a center frequency of, say, 500 Hz and a Q of 2 as follows. The normalized filter has a center

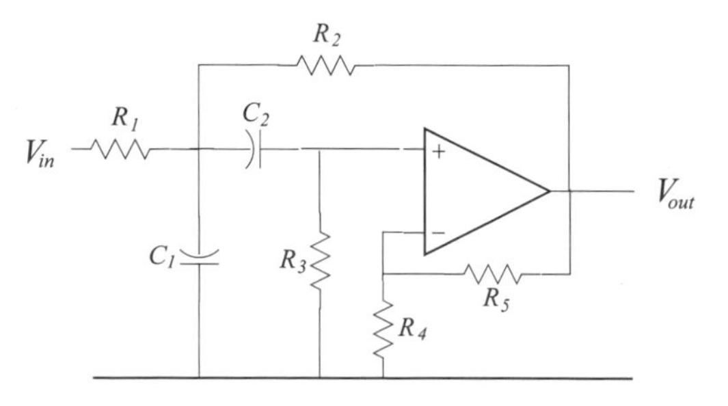

FIGURE 8.29. Second-order Sallen and Key bandpass filter.

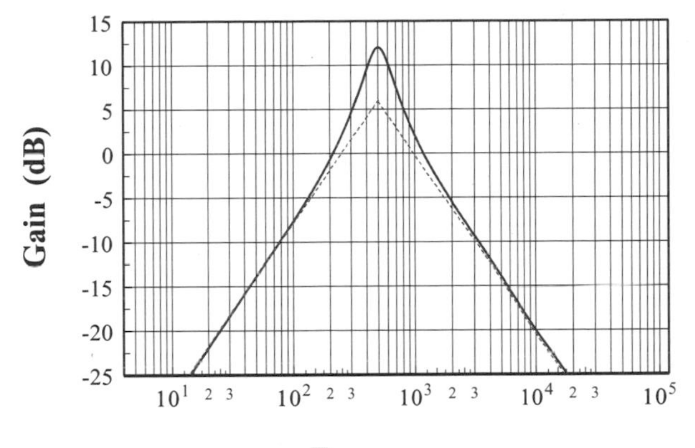

FIGURE 8.30. Second-order active band-pass characteristic with center frequency at $f_c = 500$ Hz.

frequency of $f_c = \frac{\omega_c}{2\pi}$ with $\omega_c = 1$. Thus, to raise the center frequency to 500, the capacitors should be scaled by $\frac{1}{2\pi \times 500}$ so $C_1 = C_2 = 318.31\ \mu$F. To maintain the new center frequency but adjust the capacitors to somewhat more realistic values, let us divide the C's by 1000 and at the same time multiply the resistors by 1000. The filter now is specified by $C_1 = C_2 = 0.31831\ \mu$F, $R_1 = 1$ kΩ, $R_2 = 740\ \Omega$, $R_3 = 2.351$ kΩ. Only the ratio of the gain-setting resistors matters, so choose $R_4 = R_5 = 1$ kΩ. The resulting filter characteristics are plotted in Fig. 8.30.

The dashed lines indicate the linear 20 dB per decade rolloff of the filter at low and high frequencies. For this design, $G = 2$, which is 6.02 dB. Note that the two linear extrapolations intersect at the point (500 Hz, 6.02 dB). The actual filter characteristic rises above this in a $Q = 2$ peak.

The effect of higher Q filters is illustrated in Fig. 8.31, where progressively narrower pass-bands appear. A linear vertical scale has been chosen here so that the peak effect is clearly revealed. These frequency response characteristics could appropriately be considered as defining "slot" filters (the inverse of notch filters), which are transparent to only a very restricted range of frequencies. Such a property can in fact be useful in particular applications where a nearly monochromatic "signal" is surrounded by extraneous noise.

Third-Order Filters

A third-order Sallen and Key low-pass filter is illustrated in Fig. 8.32. Adopting, as before, a normalized filter (corner frequency $\omega_c = 1.0$) with $R_1 = R_2 = R_3 = 1.0\ \Omega$, the required capacitor values are given in the following table.

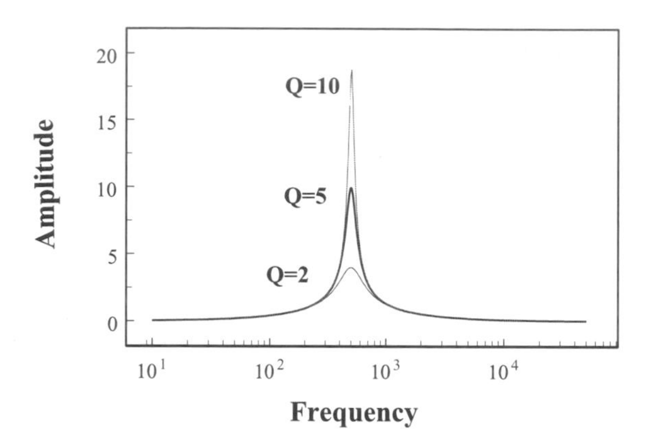

FIGURE 8.31. Second-order active band-pass filters with center frequency of 500 Hz and Q values of 2, 5, and 10.

	C_1(F)	C_2(F)	C_3(F)
Butterworth	0.20245	3.5465	1.3926
Chebyshev 1 dB	0.05872	14.784	2.3444

PSpice simulations for both Butterworth and 1.0 dB Chebyshev filters using component values as specified in the table produce the results plotted in Fig. 8.33. As expected for a third-order low-pass filter, the high-frequency rolloff is quite steep at 60 dB per decade (18 dB per octave).

As might be anticipated, a third-order high-pass filter can be derived from the low-pass circuit (Fig. 8.32) by interchanging the roles of resistor and capacitor. Hence, with $C_1 = C_2 = C_3 = 1.0$ F, the design table for normalized characteristics is

	$R_1(\Omega)$	$R_2(\Omega)$	$R_3(\Omega)$
Butterworth	4.93949	0.28194	0.71808
Chebyshev 1 dB	17.0299	0.06764	0.42655

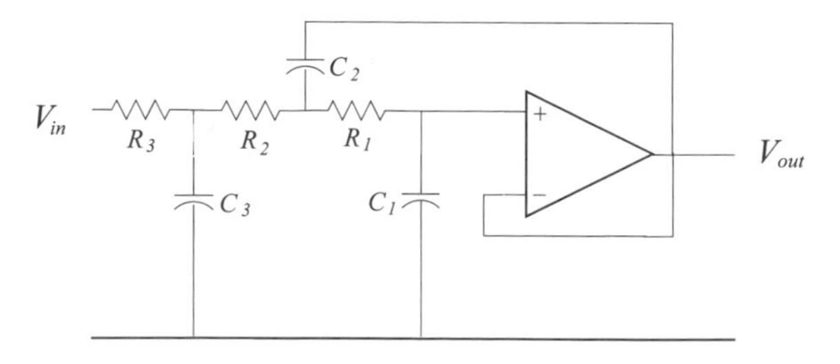

FIGURE 8.32. Third-order active Sallen and Key low-pass filter.

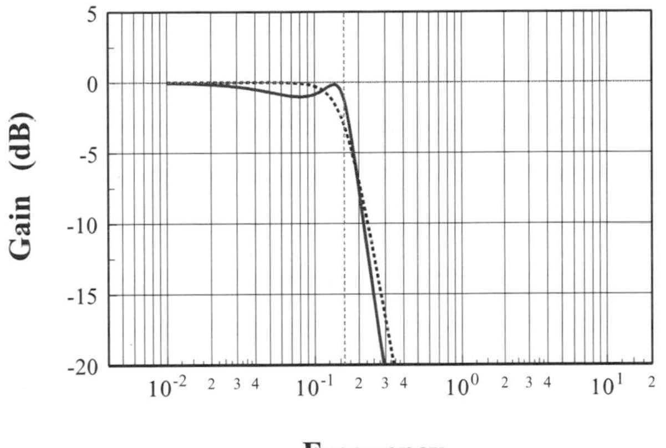

FIGURE 8.33. Frequency response of third-order active low-pass filter. Both Butterworth (dashed) and 1 dB Chebyshev (solid) characteristics are shown.

8.3 REMARKS

This chapter has provided an introduction to the basic concepts of analog filters. For all but the simplest applications, active filters are probably the best choice. Second- or third-order designs are generally sufficient.

The decision regarding tradeoffs is not always cut-and-dried, although in many instances the instrumentation application itself will dictate some choices—low ripple versus narrow transition region, low-order filter design versus sharply defined bands.

Active filter circuits are typically overdetermined in the sense that there are more variables (component values) than constraints. This allows some components to be preset to convenient values while the remaining ones are determined so that the real filter behaves as much as possible like the ideal target. The component values presented in the previous tables were specific to a particular normalizing scheme. The appropriate values were simply quoted (see [3]) without proof.

The actual method of determining correct component values is a complicated business, and the reader is referred to specialized texts for details. However, a simple example will at least illustrate the process.

Consider the general second-order active low-pass filter, as illustrated in Fig. 8.23. From Eq. (8.39) the circuit response can be written in the form

$$\frac{V_{\text{out}}}{V_{\text{in}}} = \frac{G}{[R_1 R_2 C_1 C_2]\ s^2 + [R_1 C_1 (1 - G) + (R_1 + R_2) C_2]\ s + 1} \tag{8.41}$$

with $s \equiv j\omega$. In more compact form,

$$\frac{V_{\text{out}}}{V_{\text{in}}} = \frac{Gb_0}{s^2 + b_1 s + b_0}, \tag{8.42}$$

where

$$b_0 = \frac{1}{R_1 R_2 C_1 C_2} \tag{8.43}$$

and

$$b_1 = \frac{R_1 C_1 (1 - G) + (R_1 + R_2) C_2}{R_1 R_2 C_1 C_2}. \tag{8.44}$$

A Butterworth polynomial of nth order can be expressed

$$T(s) = \frac{1}{a_n s^n + a_{n-1} s^{n-1} + \cdots + a_2 s^2 + a_1 s^1 + 1}. \tag{8.45}$$

The coefficients of these polynomials have been tabulated (see [5], p. 69) and are, for the case $n = 2$, $a_1 = \sqrt{2}$ and $a_2 = 1$. To produce a Butterworth filter, the circuit response function [Eq. (8.41) or Eq. (8.42)] must be made to match the polynomial, term by term.

For a normalized filter with $G = 1$, we wish to have $\omega_0 = 1$, and this just requires $b_0 = 1$ or

$$R_1 R_2 C_1 C_2 = 1. \tag{8.46}$$

The remaining condition needed to guarantee a match with the Butterworth polynomial is $a_1 = b_1 = \sqrt{2}$, which forces

$$(R_1 + R_2) C_2 = \sqrt{2}. \tag{8.47}$$

Notice that there are only two constraints—Eqs. (8.46) and (8.47)—but four components to be determined. In the earlier discussion of this second-order low-pass filter, an additional arbitrary but convenient choice was made: $R_1 = R_2 = 1.0\ \Omega$. With the resistors fixed, the capacitors must then be $C_2 = \frac{\sqrt{2}}{2} = 0.70711$ and $C_1 = \frac{1}{C_2} = 1.41421$. These are precisely the component values listed in the table accompanying the earlier discussion of the low-pass Butterworth filter.

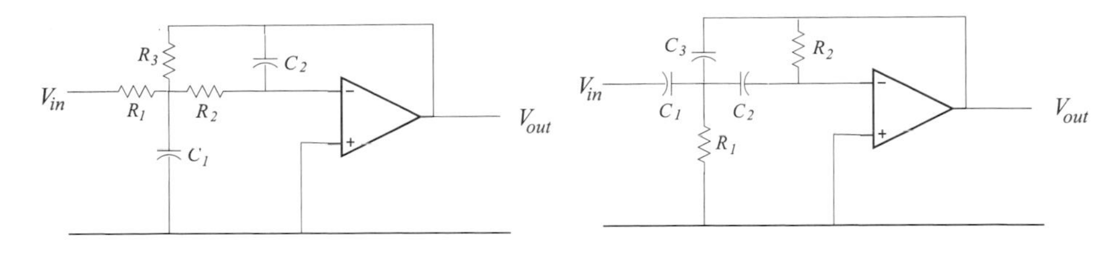

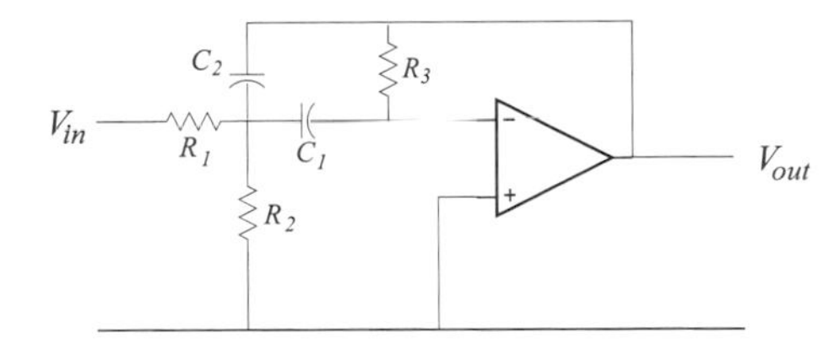

FIGURE 8.34. Configurations for second-order multiple feedback (MFB) active filters.

A third-order filter can be analyzed in a similar fashion knowing that the Butterworth coefficients are $a_1 = 2$, $a_2 = 2$, and $a_3 = 1$ (see [5], Table 8-3, p. 69).

Higher-order filters, as well as other types (Chebyshev, Elliptic, Bessel, etc.) based on different polynomials, require much more algebraic effort. These tasks are normally carried out on computers, and specialized software is now available to facilitate interactive filter design. Although such methods may be necessary for demanding applications, standard reference texts which are available in most engineering libraries provide concise summaries of design data that are usually adequate for most situations.

Although the discussion has been restricted to Sallen and Key (VCVS) circuits, it should be noted that other active filter designs are possible. The most common alternative to VCVS is the so-called Multiple Feedback (MFB) arrangement shown in Fig. 8.34. These filters are discussed in [3, 4, 5, 2, 7].

PROBLEMS

Problem 8.1. The discussion of passive first-order filters included low-pass RC, low-pass RL, and high-pass RC. A fourth possibility would be a high-pass RL filter. Design such a filter with the following characteristics: $f_c = 2$ kHz and

$R = 1$ kΩ. If possible, verify the final design with PSpice or a similar simulation package.

Problem 8.2. Design an active second-order, low-pass 1.0 dB Chebyshev filter using 5 kΩ resistors such that the corner frequency is 2 kHz. If possible, verify the final design with PSpice or a similar simulation package.

BIBLIOGRAPHY

[1] R.P. Sallen and E.L. Key, "A practical method of designing RC active filters," IRE Trans. Circuit Theory 2, 74–85, March (1955).
[2] Martin Hartley Jones, *A Practical Introduction to Electronic Circuits* (Cambridge University Press, Cambridge, U.K., 1977), pp. 162–170.
[3] Z.H. Meiksin, *Complete Guide to Active Filter Design, Op Amps, and Passive Components* (Prentice-Hall, Englewood Cliffs, NJ 1990); see in particular Table 3-1.
[4] David E. Johnson and John L. Hilburn, *Rapid Practical Designs of Active Filters* (John Wiley & Sons, New York 1975); see in particular Tables 2-2 through 2-7.
[5] Carson Chen, *Active Filter Design* (Hayden Book Company, 1982).
[6] William D. Stanley, *Operational Amplifiers with Linear Integrated Circuits*, third edition (Merrill/Macmillan College Publishing Company, New York 1994), Chapter 8.
[7] Thomas L. Floyd, *Fundamentals of Linear Circuits* (Merrill/Macmillan College Publishing Company, New York 1992), p. 388.
[8] J.D. Turner, *Instrumentation for Engineers* (Macmillan Education Ltd., New York 1988), Chapter 4.
[9] Stephen E. Derenzo, *Interfacing: A Laboratory Approach Using the Microcomputer for Instrumentation, Data Analysis, and Control* (Prentice-Hall, Englewood Cliffs, NJ 1990), pp. 62–71.

PART II

Sensors

CHAPTER 9

Temperature

Perhaps the most commonly monitored physical parameter in experimental situations is temperature. Before proceeding to the question of how temperature can be sensed, some preliminary facts and definitions are in order.

The three common temperature scales are: Fahrenheit, Celsius (or centigrade), and Kelvin (or absolute). The boiling point of water is 212 degrees Fahrenheit (°F) or, equivalently, 100 degrees Celsius (°C). The freezing point of water defines 32°F or, equivalently, 0°C. It is evident in either system that a pair of points capable of precise physical specification (freezing point and boiling point) are assigned rather arbitrary numerical "temperatures," but also that these two values are sufficient to define a complete temperature scale. Between freezing and boiling, there are 180 Fahrenheit degree increments, whereas the same range contains exactly 100 Celsius degrees. Thus, as a temperature interval, one Celsius degree is the same as 1.8 Fahrenheit degrees. Converting back and forth between the two scales is easy:

$$^\circ F = \left[\frac{9}{5}\right] {}^\circ C + 32,$$

$$^\circ C = \frac{5}{9}\,[^\circ F - 32].$$

The absolute temperature scale has intervals that are the same size as in the Celsius system, but its zero is relocated from the freezing point of water to the more universal temperature at which all thermal molecular motion theoretically ceases—"absolute zero." The conversion between Celsius and absolute is just

$$^\circ K = {}^\circ C + 273.15.$$

Absolute zero is thus -273.15°C or -459.67°F.

The product of Boltzmann's constant k (introduced earlier in the discussion of log and antilog amplifiers) and absolute temperature T represents an amount of energy typical of thermally excited molecular motion. Because room temperature is often approximated as 300°K, this energy is $kT = (1.38 \times 10^{-23})(300) = 4.14 \times 10^{-21}$ joules. Measured in units of electron volts, where $1 \text{ eV} = 1.602 \times 10^{-19}$ joules, this is equivalently 0.0258 eV.

The most familiar temperature sensor is the common thermometer in which thermal expansion drives a liquid up the fine bore of a glass tube. It is inexpensive and easy to use, but slow, not usually very accurate, applicable to limited temperature ranges, and impossible to interface with instrumentation apparatus. Fortunately, many specialized sensors are now available which provide enhanced performance and relative ease of integration within acquisition systems.

9.1 THERMISTORS

In outward appearance, a thermistor is normally a small (about the size of a match head) epoxy coated element from which a pair of electrical leads emerge. The encapsulation serves as a protection for the sensor element, which is a mass made up of a compressed blend of sintered oxides of manganese, nickel, copper, and/or cobalt. The resulting material is essentially a semiconductor with an effective energy gap E_g. Typical energy gaps are on the order of an electron volt. The resulting properties of the blend depend on the exact recipe for the mix, so infinitely many possibilities exist. However, certain compositions have been accepted as "standards" so that thermistors can be selected from manufacturers' catalogs and integrated into reproducible electronic thermometers. In essence, a thermistor is simply a temperature-dependent resistor.

The origin of this temperature dependence lies in certain properties of the semiconductor material from which the thermistor is fabricated. For an intrinsic (undoped) semiconductor, the density of conduction band electrons (and equally, valence band holes) is given by the standard expression [1]

$$n_i(T) = p_i(T) = 2\left(\frac{2\pi kT}{h^2}\right)^{\frac{3}{2}} (m_n^* \, m_p^*)^{\frac{3}{4}} \, e^{-\frac{E_g}{2kT}}, \tag{9.1}$$

where k is Boltzmann's constant, h is Planck's constant, and m_n^* and m_p^* are the so-called effective masses of electrons in the conduction band and holes in the valence band, respectively. The key point here from the perspective of thermistor behavior is the temperature dependence, which appears in two places, a power term $T^{\frac{3}{2}}$ and an exponential $e^{-\frac{E_g}{2kT}}$. Practically speaking, the exponential dependence dominates almost completely so that a plot of $\ln(n_i)$ versus T^{-1} is close to a linear relationship (see [1], p. 79, Fig. 3-17). The conductivity of the thermistor element depends on both carrier concentrations and their mobilities.

$$\sigma = n_i q \mu_e + p_i q \mu_h. \tag{9.2}$$

Because resistance is proportional to the reciprocal of conductivity, the preceding points lead to the conclusion that

$$R \propto e^{\frac{E_g}{2kT}} \tag{9.3}$$

captures the essential temperature dependence of a thermistor. Notice that with increasing temperature, the resistance will decrease—a so-called negative temperature coefficient (NTC) dependence. This is opposite to the response of a conventional metal, where increased scattering of electrons by the material lattice leads to an increase of resistance with rising temperature.

If we define

$$\beta \equiv \frac{E_g}{2k}, \tag{9.4}$$

then Eq. (9.3) becomes

$$R = A\, e^{\frac{\beta}{T}}. \tag{9.5}$$

Note that the appropriate numerical value of β will be determined by the specific choice of semiconductor compound from which the thermistor element is fabricated. Because energy gaps are of the order of 1 eV and Boltzmann's constant is 8.6×10^{-5} eV/K, the parameter β may be estimated to have a value $\sim$6000 K. Actually, β is more typically in the range 3000 to 4000 K.

Now, suppose that at some particular temperature, T_0, the thermistor resistance happens to be R_0. Then,

$$R_0 = A\, e^{\frac{\beta}{T_0}}. \tag{9.6}$$

Combining Eqs. (9.5) and (9.6) gives

$$\boxed{R = R_0 \exp\left[\frac{\beta}{T} - \frac{\beta}{T_0}\right]}. \tag{9.7}$$

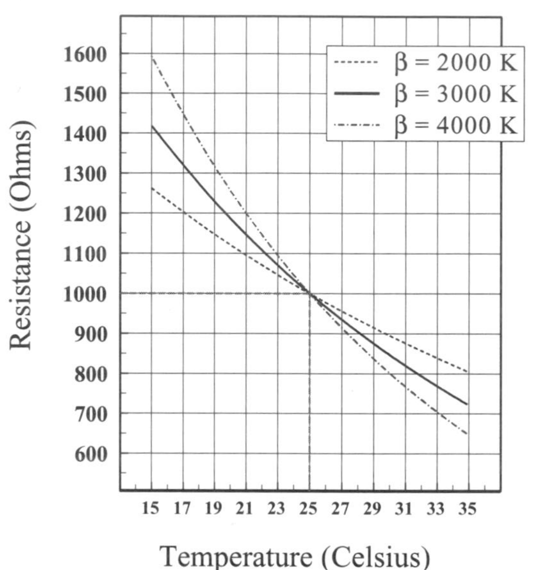

FIGURE 9.1. Temperature dependence of thermistors made from three slightly different semiconductor compounds. In this example, all the sensors have a resistance of 1000 Ω at the nominal temperature of 25°C.

As an example, suppose a thermistor is specified as having a resistance $R_0 = 1000\ \Omega$ at $T_0 = 25°\text{C}$. Figure 9.1 illustrates the resistance profile as a function of temperature for several different choices of β. An informative reference for this and subsequent topics on thermistor theory and applications is a handbook from Thermometrics [2].

Another perspective is gained by noting the implication in Eq. (9.7) that a plot of the natural logarithm of the resistance should be a linear function of reciprocal absolute temperature. In other words,

$$\ln(R) = \frac{\beta}{T} + \left[\ln(R_0) - \frac{\beta}{T_0}\right]. \tag{9.8}$$

This is apparent in Fig. 9.2.

Commercial thermistors are available in standard resistances usually specified at 25°C. These nominal values are most typically on the order of a few hundreds of ohms to tens of thousands of ohms.

Data sheets also frequently give values for the resistance ratio $[R@0°\text{C}]/[R@50°\text{C}]$; these dimensionless numbers often lie in the range 5 to 10. From Eq. (9.7) it is apparent that by setting $T_0 = 50°\text{C} = 323$ K and $T = 0°\text{C} = 273$ K,

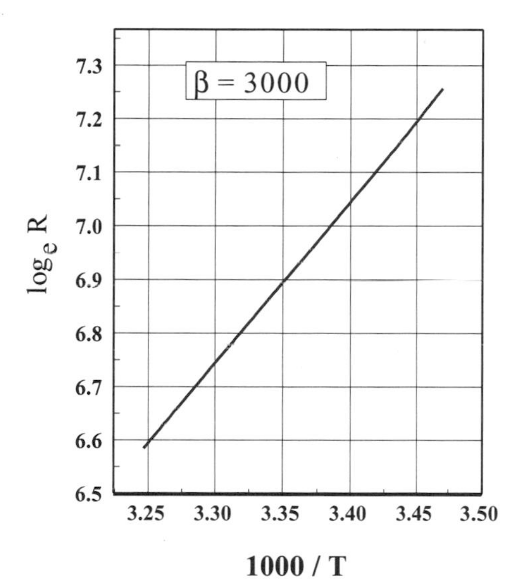

FIGURE 9.2. The same data as in the previous figure, for the case $\beta = 3000$, plotted such that the reciprocal temperature dependence of $\ln(R)$ is emphasized.

we obtain

$$\beta = \frac{273 \times 323}{323 - 273} \ln\left(\frac{R_{@0}}{R_{@50}}\right). \tag{9.9}$$

As an example, suppose a thermistor is characterized in a data sheet as having a resistance ratio of 7.04. Then, from Eq. (9.9), the material constant β may be calculated; the result in this case is $\beta = 3441$ K. Thus, a resistance ratio is simply another way of declaring the essential quantity β.

Sensitivity

The resistance-temperature characteristic is in general very nonlinear, but it may be viewed as approximately linear in the near vicinity of any given operating point. Suppose a thermistor is to be used for temperature monitoring on a system that nominally is at T_0. At T_0, the thermistor has a resistance R_0. As the temperature of the system rises above or falls below this reference value, the thermistor resistance moves to correspondingly smaller or larger values (see Fig. 9.3). The

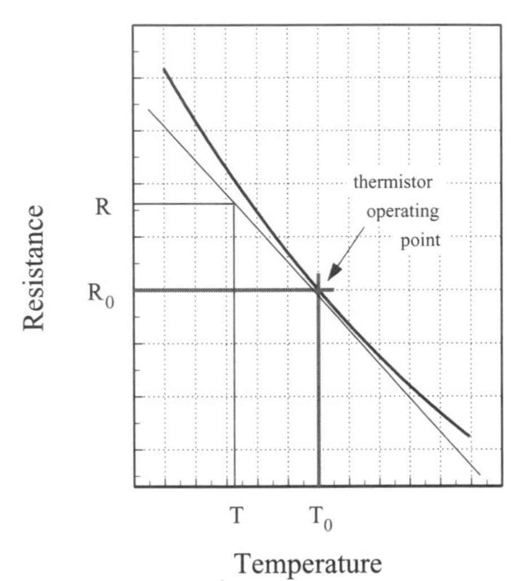

FIGURE 9.3. Thermistor operating about the point (R_0, T_0). A tangent line at this operating point provides the linear approximation to $R(T)$.

slope of the $R(T)$ characteristic may be obtained from Eq. (9.7):

$$\frac{dR}{dT} = \left[R_0 \exp\left(\frac{\beta}{T} - \frac{\beta}{T_0}\right)\right]\left(\frac{-\beta}{T^2}\right). \tag{9.10}$$

Evaluated at the operating point, this gives

$$\left.\frac{dR}{dT}\right|_{T=T_0} = -\beta \frac{R_0}{T_0^2}. \tag{9.11}$$

The slope is of course negative, as expected of an NTC device. As a general measure of sensitivity of resistance to temperature, we define

$$\boxed{\alpha = \frac{1}{R}\frac{dR}{dT}}, \tag{9.12}$$

which in this case is

$$\alpha = -\frac{\beta}{T_0^2}. \tag{9.13}$$

Writing Eq. (9.12) in the form

$$\alpha = \frac{\Delta R}{R} \frac{1}{\Delta T}$$

points to the interpretation of the coefficient α as the relative change in resistance per degree change in temperature. The magnitude of the temperature coefficient α is obviously the local sensitivity (at the operating point) of the thermistor; when multiplied by 100, it represents the percentage change in resistance per degree Kelvin or, since an interval of one degree absolute is exactly the same as an interval of one degree Celsius, it is also correct to say that the sensitivity is per degree Celsius.

As an example, suppose a thermistor has a resistance of 5000 Ω at 25°C and $\beta = 3000$ K. Then,

$$\alpha|_{T=25^\circ\mathrm{C}} = -\frac{3000}{(298)^2} = -0.034,$$

so the sensitivity could be stated as 3.4% per Kelvin or Celsius. Note that the temperature in Eq. (9.13) must be expressed in absolute degrees.

The equation of the linear approximation to $R(T)$ is thus

$$(R - R_0) = \beta \frac{R_0}{T_0^2} (T_0 - T), \tag{9.14}$$

from which a measured thermistor resistance value R quickly gives a temperature estimate T, provided R_0, T_0, and β are already known.

Thermal Dynamics

The temperature-dependence of the thermistor's resistance obviously can serve as the basis of a temperature measuring scheme. When employed as a thermometer, a thermistor must, ideally, be at exactly the same temperature as the object or medium whose temperature is to be measured. That is, it should be in intimate thermal contact with the object or medium, say by cementing it to a solid surface or by immersing it in a liquid or gas. However, complications can arise because of two effects: (1) the self-heating caused by necessary bias currents flowing through the thermistor; (2) the finite response-time of the device caused by its thermal mass. Precise temperature sensing requires an appreciation of these factors.

To measure an unknown temperature with a thermistor, one equivalently measures the thermistor resistance at that temperature and then deduces the temperature from a calibration curve of $R(T)$. In reality, measuring a resistance means feeding a known bias current I_b through the element and then measuring the voltage V across it. Then, simply, $R = V/I_b$.

Note that while the temperature to be measured is the ambient value T_A, only the actual temperature of the sensor itself, T, will determine the thermistor resistance, R. Hence T, not T_A, will be monitored. In the event that $T \simeq T_A$, measurements would yield accurate values, but it may happen that the thermistor temperature deviates significantly from the target value T_A. This will occur if the bias current supplies enough energy to the sensor element to raise its temperature somewhat above its immediate surroundings. Clearly, small bias currents are desirable, but the necessity of producing measurable voltages forbids endless reductions.

Suppose the thermistor element is characterized by its mass m and specific heat capacity c (with units of calories per gram and degrees Celsius). In a time increment Δt, the temperature will rise from T to $T + \Delta T$.

Conservation of heat energy during the interval Δt can be stated in the form

$$\Delta H_i = \Delta H_\ell + \Delta H_a, \tag{9.15}$$

which just expresses the requirement: energy flow into sensor element = energy lost to surroundings + energy absorbed. The input term is due simply to Joule heating:

$$\Delta H_i = \left(I_b^2\, R\right) \Delta t = (I_b V)\, \Delta t. \tag{9.16}$$

The loss term can be expressed as

$$\Delta H_\ell = [\delta(T - T_A)]\, \Delta t, \tag{9.17}$$

where δ is a so-called dissipation constant. This equation is also known as Newton's Law of Cooling, and it simply says that the rate of flow of heat from a hot object to a cooler one is proportional to the temperature difference between the two objects. This is reminiscent of Ohm's law in circuits, $I = V/R$, which declares that the rate of flow of electrical charge, I, equals a proportionality constant, R^{-1}, times the potential difference V. From this perspective, the dissipation constant clearly functions as an inverse thermal resistance.

When an amount of heat ΔQ is absorbed by an object whose mass is m, its temperature will rise by an amount ΔT according to $\Delta Q = mc\, \Delta T$, where c is the specific heat capacity of the object. Consequently,

$$\Delta H_a = mc\, \Delta T \tag{9.18}$$

Substituting Eqs. (9.16), (9.17), and (9.18) in Eq. (9.15) yields

$$(I_b V)\,\Delta t = [\delta\,(T - T_A)]\,\Delta t + mc\Delta T \tag{9.19}$$

or

$$mc\frac{\Delta T}{\Delta t} + \delta\,(T - T_A) = I_b V, \tag{9.20}$$

which in the limit $\Delta t \to 0$ becomes the differential equation

$$mc\frac{dT}{dt} + \delta\,(T - T_A) = I_b V. \tag{9.21}$$

A typical initial condition for this first-order differential equation is that the sensor element is at ambient temperature T_A at time 0. The solution in this case is

$$T - T_A = \frac{I_b V}{\delta}\left[1 - e^{-\frac{\delta}{mc}t}\right]. \tag{9.22}$$

This equation describes the way in which self-heating induced by a finite bias current will cause the sensor temperature to rise above the ambient value. The time constant for this exponential process is

$$\tau = \frac{mc}{\delta}. \tag{9.23}$$

Note that the sensor mass ultimately approaches a temperature that is $\frac{I_b V}{\delta}$ above T_A. Thus, by measuring the thermistor resistance via a measurement of its terminal potential V, $[T_A + \frac{I_b V}{\delta}]$ rather than T_A is determined. As noted earlier, this self-heating error can be reduced by employing small-bias currents. It is also apparent that self-heating effects can be reduced by improving the thermal path between thermistor and sample (i.e., by decreasing the thermal resistance δ^{-1}). Practical means of accomplishing this depend upon particulars of the measurement problem. For example, stirring a fluid when a liquid temperature is being determined, or using heat-sink pastes to anchor a thermistor to a solid surface, can improve matters.

Example

As an example, suppose a data sheet gives the following values for a glass-coated bead thermistor with a nominal resistance of 5000 Ω at 25°C: free-air dissipation

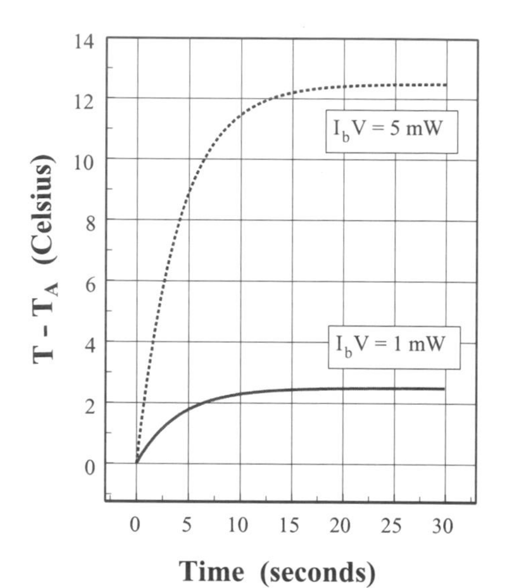

FIGURE 9.4. Self-heating in a thermistor: the thermistor temperature T rises to a value $I_b V/\delta$ above the ambient value T_A. The dissipation constant used here was $\delta = 0.4$ mW/°C.

constant 0.4 mW/°C; time constant 4 sec. Two self-heating curves are illustrated in Fig. 9.4, each specified by the particular power assumed to be dissipated in the thermistor element. As expected from Eq. (9.22), the thermistor temperature rises to a value $\frac{I_b V}{\delta}$ above ambient. For even 1 milliwatt (mW) of dissipation, the temperature shift is 2.5°C—a potentially large measurement error in trying to determine T_A.

The power dissipation can also be expressed [Eq. (9.16)] $P = I_b V = (I_b)^2 R$. For measurements with this thermistor in the neighborhood of 25°C, for example, 1 mW corresponds to a bias current of $I_b = \sqrt{[\frac{P}{R}]} = \sqrt{[\frac{10^{-3}}{5000}]} = 0.45$ mA. The voltage to be measured would in this case be of order $I_b R = 2.25$ V. Clearly, an improvement would be achieved by the use of a smaller bias current. Because of the squared dependence in $P = (I_b)^2 R$, dropping I_b by a factor of 10 would result in a temperature-shift reduction by a factor of 100 to only 0.025°C while at the same time preserving an easily measured thermistor voltage of 0.225 V.

This example illustrates one of the possible shortcomings of thermistors—their slow response time. Of course, in many applications where changes occur at rates slower than a few seconds, such time constants do not represent a liability.

Voltage-Current Characteristic

When a thermistor is in equilibrium, Eq. (9.21) becomes

$$I_b V = \delta (T - T_A), \tag{9.24}$$

so

$$T = \frac{I_b V}{\delta} + T_A. \tag{9.25}$$

A voltage-current characteristic can be constructed from the iterative process outlined in Fig. 9.5.

As an example, consider a thermistor whose parameters are: $R_0 = 5000\,\Omega$ @ 25°C, $\beta = 3000\,\mathrm{K}$, and $\delta = 0.3\,\mathrm{mW/^\circ C}$. The sequence of calculations depicted in Fig. 9.5 was executed for two different possible ambient temperatures: $T_A = 10^\circ\mathrm{C}$ and 40°C. Power values ranging from 0.2 mW to 40 mW were employed to generate the curves. The results are plotted in Fig. 9.6.

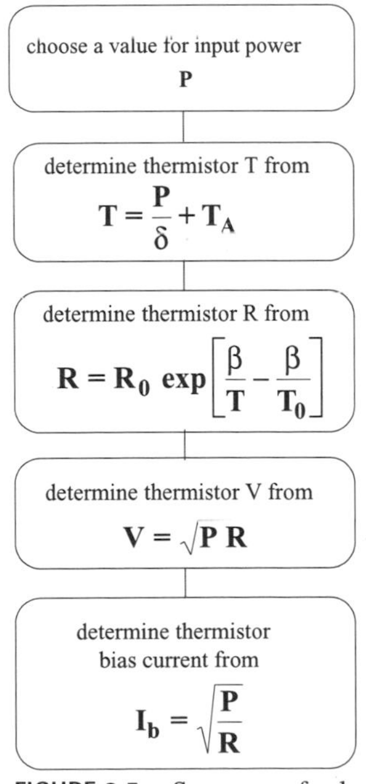

FIGURE 9.5. Sequence of calculations for computing voltage-current characteristics.

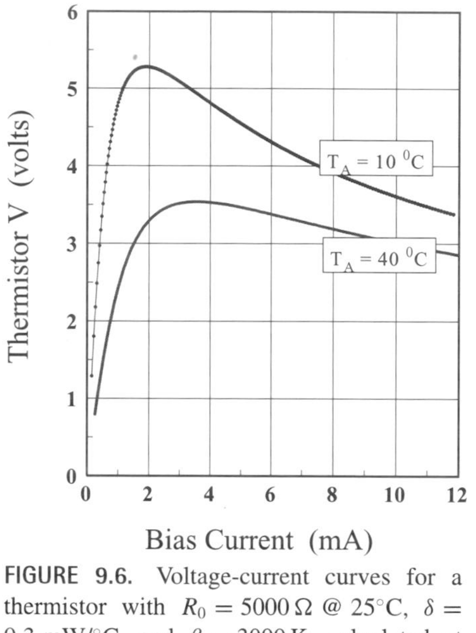

FIGURE 9.6. Voltage-current curves for a thermistor with $R_0 = 5000\,\Omega$ @ 25°C, $\delta = 0.3$ mW/°C, and $\beta = 3000$ K, calculated at two ambient temperatures.

A Simple Thermometer

The temperature dependence of a thermistor's R is the basis of thermometry with these devices. However, apart from issues of nonlinearity, the disadvantage of directly monitoring resistance is that this requires either two measurements (I and V) or a single voltage measurement when a constant current source feeds the thermistor. These approaches are certainly practical, but a simpler circuit is also worth considering.

Suppose a thermistor is connected in series with a fixed resistor R_S and a battery V_S, as shown in Fig. 9.7.

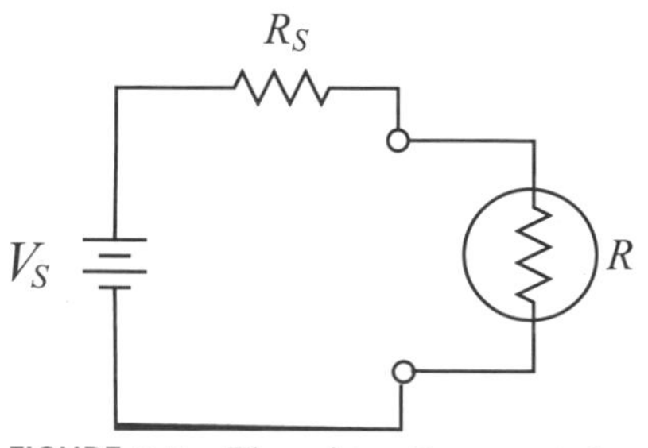

FIGURE 9.7. Thermistor R connected to the terminals of a bias circuit consisting of battery V_S in series with resistor R_S.

The combination of battery plus series resistor may be viewed as a two-terminal bias circuit, which drives a load (a thermistor in this case). Let the load resistance be R. Ohm's law then says

$$V_S = IR_S + V$$

or

$$V = -IR_S + V_S. \tag{9.26}$$

This is the equation of a straight line ($y = mx + b$) when voltage is assigned to the vertical axis and current is assigned to the horizontal axis; the slope is $-R_S$. This is known as the *load line* and it is the locus of all possible combinations of current and voltage that a two-terminal bias circuit is capable of supplying to any load. In the event that $R = 0$, that is, a shorted load, clearly $V = 0$ and $I_{\text{short}} = \frac{V_S}{R_S}$. Similarly, in the event that the load terminals are open circuited, then $I_{\text{open}} = 0$ and $V = V_S$. These two special points define the ends of the load line.

The fact that the two connecting terminals link the bias circuit to the load circuit means that the resulting current flowing through the terminals, and the voltage across the terminals, must simultaneously lie on both the load line (the locus of possible bias points) and the characteristic of the load itself (locus of all possible points for the thermistor). This requirement is easily visualized by plotting both curves and noting the points at which they intersect.

To continue the previous example, the load line and thermistor $V - I$ characteristics are combined in Fig. 9.8. Here, the bias circuit was taken to have $V_S = 5$ V and $R_S = 500\ \Omega$. Clearly, a change in ambient temperature produces a change in output voltage ΔV, which in this case is about 1 V for $\Delta T_A = 30°$C. Thus, if measurements in the vicinity of 25°C were being made with this circuit, the nominal output voltage would be around 4 V, so the sensitivity would be approximately 1 in 4, or 25% in an interval of 30°C. Equivalently, this is ~0.8% per °C.

Power Rating

Circuit elements such as resistors, transistors, or thermistors dissipate power internally when driven by bias currents. This dissipation takes the form of heat. Obviously, excessive internal heating may be harmful to the physical structure of a device and can lead in extreme scenarios to damage or even self-destruction. Proper circuit design can assure that such events will not occur.

Any power-sensitive component will be rated for some safe maximum such as 5 W, 50 mW, etc. As a design example, suppose a thermistor has the parameters

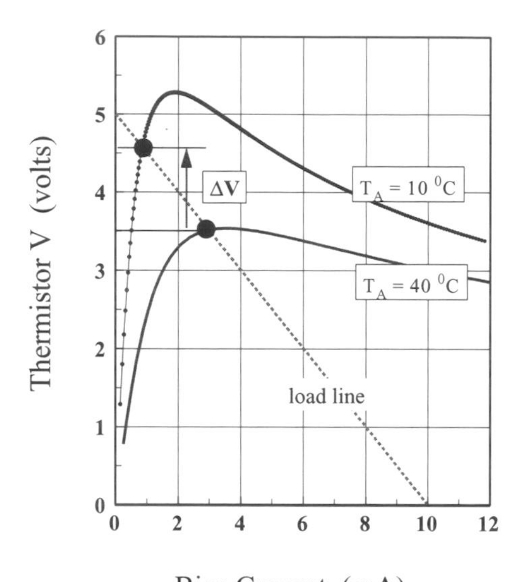

FIGURE 9.8. Thermistor characteristics at two ambient temperatures combined with a bias-circuit load line. This shows how the output voltage changes by ΔV as a result of the change in temperature.

employed in the previous two sections. The voltage–current characteristics presented in Fig. 9.8 are replotted in Fig. 9.9, but with the two axes reversed so that the representation is now current–voltage.

Internal dissipation is given by the product of the current flowing through the thermistor with the voltage appearing across the terminals of the thermistor (i.e., $P = IV$). Constant-power contours are also plotted in Fig. 9.9 at levels of 1 mW, 3 mW, and 5 mW. The load line from Fig. 9.8 corresponding to the previous source resistance of 500 Ω also appears. As the ambient temperature shifts among values within the range 10°C to 40°C—for this example—the operating point must slide along the load line. From the figure, it is clear that over this entire temperature range the power dissipation will exceed approximately 4 mW.

Suppose, however, that the thermistor has a power rating of only 1 mW. Damage to the thermistor would be very likely. The bias cicuit, in this case shown in Fig. 9.7, must therefore be changed.

The objective in achieving a safe design is to find a load line that is approximately tangent to the maximum power contour. Any source resistance greater than, or equal to, this critical value will guarantee that no point on the load line will ever correspond to an excessive power.

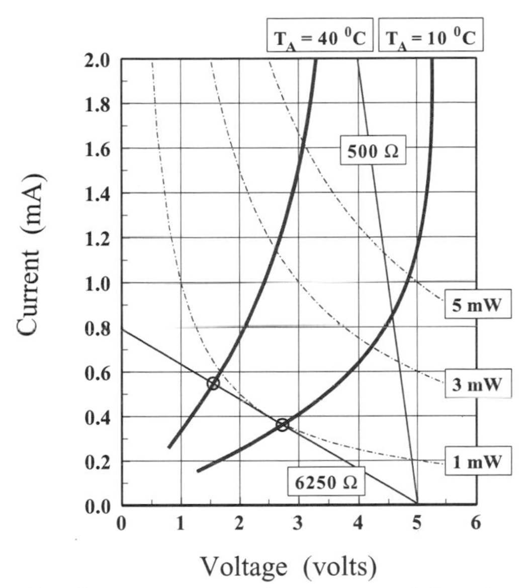

FIGURE 9.9. Current–voltage characteristics of a thermistor for two different ambient temperatures. Three constant-power contours are plotted as well as two possible load lines. Assuming the thermistor is rated at 1 mW, the 6250 Ω source resistance would assure safe operation.

For the case at hand, a load line of 6250 Ω is essentially tangent to the 1 mW contour and so represents the minimum safe source resistance for this design problem. Examining Fig. 9.9, it is apparent that the output voltage now will range from about 2.8 V at $T_A = 10°\text{C}$ down to about 1.5 V at $T_A = 40°\text{C}$.

Linearization with Parallel Resistor

The resistance-temperature characteristic of a thermistor is very nonlinear, as was evident in Eq. (9.7) and Fig. 9.1. This nonlinearity carries over into the temperature dependence of the output signal from a circuit such as Fig. 9.7. In other words, the temperature being sensed cannot be stated simply as a constant times the output voltage.

One approach to this situation is to first calibrate the "thermometer" by recording the output at a number of known temperatures and then interpolate any unknown temperature from the measured voltage. Calibration data storage and interpolation procedures can be implemented with memory and a microprocessor, so that an easy-to-use instrument results.

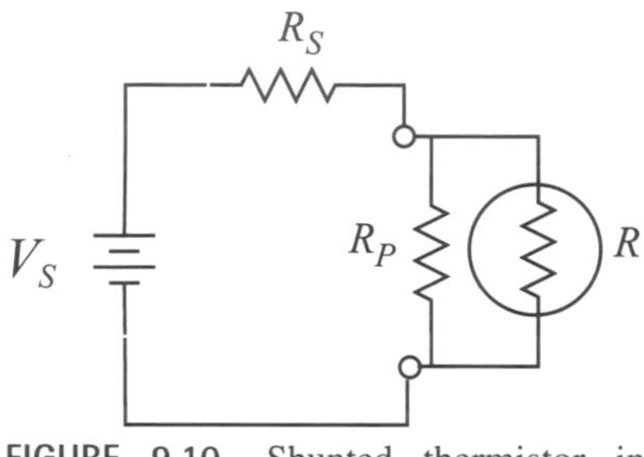

FIGURE 9.10. Shunted thermistor in simple bias circuit.

It is also possible to slightly modify the original circuit in such a way that a more nearly linear response occurs. If the resulting temperature dependence is linear enough, then the complexity of data sets and microprocessors can be avoided.

To begin, suppose that a thermistor is shunted with a fixed resistor R_P, as indicated in Fig. 9.10.

The single resistance which is equivalent to the parallel pair R_P and R is

$$R_{\text{eq}} = \frac{RR_P}{R + R_P}. \tag{9.27}$$

R is of course just the thermistor resistance expressed in Eq. (9.7). Suppose the temperature region of interest for this thermometer is centered around T_0. Then, from Eqs. (9.7) and (9.27), with R_P chosen to equal R_0,

$$R_{\text{eq}} = \frac{R_0}{1 + \exp\left[-\beta\left(\frac{1}{T} - \frac{1}{T_0}\right)\right]}, \tag{9.28}$$

which may be written

$$R_{\text{eq}} = \frac{R_0}{1 + e^{\varepsilon}} \tag{9.29}$$

with

$$\varepsilon = -\beta\left[\frac{1}{T} - \frac{1}{T_0}\right] = \beta\frac{T - T_0}{TT_0}. \tag{9.30}$$

Recall the expansion

$$e^{\varepsilon} = 1 + \varepsilon + \frac{\varepsilon^2}{2!} + \frac{\varepsilon^3}{3!} + \cdots \tag{9.31}$$

If the temperature T remains close to T_0 such that $T - T_0$ is small and $T\ T_0 \gg \beta$, then clearly $\varepsilon \ll 1$, in which case

$$R_{eq} \approx \frac{R_0}{2 + \varepsilon + \frac{\varepsilon^2}{2}}. \tag{9.32}$$

Higher powers of the small quantity ε have been dropped. This expression can further be simplified to

$$R_{eq} \approx R_0 \left[\frac{1}{2} - \frac{\varepsilon}{4}\right], \tag{9.33}$$

again dropping cubic and higher powers of the small quantity ε.

Returning to the definition given in Eq. (9.30), noting that $T\ T_0 \approx T_0^2$, and using $\Delta T \equiv (T - T_0)$,

$$\boxed{R_{eq} \approx R_0 \left[\frac{1}{2} - \frac{\beta \Delta T}{4\ T_0^2}\right]}. \tag{9.34}$$

The equivalent resistance is seen to be approximately a linear function of the temperature shift ΔT. The slope of this new characteristic $R_{eq}(T)$ is

$$\frac{dR_{eq}}{dT} = -\frac{R_0\beta}{4\ T_0^2}\frac{d\Delta T}{dT} = -\frac{R_0\beta}{4T_0^2}. \tag{9.35}$$

From the earlier definition of the temperature coefficient α [Eq. (9.12)],

$$\alpha = -\frac{\beta}{4\ T_0^2}. \tag{9.36}$$

Comparing this with Eq. (9.13), it is apparent that the tradeoff with this technique of linearizing the temperature dependence is that the sensitivity as expressed by the coefficient α has dropped to one-quarter of its original value.

Linearization with Series Resistor

A different method for generating a voltage that is a quasilinear function of temperature occurs when the output signal is taken from the series resistance instead of from the thermistor, as previously. The arrangement is shown in Fig. 9.11.

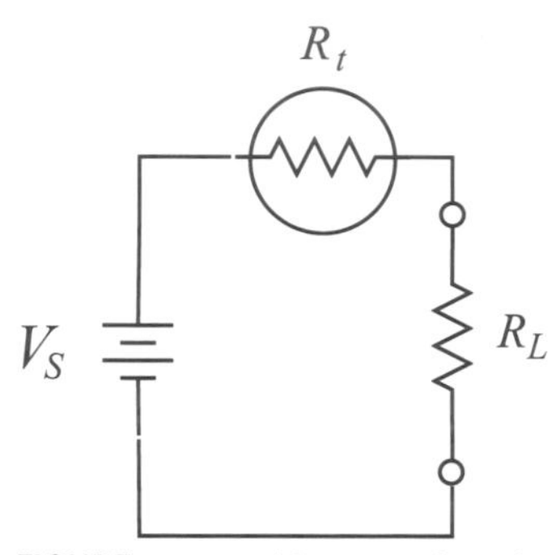

FIGURE 9.11. Alternate thermistor circuit in which the output is taken across the load resistor.

The voltage across the load R_L is

$$V_O = V_S \frac{R_L}{R_t + R_L}, \tag{9.37}$$

or

$$V_O = V_S \frac{1}{1 + \frac{R_t}{R_L}}. \tag{9.38}$$

The thermistor resistance is of course temperature-dependent. Suppose its value at a particular temperature is known. To make the process explicit, let us designate by R_{25} the sensor resistance at 25°C (298°K). Note that this value is a single fixed number, which is set once a specific thermistor is chosen. Then,

$$V_O = V_S \frac{1}{1 + \frac{R_t}{R_{25}} \frac{R_{25}}{R_L}}. \tag{9.39}$$

Consider the two factors that form the product in the denominator. The first of these ratios is temperature-dependent by reason of the thermistor resistance R_t. The second ratio may be viewed as an adjustable parameter in the design process, the adjustment being made by varying the load resistor R_L. For ease of notation, let

$$\eta = \frac{R_{25}}{R_L} \tag{9.40}$$

so

$$\boxed{\frac{V_O}{V_S} = \frac{1}{1 + \eta \frac{R_t}{R_{25}}}}. \tag{9.41}$$

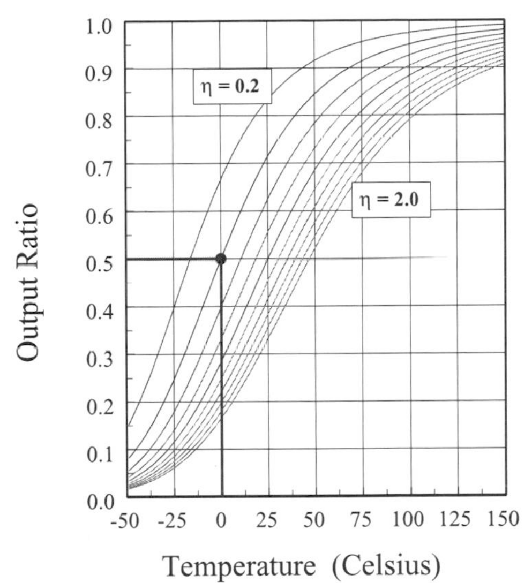

FIGURE 9.12. Plot of output ratio as a function of temperature. Each curve is specified by the parameter η, which ranges from 0.2 to 2.0 in ten equal steps of 0.2.

Again, to illustrate this process by way of example, suppose the thermistor is specified on a data sheet as having $R_{25} = 5000\ \Omega$ and $\beta = 3000$ K. Then, from Eq. (9.7),

$$\frac{R_t}{R_{25}} = \exp\left[\frac{3000}{T} - \frac{3000}{298}\right]. \tag{9.42}$$

Using Eq. (9.42) in Eq. (9.41), it is possible to construct a plot, for any given choice of η, of the output ratio $\frac{V_O}{V_S}$ as a function of temperature T. A series of such plots is shown in Fig. 9.12.

These curves are S-shaped, asymptotically limited by a ratio of unity at high temperature, and zero at low temperature. On any particular curve defined by a specific value of η, there is an inflection point. Above the inflection point, the curvature is positive; below the inflection point, the curvature is negative. The neighborhood of the inflection point is the most linear portion of any output ratio-temperature characteristic.

Once more adopting specific numbers for the purpose of illustration, suppose this thermometer is to operate about an ambient temperature of 0°C. As suggested in Fig. 9.12, the characteristic for $\eta = 0.4$ seems to possess an inflection very near this temperature, but bear in mind that this is just a visual estimate. The figure also suggests that the output ratio will be an approximately quasilinear function of temperature, varying between ~ 0.25 and ~ 0.71 over the range $-25 < T < +25$.

Inflection Point Optimization

The preceding ideas can be made more precise by treating the inflection point properties in a more analytical fashion. Begin with Eq. (9.41):

$$\frac{V_O}{V_S} = F(T) = \frac{1}{1 + \eta\, g(T)}. \tag{9.43}$$

For notational convenience, the output ratio has been redesignated as the function $F(T)$, and

$$g(T) = \frac{R_t(T)}{R_{25}} \tag{9.44}$$

contains the temperature dependence which enters through the thermistor resistance $R_t(T)$.

The inflection point in $F(T)$ is defined by the condition $\frac{d^2F}{dT^2} = 0$. Using Eq. (9.43),

$$\frac{dF}{dT} = -\frac{\eta \frac{dg}{dT}}{[1 + \eta g]^2}.$$

Then, at the inflection point,

$$\frac{d^2F}{dT^2} = -\frac{[1 + \eta g]^2\, \eta \frac{d^2 g}{d\,T^2} - 2\left(\eta \frac{dg}{dT}\right)^2 [1 + \eta g]}{[1 + \eta g]^4} = 0,$$

so

$$[1 + \eta g]\, \frac{d^2 g}{d\,T^2} - 2\eta \left(\frac{dg}{dT}\right)^2 = 0,$$

from which we obtain

$$\eta = \frac{\left(\frac{d^2 g}{d\,T^2}\right)}{\left[2\left(\frac{dg}{dT}\right)^2 - g\left(\frac{d^2 g}{d\,T^2}\right)\right]}. \tag{9.45}$$

If the terms on the right-hand side of this equation are evaluated for a chosen thermistor at some desired mean operating temperature T_0 (0°C in the example discussed previously), then the exact optimal value of the parameter η will be found. From earlier expressions for thermistor characteristics,

$$g(T) = \exp\left[\frac{\beta}{T} - \frac{\beta}{298}\right]. \tag{9.46}$$

Therefore,

$$\left.\frac{dg}{dT}\right|_{T=T_0} = -\frac{\beta}{T_0^2}\, g(T_0) \tag{9.47}$$

and

$$\frac{d^2 g}{dT^2} = -\frac{\beta}{T^2}\left[-\frac{\beta}{T^2}\, g\right] + \frac{2\beta g}{T^3}$$

or

$$\left.\frac{d^2 g}{dT^2}\right|_{T=T_0} = \left[\frac{\beta^2}{T_0^4} + \frac{2\beta}{T_0^3}\right] g\,(T_0). \tag{9.48}$$

Equations (9.47) and (9.48) can now be inserted into Eq. (9.45) with the result

$$\boxed{\eta = \left[\frac{\beta + 2T_0}{\beta - 2T_0}\right] \frac{1}{g\,(T_0)}}. \tag{9.49}$$

Returning to the previous example, which had $\beta = 3000\,\mathrm{K}$ and a desired center temperature of $T_0 = 0$°C (273°K), from Eq. (9.46)

$$g\,(@0°\mathrm{C}) = 2.514,$$

and hence from Eq. (9.49)

$$\eta = 0.575.$$

This value is close to the earlier estimate of $\eta \approx 0.4$, which was based only on a visual examination of the $F(T)$ characteristics presented in Fig. 9.12. The new value is of course more precise. The parameter η is just the ratio $\frac{R_{25}}{R_L}$, and thus the optimum load resistor can be determined.

Sensitivity

The sensitivity of the arrangement illustrated in Fig. 9.11 may be defined in the usual fashion as

$$S = \frac{1}{V_0} \frac{dV_0}{dT}, \tag{9.50}$$

or equivalently

$$S = \frac{1}{F} \frac{dF}{dT}. \tag{9.51}$$

In the last section, we had

$$\frac{dF}{dT} = -\frac{\eta \frac{dg}{dT}}{[1 + \eta g]^2} = -\eta\, F^2 \frac{dg}{dT},$$

so

$$S = -\eta\, F \frac{dg}{dT} = -\frac{1}{\frac{1}{\eta} + g} \frac{dg}{dT}. \tag{9.52}$$

Suppose the desired operating point for the thermometer is T_0. Equation (9.49) gives the optimum value of the parameter η. Using the expression for dg/dT given by Eq. (9.47), the sensitivity at this optimum choice is seen to be

$$S|_{T_0} = \left[\frac{\beta - 2T_0}{\beta + 2T_0} g + g \right]^{-1} \frac{\beta}{T_0^2} g. \tag{9.53}$$

Finally,

$$S|_{T_0} = \frac{\beta + 2T_0}{2T_0^2}. \tag{9.54}$$

This result for the sensitivity may be compared with the magnitude $\frac{\beta}{T_0^2}$ [Eq. (9.13)] that arises from considering the thermistor as a temperature-dependent resistor, and $\frac{\beta}{4T_0^2}$ [Eq. (9.36)], which is applicable when a parallel fixed resistor is added for linearization.

9.2 RESISTANCE TEMPERATURE DETECTORS

A resistance temperature detector (RTD) is simply a resistor fabricated from a nearly pure metal such as platinum, copper, or nickel. The conductivity of a sample is expressed by Eq. (9.2), modified to the extent that a metal, in contrast to a semiconductor, does not have an energy gap and so the free carriers are only electrons at a concentration n.

$$\sigma = nq\mu. \tag{9.55}$$

The electron density in metals is nearly independent of temperature over a considerable range because the outer conduction electron(s) in each atom are easily set free by even weak thermal activation so ionization is virtually complete beginning at relatively low temperatures, while at the same time the remaining inner atomic electrons are tightly bound and remain so to very high temperatures. The temperature dependence of the conductivity thus is due almost entirely to the temperature dependence of the electron mobility μ. Increasing temperature brings increasing electron scattering by the host lattice of the metal and a decrease in mobility.

Hence, overall, conductivity will decrease with temperature; or equivalently, resistivity will increase with temperature. This implies a positive temperature coefficient (PTC) in contrast with thermistors, which are characterized by negative temperature coefficients (NTC).

The $R(T)$ behavior of platinum is shown in Fig. 9.13. Such a quasilinear characteristic is easily represented in the form of a power series

$$R = R_0\,[1 + \alpha T + \beta T^2 + \gamma T^3 + \cdots], \tag{9.56}$$

where T is the temperature in °C and R_0 is the resistance at zero degrees Celsius. In this example, the linear approximation is very good over the range $-100 < T < +200$, with a temperature coefficient for platinum of $\alpha = 0.00392$, which represents a proportional change of 0.39% per degree Celsius. This is a much smaller numerical value than a typical coefficient for thermistors; in an earlier example, α was shown to be ~ -0.034. Hence, the RTD is about 10 times less

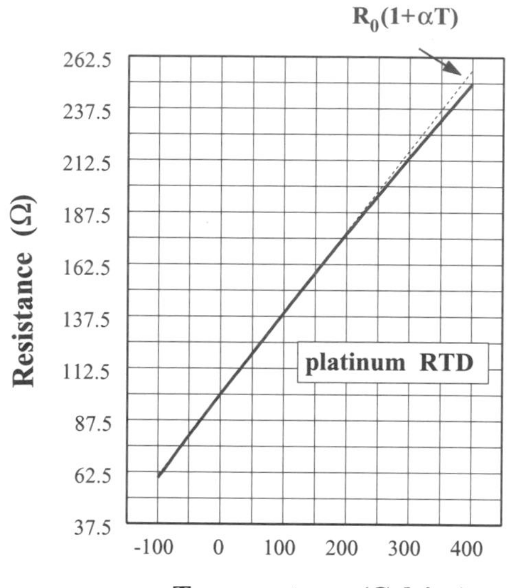

FIGURE 9.13. Resistance temperature characteristic of a platinum RTD. The resistance at 0°C is 100 Ω. Also shown is the linear approximation: $R = R_0(1 + \alpha\, T)$ with $\alpha = 0.00392$.

sensitive than a thermistor, but it is very linear and can be operated over a much larger temperature interval.

An improved mathematical fit for platinum devices over a wider range in T than the linear approximation provides is achieved with the quadratic

$$R = R_0\,[1 + 0.00392\, T - 5.8 \times 10^{-7}\, T^2]. \tag{9.57}$$

There are, in fact, two international standards for platinum RTDs—the "American" characteristic with $\alpha = 0.00392$, and the "European" characteristic with $\alpha = 0.00385$ [3].

RTDs are also available that employ resistive elements made from either copper or nickel; the temperature coefficients are $\alpha = 0.00427$ and $\alpha = 0.00672$, respectively. Platinum devices can be used over a temperature range of about (−200°C, +800°C), but the less expensive copper and nickel RTDs are suited to somewhat narrower intervals, the most limited being copper. Note that the melting temperatures of the components of common solder are: lead 327°C and tin 232°C.

Structurally, RTDs are generally made with the resistive wire element coiled on a ceramic form, or as metal films deposited on a rigid ceramic substrate. A newer version of the platinum RTD consists of a metal-film element embedded in a flexible carrier, which can be bonded to even curved test surfaces with appropriate adhesives. The operating range for these sensors is typically reduced to (−20°C, +180°C).

9.3 THERMOCOUPLES

In 1822, Thomas Seebeck discovered that current would circulate in a loop formed by joining two segments of dissimilar wires if the two joining points (junctions) were at different temperatures. Only if the metals are different and only if $T_1 \neq T_2$ does the effect exist. The fundamental property is actually associated with every single junction made of two dissimilar metals. When the junction is at a finite temperature, a thermal electromotive force (emf) is generated at the junction. The magnitude and polarity of this emf depends on the temperature and the specific metals involved.

For the composite loop in Fig. 9.14, unequal temperatures results in one emf being larger than the other, and hence a net emf is found for a complete path around the loop. A circulating current will be maintained.

If the loop is broken, as indicated in the lower schematic of Fig. 9.14, no current will circulate, but a potential difference will be present across the free ends of the wire.

Suppose metal A is copper. Then, a voltmeter (with internal copper wiring) installed across the free ends will correctly read the circuit emf because this arrangement has only introduced two additional copper–copper connections, which do not themselves generate thermal emf's.

But next suppose that metal A is not copper. As shown in Fig. 9.15, the new connecting junctions at the meter involve dissimilar metals, so additional thermal

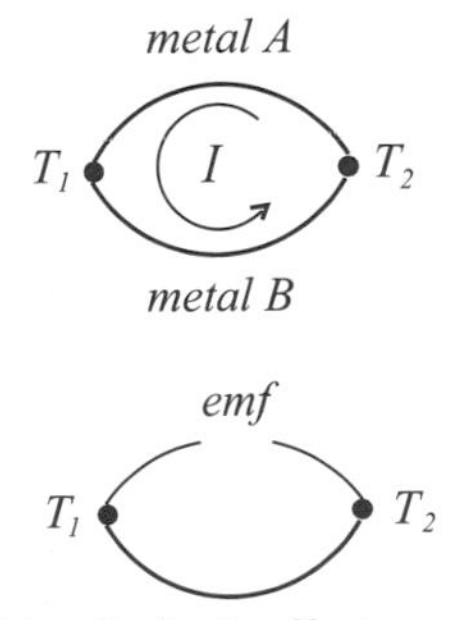

FIGURE 9.14. Seebeck effect as a circulating current in a closed loop with junctions at different temperatures, or as an emf appearing across the terminals in a broken loop.

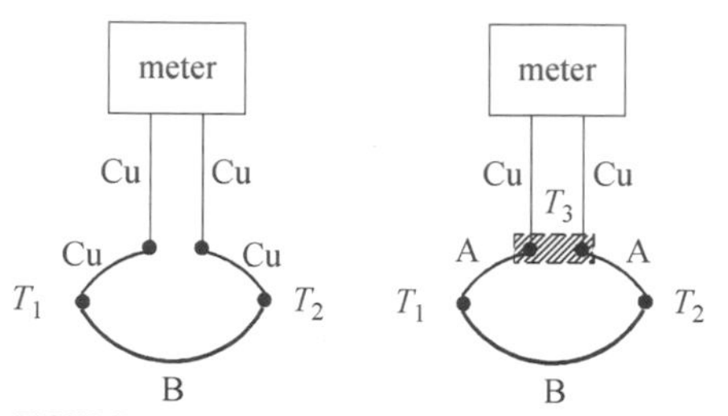

FIGURE 9.15. Two possible meter connections to the thermocouple defined by the two junctions at temperatures T_1 and T_2. In the first case, metal A is assumed to be copper, whereas in the second case neither A nor B is copper. The terminal block (shaded) is at temperature T_3.

emf's will be generated. However, if both connecting junctions are at exactly the same temperature T_3 by being thermally (but not electrically) anchored to a common point, then the associated emf's will be identical in magnitude and opposite in sign and hence will cancel. That is, from the diagram

$$V_{\text{meter}} = V_{\text{Cu}-A}(T_3) + V_{A-B}(T_2) + V_{B-A}(T_1) + V_{A-\text{Cu}}(T_3), \tag{9.58}$$

and because, trivially,

$$V_{\text{Cu}-A}(T_3) = -V_{A-\text{Cu}}(T_3), \tag{9.59}$$

then

$$V_{\text{meter}} = V_{A-B}(T_2) - V_{A-B}(T_1). \tag{9.60}$$

This can be the basis of a thermometer if the properties of a metal A/metal B junction have been experimentally determined in advance, and if one of the two junctions is at a known temperature, the usual choice being 0°C. The second junction then functions as a probe whose temperature can be determined from Eq. (9.60). This final arrangement is depicted in Fig. 9.16.

Thermocouples are manufactured using combinations of a few standard metals and alloys. The compositions are: iron, copper, 55% copper–45% nickel (Constantan), 95% nickel–2% manganese–2% aluminum (Alumel), and 90% nickel–10% chromium (Chromel). Conventionally, a thermocouple is a single-junction device formed from a pair of wires. Thermocouples are classified by type as in the following table:

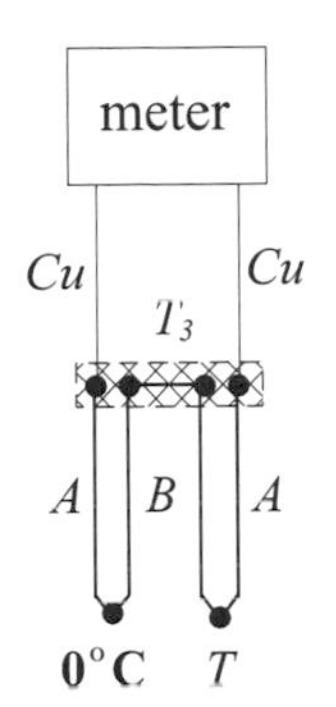

FIGURE 9.16. Final design for a thermometer based on two thermocouple probes, one at the reference temperature 0°C.

Type	Metal A – Metal B	Temperature Range (°C)
Type E	Chromel–Constantan	−200 to +900
Type J	Iron–Constantan	0 to +750
Type K	Chromel–Alumel	−200 to +1250
Type T	Copper–Constantan	−200 to +350

Plots of thermocouple output (as compared to a reference output for a similar junction at 0°C) versus temperature for these thermocouples are shown in Fig. 9.17. From the slopes of these characteristics, the approximate sensitivities

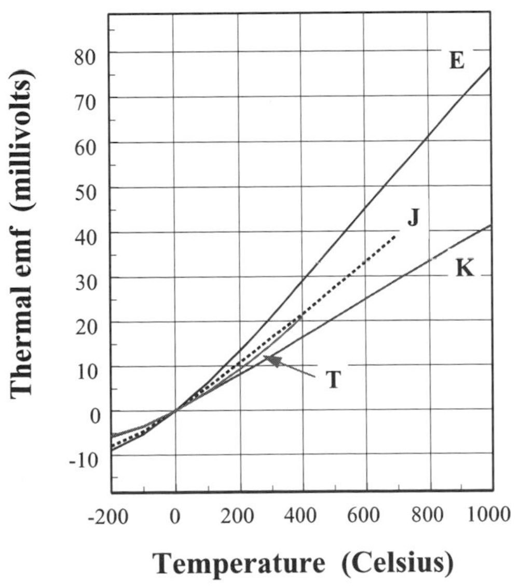

FIGURE 9.17. Thermocouple characteristics with reference junctions at 0°C.

are: Type E—70 μV/°C, Type J—55 μV/°C, Type T—50 μV/°C, and Type K—40 μV/°C.

There is clearly significant curvature, so accurate temperature readings based on measurements of the thermal emf's require table look-up techniques or polynomial representations of the characteristics. In microprocessor-based instrumentation, the polynomial approach is feasible.

The general expression to be used is of the form

$$T = a_0 + a_1 V + a_2 V^2 + a_3 V^3 + \cdots + a_n V^n, \qquad (9.61)$$

where V is the observed thermocouple emf in volts, T is the junction temperature in Celsius, and the coefficients appropriate for the particular types of thermocouple are given in the following table.

	Type E	Type J	Type K	Type T
	−100 to 1000°C	0 to 760°C	0 to 1370°C	−160 to 400°C
a_0	0.104967248	−0.048868252	0.226584602	0.100860910
a_1	17189.45282	19873.14503	24152.10900	25727.94369
a_2	−282639.0850	−218614.5353	67233.4248	−767345.8295
a_3	12695339.5	11569199.78	2210340.682	78025595.81
a_4	−448703084.6	−264917531.4	−860963914.9	−9247486589
a_5	1.10866E+10	2018441314	4.83506E+10	6.97688E+11
a_6	−1.76807E+11		−1.18452E+12	−2.66192E+13
a_7	1.71842E+12		1.38690E+13	3.94078E+14
a_8	−9.19278E+12		−6.33708E+13	
a_9	2.06132E+13			

These expansions yield junction temperatures from observed thermocouple emf's to better than 1% accuracy over the ranges indicated. As an example, suppose a Type J thermocouple is observed to have an output of 10.0 mV relative to a reference junction at 0°C. Then, from Eq. (9.61) and the tabulated coefficients, $T \simeq 186$°C. As a cross-check, thermocouple tables [3] for Type J devices do indeed show 10.0 mV output at this temperature.

Single Probe Configuration

As already noted, thermocouples are generally supplied in the form of two-wire, single-junction probes, whereas the thermometers discussed so far have employed two junctions, one of them held at a known reference temperature. Clearly, it would be a considerable simplification if this requirement for two probes could be relaxed.

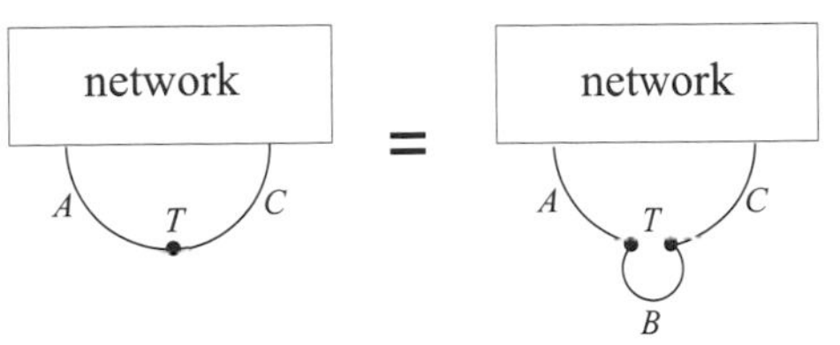

FIGURE 9.18. Illustration of the Law of Intermediate Metals.

As a first step in the simplification process, let us consider the so-called "Law of Intermediate Metals." This states that if an existing node at which two metals, say A and C, are connected is broken and a third metal, B, is inserted between A and C, then the original circuit and the new circuit will be thermally equivalent in their role within a larger network, provided both new junctions remain at the original temperature T. This is illustrated in Fig. 9.18, which is a graphical representation of the statement

$$V_{A-C}(T) = V_{A-B}(T) + V_{B-C}(T). \tag{9.62}$$

This idea can be utilized by revising the earlier schematic Fig. 9.15, first by combining the meter connectors and the reference junction onto a single thermal anchor at temperature T_{ref} and then eliminating the "intermediate metal," which in this case is the A segment. The two steps are illustrated in Fig. 9.19. Referring back to Eq. (9.60), the equation for the final configuration is now seen to be

$$V_{\text{meter}} = V_{A-B}(T) - V_{A-B}(T_{\text{ref}}). \tag{9.63}$$

Thus,

$$V_{A-B}(T) = V_{A-B}(T_{\text{ref}}) + V_{\text{meter}}, \tag{9.64}$$

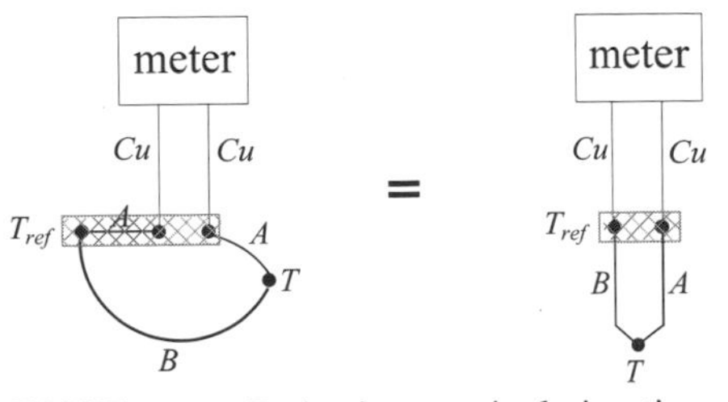

FIGURE 9.19. Reduction to a single-junction thermistor probe with a thermal anchor at T_{ref}.

which can also be expressed in the form

$$[V_{A-B}(T) - V_{A-B}(0°\text{C})] = [V_{A-B}(T_{\text{ref}}) - V_{A-B}(0°\text{C})] + V_{\text{meter}}. \tag{9.65}$$

The left-hand side represents the voltage difference that would have been generated by a pair of A–B thermocouple junctions, one of which is at a standard reference temperature of 0°C. This difference is precisely the output voltage referred to as V in the polynomial expansion: Eq. (9.61). Hence, evaluating the right-hand side of Eq. (9.65) would yield V, and from it, using the appropriate expansion coefficients in the earlier table, the desired temperature T.

The right-hand expression in Eq. (9.65) requires, in addition to the meter reading, a value for the factor

$$V_{A-B}(T_{\text{ref}}) - V_{A-B}(0°\text{C}). \tag{9.66}$$

Suppose that the temperature T_{ref} is measured by means of a sensor such as a thermistor or RTD that is bonded to the isothermal block indicated in the right-hand diagram of Fig. 9.19. This temperature can then be located in the standard table for the specific type of thermistor in use, and the table entry will represent precisely the factor in Eq. (9.66). All items on the right-hand side of expression (9.65) are thus determined, so by the process just outlined, the unknown temperature T is found.

Note that it has been possible with this scheme to eliminate the requirement of a second thermocouple immersed in an ice bath (0°C), but to accomplish this a second sensor must be employed to read the temperature T_{ref}. Such a tradeoff is frequently desirable.

9.4 TEMPERATURE-SENSITIVE DIODES

A diode formed from p- and n-type semiconductors obeys a current–voltage relationship of the form

$$I = I_0 \left[e^{\frac{qV}{kT}} - 1 \right], \tag{9.67}$$

where k is Boltzmann's constant, q is the electron charge, I and V are the diode current and voltage, respectively, and T is the absolute temperature of the device.

The reverse-bias saturation current is given by

$$I_0 = Aq\left[\frac{D_p}{L_p}\,p_n + \frac{D_n}{L_n}\,n_p\right]. \tag{9.68}$$

In this expression (this is a standard result; see, for example, Eq. 5-36 in [1]), A is the area of the diode structure, D_p and D_n are diffusion coefficients for holes and electrons, L_p and L_n are diffusion lengths for holes and electrons, p_n is the hole concentration in the n-material, and n_p is the electron concentration in the p-material.

Suppose the semiconductors are doped such that there are N_d donors per cm^3 in the n-side and N_a acceptors per cm^3 in the p-side. Essentially all donors and acceptors are ionized at temperatures above ~100 K (see, for example, [1], Fig. 3-18). Beyond about 400 K additional carriers are created by thermal excitation across the semiconductor energy gap. Thus, the majority carrier concentrations are relatively constant at the doping levels within the temperature range (100 K, 400 K).

$$\begin{aligned} n_n &= N_d, \\ p_p &= N_a. \end{aligned} \tag{9.69}$$

A fundamental result known as the Carrier-Product Equation (see, for example, [1], Eq. 3-24) stipulates

$$n_n\,p_n = n_i^2 \tag{9.70}$$

for the n-material and

$$n_p\,p_p = n_i^2 \tag{9.71}$$

for the p-material; n_i is the concentration of free carriers in intrinsic (undoped) material. Combining these expressions gives

$$I_0 = Aq\left[\frac{D_p}{L_p N_d} + \frac{D_n}{L_n N_a}\right] n_i^2. \tag{9.72}$$

The concentration of electrons (and equally, holes) in undoped material is itself temperature dependent. A standard expression (see, for example,

[1], Eq. 3-26) is

$$n_i(T) = 2\left[\frac{2\pi kT}{h^2}\right]^{\frac{3}{2}} (m_n^* m_p^*)^{\frac{3}{4}}\, e^{-\frac{E_g}{2kT}}, \tag{9.73}$$

where h is Planck's constant, E_g is the energy gap of the diode material, and m_n^*, m_p^* are the electron and hole effective masses, respectively. Hence,

$$\boxed{I(T) \simeq G\, T^3\, e^{-\frac{E_g}{kT}}\, e^{\frac{qV}{kT}}}. \tag{9.74}$$

In arriving at Eq. (9.74), all nontemperature-dependent terms have been grouped into the factor G and, in addition, the approximation

$$e^{\frac{qV}{kT}} \gg 1$$

has been made. This is usually valid for all but the smallest diode voltages or highest temperatures.

Constant Bias

Now suppose that the diode is supplied with a fixed dc current, I_{bias}. Then, obviously $dI/dT = 0$. Differentiating Eq. (9.74) gives, after some algebraic manipulation,

$$\boxed{\frac{dV}{dT} = \frac{V}{T} - \frac{V_g}{T} - \frac{3k}{q}}. \tag{9.75}$$

In this expression, $V_g = qE_g$ and V is the diode voltage. It is immediately apparent that in this constant bias mode, the diode voltage is temperature dependent. This, then, is the basis of a thermal sensor that uses a solid-state diode.

The sensitivity of the diode temperature sens or is just $\frac{dV}{dT}$ as defined in Eq. (9.75). This can be estimated as follows.

Consider a silicon diode biased at a current that results in a voltage of, say, 0.6 V. Suppose the diode is operating in the neighborhood of room temperature ($T = 300$ K). For silicon, $V_g = 1.12$ V, so

$$\frac{dV}{dT} = \frac{0.6}{300} - \frac{1.12}{300} - \frac{3 \times 1.381 \times 10^{-23}}{1.602 \times 10^{-19}} \approx -0.002 \text{ V/K}, \tag{9.76}$$

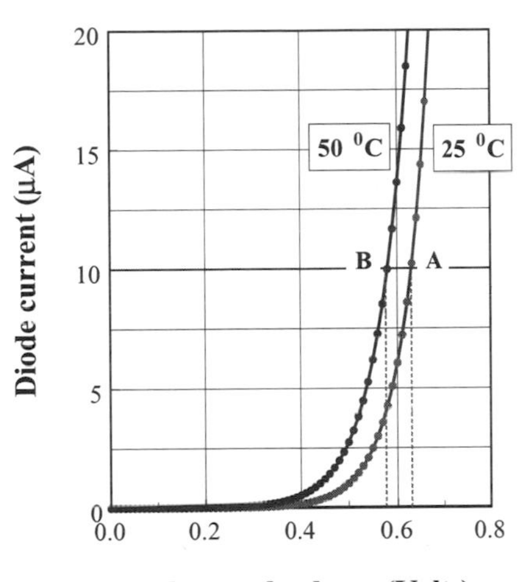

FIGURE 9.20. Current–voltage characteristics for a type 1N914 silicon diode (PSpice simulation) at two temperatures. If a constant bias of 10 μA is applied, the diode voltage changes over the interval indicated.

which is also -0.002 V/°C. Since the diode was presumed to be nominally at 0.6 V, this sensitivity could equally be expressed as about one part in three hundred per degree Celsius, or ~0.3% per °C. Interestingly, this numerical value is comparable to the sensitivity of platinum RTDs.

The action of the fixed-bias diode as temperature varies can be visualized with the aid of Fig. 9.20. At a fixed bias of 10 μA, the voltage across the device changes from 0.577 V to 0.629 V as the temperature changes by 25°C. Therefore, in this example, the sensitivity $\Delta V/\Delta T$ is approximately 0.00208, in good agreement with the earlier prediction in Eq. (9.76).

PROBLEMS

Problem 9.1. A resistive sensor is known to have a temperature dependence given by $R(T) = aT^2$, where T is in °C. Suppose this sensor is shunted with a parallel fixed resistance whose value is $R_0 = aT_0^2$. Show that this parallel combination has an equivalent resistance that is approximately a linear function of the distance from the reference temperature T_0. [Ans. $R_{\text{eq}} \approx \frac{aT_0^2}{2}\left[1 + (T - T_0)/T_0\right]$].

Problem 9.2. The data sheet for a particular thermistor gives the following information: resistance ratio (0 degrees C/50 degrees C) of 8.00 and a resistance at 25°C of 5000 Ω.

1. What is β? [Ans. 3667 K].
2. What is the thermistor resistance at 15°C? [Ans. 7665 Ω].
3. What is the sensitivity α at a temperature of 15°C? [Ans. −0.044].
4. What percentage change in thermistor resistance would occur if the temperature changed from 15°C to 16°C? [Ans. 4.4%].

Problem 9.3. What is the sensitivity dV/dT of a germanium diode (energy gap 0.66 eV) at an operating temperature of 300 K if it is biased at 0.400 V? [Ans. −0.00113 V/K].

BIBLIOGRAPHY

[1] Ben G. Streetman, *Solid State Electronic Devices*, fourth edition (Prentice Hall, Englewood Cliffs, NJ, 1995).
[2] *Thermistor Sensor Handbook* (Thermometrics Inc., Edison, NJ).
[3] *The Temperature Handbook* (Omega Engineering, Inc., Stamford, CT).

CHAPTER 10

Light

Visible light is just that small portion of the complete electromagnetic spectrum that the human eye, as a specific sensor, is capable of detecting. Because we can "see" this radiation, we have a special interest in it. But from a scientific or engineering point of view, other portions of the spectrum are also potentially important.

A schematic representation of the spectrum is shown in Fig. 10.1. The scale is broken at the marker to allow the longer wavelengths of television and radio to be included. Frequency is in hertz (Hz), which is the same as cycles-per-second. Wavelength is expressed typically in meters (m), or in angstroms (Å), where $1\,\text{Å} = 10^{-10}$ m. These two quantities are related by

$$c = f\lambda, \tag{10.1}$$

with c being the speed of light:

$$c = 2.9979245 \times 10^{8}\ \text{m/sec}$$

in a vacuum. In a transparent medium such as glass or air, light travels at a speed v that is less than c, and the wavelength is proportionately shortened (the frequency remaining unchanged). The ratio c/v is the *index of refraction* of a given medium.

Since Einstein's explanation of the photoelectric effect in 1905, it has been appreciated that there is a fundamentally quantum nature to light. An electromagnetic wave carries energy, and from the quantum point of view this energy is packaged in the form of discrete entities called *photons*. Every photon associated with radiation of frequency f carries exactly the same energy:

$$E = hf, \tag{10.2}$$

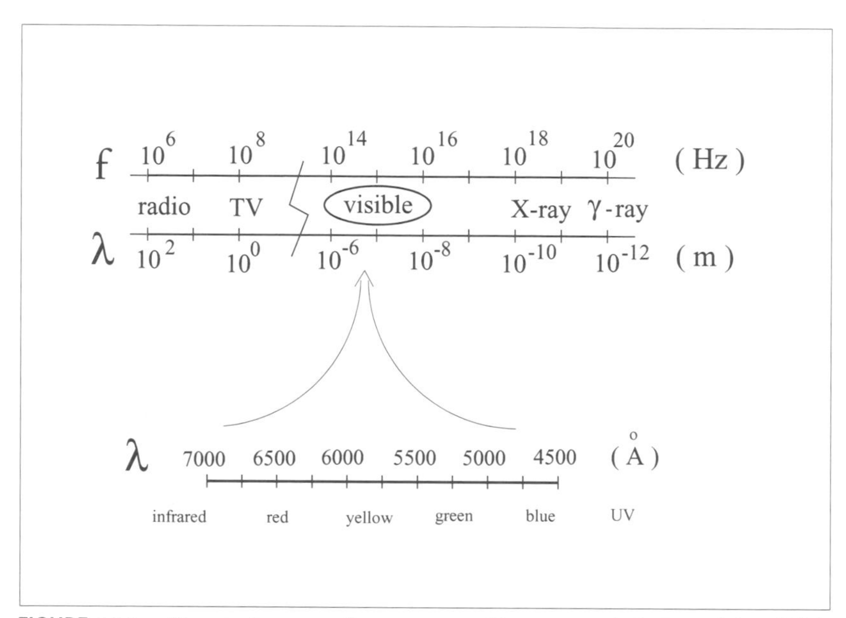

FIGURE 10.1. The electromagnetic spectrum with an expanded view of the visible portion. λ and f are the wavelength and frequency of the radiation.

where h is Planck's constant

$$h = 6.626075 \times 10^{-34} \text{ Joule–sec.}$$

Maxwell's classical theory is a description of propagating electromagnetic waves comprised of oscillating electric and magnetic fields. The quantum theory pictures radiation in terms of streams of photons, each propagating at the velocity of light, having zero rest mass, and carrying an exact quantity of energy, as specified in Eq. (10.2). These two descriptions form a duality in the sense that light may manifest itself in either form. Hence, the photoelectric effect can be explained only from a quantum perspective, whereas the operation of a lens requires a wave description.

Clearly, the operation of light sensors will be intimately bound up in this duality. Some types of detector respond to photons, whereas others react to the wave properties. Photodiodes belong to the first category; bolometers belong to the second. Whatever the operating principle, all photodetectors must be packaged such that radiation can enter the otherwise sealed unit and fall upon an active surface of the sensitive material. This entails providing a small window which is either flat or in the form of a minilens. The window type determines the field of view of the sensor.

In the previous chapter, sensors were discussed that were appropriate for temperatures generally limited to the range (−150°C, 1000°C). This range was deemed to encompass the normal working environment of most scientists. In a similar vein, here we limit the discussion to the visible or near-visible portion of the spectrum. Sensors for radio, television, x-rays, γ-rays, etc., will not be discussed.

10.1 PHOTOCONDUCTIVE SENSORS

A photoconductive sensor is simply a light-dependent resistor. As in the case of a thermistor [Eq. (9.2)], the conductivity of an intrinsic semiconductor is

$$\sigma = n_i\, q\, \mu_e + p_i\, q\, \mu_h, \tag{10.3}$$

but now the carrier density is controlled by photoexcitation rather than temperature, as before. μ_e and μ_h are the electron and hole mobilities.

The excitation process is depicted in Fig. 10.2. If the incoming photon has sufficient energy, it will be capable of liberating a valence electron from its parent atom. The free electron then is added to the conduction band. At the same time, the empty site of the original valence electron becomes a hole in the valence band. Both electron and hole constitute additional mobile charges that can contribute to current flow in the semiconductor.

The change in conductivity due to photoinduced carriers is

$$\Delta\sigma = q\,(\mu_e \Delta n + \mu_h \Delta p)\,. \tag{10.4}$$

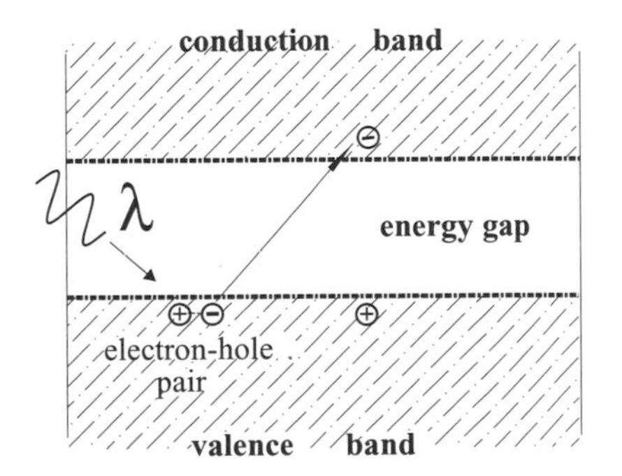

FIGURE 10.2. Schematic illustration of incoming photon (λ) exciting an electron from the valence band into the conduction band. This results in the creation of two carriers: the electron in the conduction band and a hole in the valence band.

Let g_{op} represent the number of photoexcitations per unit volume per unit time, and let τ_e and τ_h denote the lifetimes of electrons in the conduction band or holes in the valence band, respectively. Clearly, there are two competing processes—the creation of carriers through photoexcitation and the loss of free carriers through recombination. The net steady-state increase in density of conduction electrons is just ($g_{op}\ \tau_e$), and the net increase in density of holes is ($g_{op}\ \tau_h$). Therefore (see [1], Eq. 4-17, p. 112),

$$\Delta\sigma = q g_{op} \left(\mu_e \tau_e + \mu_h \tau_h\right). \tag{10.5}$$

Suppose the photoconductor is a sample of length ℓ and that a voltage V is applied from end to end. From the definition of mobility (drift velocity per unit electric field), the transit times for carriers down the sample are

$$t_e = \frac{\ell}{\mu_e \frac{V}{\ell}}, \tag{10.6}$$

$$t_h = \frac{\ell}{\mu_h \frac{V}{\ell}}. \tag{10.7}$$

The *gain* of the photoconductor is defined as the ratio of carrier lifetime to transit time. Thus,

$$\text{Gain} = \left(\tau_e \mu_e + \tau_h \mu_h\right) \frac{V}{\ell^2}. \tag{10.8}$$

From this expression, it is apparent that a small sample length is desirable. This condition is achieved in practical photoresistors by means of interleaved contacts, as shown in Fig. 10.3.

In effect, a very wide but short photoconductor has been folded like an accordion. The "length" is now the spacing between digits for contact A and contact

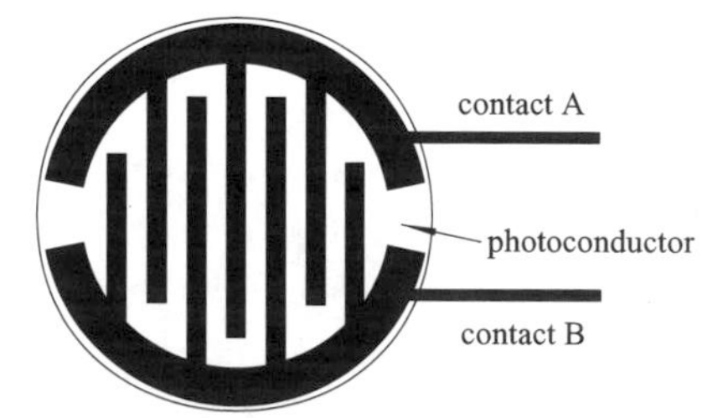

FIGURE 10.3. Layout of a photoconductive cell. The fingers of the two contacts are interleaved to produce an equivalent of a short but wide photoconductor.

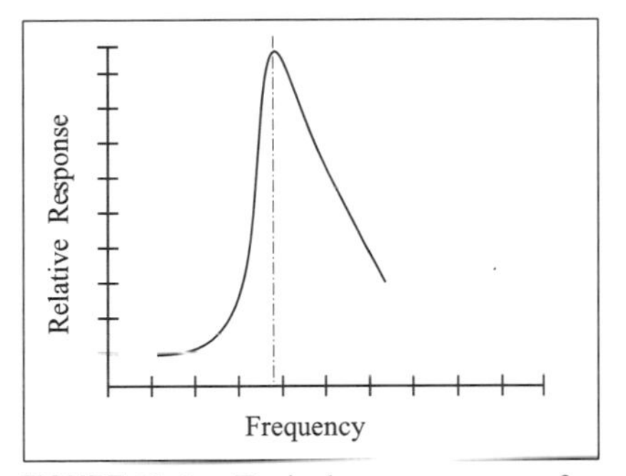

FIGURE 10.4. Typical response curve for a photoconductor. A rapid increase in absorption occurs close to the frequency for which $hf \approx E_g$. At higher frequencies, the response drops off more slowly.

B, and the "width" is the linear distance summed over the fingers of either contact.

A number of materials are in common use for photoconductive sensors. These include CdS, CdSe for the visible range, and PbS, PbTe, PbSe for the infrared. Each semiconducting compound may be characterized by an energy gap, and this gap in turn dictates the minimum photon energy which will be capable of exciting electrons into the conduction band.

Each of these materials exhibits a relative response characteristic whose shape is roughly as depicted in Fig. 10.4. Beginning at low frequencies (long wavlengths), the response is very small since the photon energy is less than the gap. When $hf \approx E_g$, the response rises rapidly because of the photo-generation of additional carriers. Slightly beyond this point, the response tails off due to the increasing surface effects. The band gaps and corresponding wavelengths of some common photoconductors are given in the following table.

Material	Energy Gap (eV)	Wavelength (Å)
CdS	2.42	5130
CdSe	1.73	7176
PbS	0.37	33554
PbTe	0.29	42811
PbSe	0.27	45982

The relationship between incident optical power and resulting resistance is very nonlinear for photoconductors. An approximate expression is

$$\log_{10} R = a - b \log_{10} P, \tag{10.9}$$

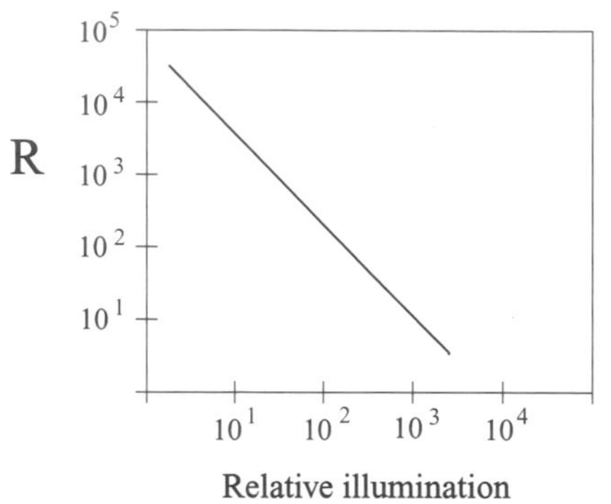

FIGURE 10.5. Illustration (hypothetical) of typical log–log dependence of photoresistance on illumination.

which yields a linear log–log plot, as illustrated in Fig. 10.5. The slope is of course negative, since increasing radiation increases conductivity and so decreases resistance. For commercial sensors, it is quite typical to find ratios of dark resistance (high) to well-illuminated resistance (low) having values of 10^4, as this figure suggests.

Photoresistors are inexpensive but very nonlinear and slow. Response times are as long as 100 msec. They are most often employed where light threshhold switching is required.

10.2 PHOTODIODES

To understand the consequences of light radiation incident on a diode, first the fundamental internal behavior of a p-n junction must be recalled (see [1], Chapter 5 and Section 6.3). As depicted in Fig. 10.6, free electrons from the n-side diffuse into the p-side of the junction, while mobile holes from the p-side diffuse into the n-side. This transfer of charge ends in an equilibrium state in which thin non-neutral regions called *depletion zones* exist immediately on either side of the original contact plane. The internal electric field set up by these shifted charges is, as indicated, directed from the n-side to the p-side.

When radiation is directed into the semiconductor, electron-hole pairs are generated, provided the incoming photon energy is sufficient. In the p-side, any electron from such a pair is a minority carrier and will diffuse toward the junction. If this electron (from the electron-hole pair) was generated within a distance L_n (minority carrier diffusion length) of the junction, then it will reach the narrow region in which the internal electric field E exists. The direction of this field, as

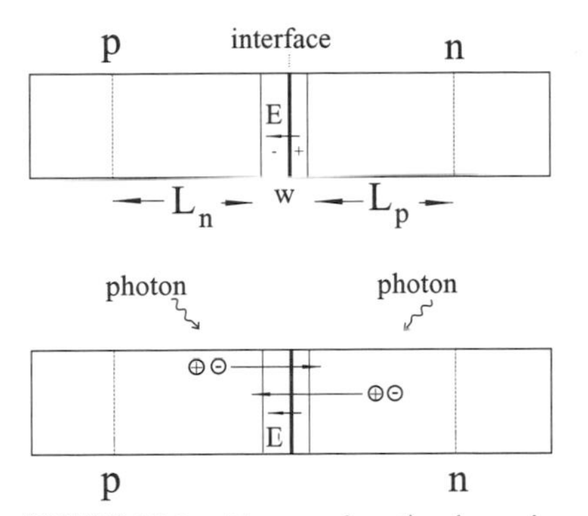

FIGURE 10.6. Top: p-n junction in equilibrium with depletion zone of total width w and internal electric field E. Bottom: electron-hole pair generation from incoming photons, and diffusion of minority carriers towards the junction, resulting in a net charge transfer.

the diagram suggests, causes the electron to be swept rapidly across the junction and into the n-side. Likewise, holes from optically generated electron-hole pairs in the n-side, that are created within a diffusion length L_p of the interface will diffuse to the region of E and then be swept into the p-side. The incident radiation thus effectively "pumps" charge across the diode junction.

To quantify the preceding picture, recall the current–voltage characteristic of a semiconductor diode, which was given in the previous chapter as Eqs. (9.67) and (9.68). The relevant expressions were

$$I = I_0 \left[e^{\frac{qV}{kT}} - 1 \right] \tag{10.10}$$

and

$$I_0 = Aq \left[\frac{D_p}{L_p} p_n + \frac{D_n}{L_n} n_p \right]. \tag{10.11}$$

But an optically generated current now exists, and this must be added to the right-hand side of Eq. (10.10).

$$I = I_0 \left[e^{\frac{qV}{kT}} - 1 \right] - I_{op}. \tag{10.12}$$

The negative sign on I_{op} properly accounts for the direction of the total photocurrent, which is from the n-side toward the p-side, and is thus opposite to the sense of positive forward current in the diode.

Let g_{op} represent the rate of photon-induced electron-hole pairs per unit volume per unit time. Then, within the diffusion region to the left, there will be $g_{op}AL_n$ electrons per second, which will all reach the p-n interface via diffusion. Similarly, there will be $g_{op}AL_p$ holes per second created in the n-side, which will all diffuse to the interface and cross over. Neglecting photon processes in the very thin depletion zone (w), the total current thus generated will be

$$I_{op} = g_{op} q A \left(L_n + L_p \right). \tag{10.13}$$

This extra term in Eq. (10.12) has the simple effect of translating the original diode $I - V$ curve downward by an amount I_{op}. A set of such displaced characteristics is shown in Fig. 10.7.

Two special points may be identified on any characteristic. One is defined by the condition $I = 0$ and is equivalent to an open-circuit condition. The forward diode voltage at this point is thus referred to as the open-circuit voltage V_{oc}. The other point is defined by the condition $V = 0$, which is equivalent to a short-circuit so the corresponding current is labeled I_{sc}.

As already noted, optical generation causes a downward displacement in the $I - V$ characteristic of the diode. As this happens, I_{sc} moves to more negative

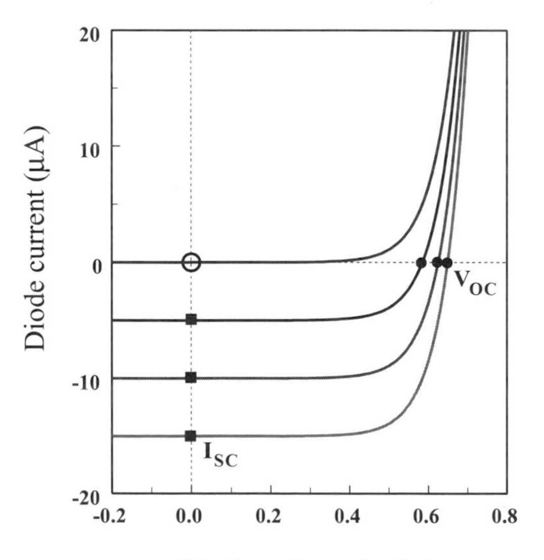

FIGURE 10.7. Current–voltage characteristics for a photodiode under a sequence of evenly increasing illumination levels. The open-circuit voltages are marked by filled circles and the short-circuit currents by filled squares.

values and V_{oc} moves to more positive values. The dependence of I_{sc} on illumination level is linear, but the variation of V_{oc} is nonlinear, as the figure makes clear.

Under short-circuit conditions ($V = 0$), Eq. (10.12) yields

$$I_{sc} = -I_{op} = -g_{op}qA(L_n + L_p). \tag{10.14}$$

As mentioned earlier, the negative sign means that current is flowing internally from the n-side of the diode to the p-side. To the external world, current would be seen to emerge from the p-side (anode), flow through the wire constituting the short, and reenter the diode at the n terminal (cathode).

Sensitivity

The sensitivity (sometimes called responsivity) of a photodiode is defined as

$$S = \frac{I_{diode}}{P}, \tag{10.15}$$

where P is the incident radiant power per unit area. Note that radiant power per unit area is just energy per unit time per unit area, and this can be viewed as the incident photon flux (number per unit area per unit time). The photodiode is of course responding specifically to the flux of photons impinging on its surface. Typical values for the sensitivity of commercial photodiodes might be $S \approx 2\ \mu\text{A/mW/cm}^2$.

Photoresistors were seen to exhibit a peaked spectral response partly because of the requirements of minimum photon energy for the excitation of electrons from the valence band into the conduction band. The same is true for photodiodes since electron-hole pairs must be created. The diode material of choice is silicon, and this sets the spectral peak at a wavelength around 900 nanometers, or equivalently 9000 Å, which is in the infrared. At $\lambda \approx 6000$ Å, silicon photodiode response is typically reduced to about 60 or 70% of the peak value.

Speed

In the earlier description of this device, the photocurrent was shown to be the result of the creation of electron-hole pairs followed by the diffusion of the resulting minority carriers toward the depletion zone at the junction. The internal electric field at the depletion zone then rapidly sweeps these carriers across the barrier. In terms of speed of response to abrupt changes in the radiation—say, for applications to fiber-optic communications—the slow portion of the process is associated with diffusion.

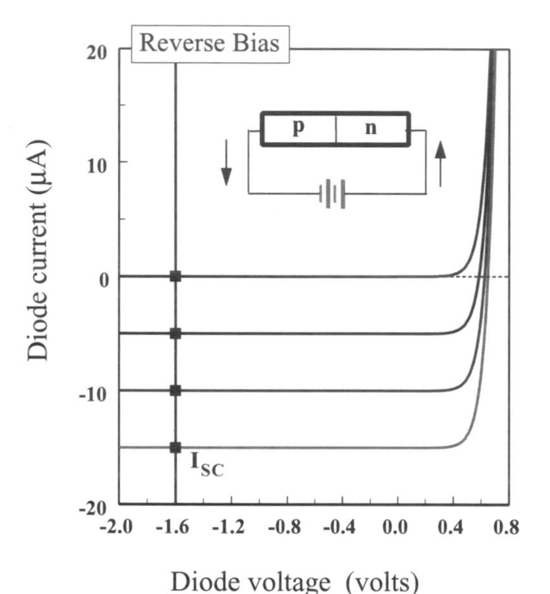

FIGURE 10.8. Photodiode operating in reverse-bias mode. The inset shows a device biased with a battery. The indicated current (arrows) corresponds to negative values in the plot, since the flow enters the diode cathode and exits the cathode.

Significant increases in speed can be achieved by operating a photodiode in a reverse-bias state rather than the pure short-circuit (zero-bias) condition. This means that a battery or power supply must be connected across the diode so that the + battery terminal is connected to the n-side (cathode) of the diode while the − battery terminal is connected to the p-side (anode) of the diode. This situation is illustrated in Fig. 10.8.

With the application of an external bias, the depletion zones in a p-n junction change according to the expression (see, for example, [1], Eq. 5-57, p. 174)

$$w = \left[\frac{2\epsilon\,(V_0 - V)}{q}\left(\frac{N_a + N_d}{N_a N_d}\right)\right]^{\frac{1}{2}}. \tag{10.16}$$

In this expression for the total width of the depletion zone, ϵ is the material permittivity, N_a and N_d are acceptor and donor doping concentrations, V is the applied bias voltage, and the contact potential is given by (see, for example, [1], Eq. 5-8, p. 142)

$$V_0 = \frac{kT}{q}\ln\frac{N_a N_d}{n_i^2}. \tag{10.17}$$

A *reverse* bias is specified by a negative applied voltage, and it is apparent from Eq. (10.16) that this results in a larger value for w. The internal electric field associated with the distribution of charges in the depletion zone now extends over a greater distance on either side of the junction, thus capturing and accelerating more photo-induced minority carriers across the barrier. Hence, an increased proportion of carriers are moved rapidly by the field, instead of more slowly by diffusion.

Recommended values of reverse bias for many commercial photodiodes are in the neighborhood of 10 to 20 V. Response times of less than 1 μsec are readily achieved.

For more demanding applications, a variation on the p-n structure is used. Reduced device capacitance and enhanced speed result when an intrinsic semiconducting layer is interposed between the p and n materials. Photon-induced electron-hole pairs within the intrinsic middle layer of such a p-i-n diode result in larger currents, which are transported more rapidly. Under reverse-bias conditions, response times of less than 1 nsec are commonly achieved.

Interface Circuits

In either the short-circuit or reverse-bias mode, the photodiode acts as a generator of photocurrent. This current comes out of the p-side and reenters the n-side. A practical light meter requires a means of sensing the photocurrent but at the same time without perturbing it. Ideally, in other words, no new resistance should be caused to appear in the diode loop.

One application of op-amps discussed in Chapter 6 (see Fig. 6.11), was the current-to-voltage converter. This circuit can be used in the present context, as indicated in Fig. 10.9.

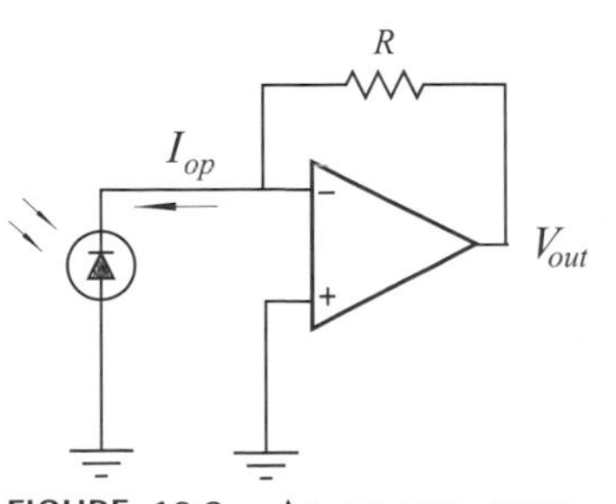

FIGURE 10.9. An op-amp arranged as a current-to-voltage converter with a photodiode at the input. The output voltage is proportional to the photocurrent and hence to the illumination.

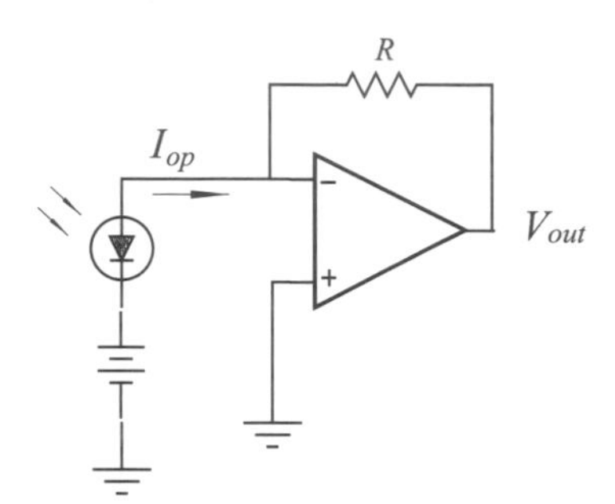

FIGURE 10.10. Modification of the circuit in Fig. 10.9 to provide reverse bias on the photodiode.

The input impedance is essentially zero, so the diode operation is unaffected. The output voltage is just

$$V_{\text{out}} = I_{op} R, \tag{10.18}$$

so the feedback resistor sets the conversion ratio.

For reverse-bias requirements, the circuit can be modified slightly, as indicated in Fig. 10.10.

The output voltage for this arrangement is

$$V_{\text{out}} = -I_{op} R. \tag{10.19}$$

The changed sign as compared with Eq. (10.18) merely reflects the changed direction of the diode and hence the photocurrent.

PROBLEMS

Problem 10.1. A p-n junction photodiode is illuminated with light whose photons have energy larger than the semiconductor gap. The diode has an effective area of 0.01 cm^2, and the diffusion lengths are $L_p = 0.04$ mm and $L_n = 0.20$ mm. How many electron-hole pairs are being created in the diode if the short-circuit current is measured to be 10 μA? [Ans. 2.60×10^{17} cm^{-3} sec^{-1}].

Problem 10.2. A photoconductor is characterized by the following data: $\tau_e = \tau_h = 3$ nsec; $\mu_e = 575$ cm^2/volt–sec; $\mu_h = 200$ cm^2/volt–sec. A bias of 30 V is applied along a length $\ell = 1$ mm of material. What is the resulting value of the gain? [Ans. 6.98×10^{-4}].

Problem 10.3. A photodiode is operated in the open-circuit mode. Suppose the observed voltage, at some light level, is 0.3 V. What will the open-circuit voltage become if the light level is now doubled? You may assume that g_{op} is relatively large and that $T = 300$ K. [Ans. 0.318 V].

Problem 10.4. A photodiode has characteristics as shown in Fig. 10.11. It is shunted by a 100 K resistor.

1. Draw the resistive load line on the figure.
2. Estimate the change in output voltage that would occur if the illumination went from 10 mW/cm^2 to 30 mW/cm^2.

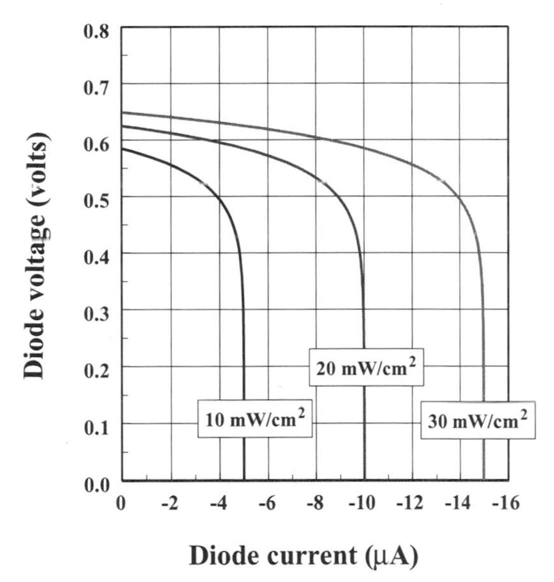

FIGURE 10.11. Problem 10.4.

BIBLIOGRAPHY

[1] Ben G. Streetman, *Solid State Electronic Devices*, fourth edition (Prentice Hall, Englewood Cliffs, NJ, 1995).

CHAPTER 11

Magnetic Fields

Static electric charge creates an electric field in space. Distributions of charge produce an overall electric field which is the vector summation of the contributions from all elements. *Moving* electric charge (i.e., an electrical current) creates a magnetic field in space. This connection was discovered by H.C. Oersted in 1819. The exact relationship between an elemental unit of current $\overrightarrow{dI}$ which is a vector quantity having both magnitude and direction, and the resulting magnetic field vector $\overrightarrow{dB}$ at location $\overrightarrow{r}$, is expressed in the Biot-Savart Law:

$$\overrightarrow{dB} = \frac{\mu_0 I}{4\pi} \frac{\overrightarrow{d\ell} \times \overrightarrow{r}}{r^3} \tag{11.1}$$

The current element has been taken to be a current of magnitude I flowing through a differential length of wire $\overrightarrow{d\ell}$. Summing such infinitesimal contributions over a macroscopic arrangement of current-carrying wires can easily become a challenging problem in integral calculus.

In expression (11.1), the constant μ_0 is called the *permeability of free space*, and its value is

$$\mu_0 = 4\pi \times 10^{-7} \text{ N amp}^{-2} \tag{11.2}$$

The (MKS) unit of magnetic field B is thus N amp^{-1}m^{-1}. But the force unit (N) is itself dimensionally equivalent to J m^{-1} and hence to volt coul m^{-1}. Consequently the magnetic field unit is volt sec m^{-2}, which is termed a tesla (T). A name has been attached to a portion of this:

$$1 \text{ volt sec} = 1 \text{ weber (Wb)}$$

So the magnitude of a magnetic field B may be expressed either in tesla or, equivalently, webers per square-meter.

An alternate unit for B is a gauss (G) and the relationship between the two is very simple

$$10^4\,\mathrm{G} = 1\,\mathrm{Wb/m^2} = 1\mathrm{T}$$

For weak magnetic fields, a unit called the *gamma* (γ) is sometimes used.

$$1\ \mathrm{gamma} = 10^{-5}\,\mathrm{G}.$$

As discussed in the section of Chapter 1 dealing with magnetism, the magnetic field of Earth—which constitutes an ever-present laboratory background unless special shielding is in place—has a magnitude of approximately 0.5 G. As another illustration, the magnetic field at a distance of 1 cm from a wire carrying a steady current of 1 amp is about 0.2 G. Biomagnetic fields are very much smaller, with nominal values ranging from $\sim 10^{-6}\,\mathrm{G} = 0.1\gamma$ for cardiograms to perhaps $10^{-10}\,\mathrm{G} = 10\,\mu\gamma$ for evoked cortical signals.

Larger fields are associated with both permanent magnets and electromagnets. Even in common devices such as loudspeakers, fields in the narrow gaps of the permanent magnets can easily reach levels of $1\,\mathrm{T} = 10^4\,\mathrm{G}$.

Notice that B is really a specification of flux density in the sense that a Weber can be viewed as a total quantity of magnetic flux that is contained within a given area. Thus, if Φ denotes a total magnetic flux, then $\Phi = BA$, where A is the area over which the magnetic field is B.

11.1 HALL-EFFECT SENSORS

Perhaps the most widely used detector of magnetic fields in use today is the Hall probe [1]. The physical principle on which its operation is based is the Hall effect, discovered in 1879. Suppose free carriers moving down a sample as a result of an applied longitudinal electric field are at the same time exposed to a magnetic field oriented as shown in Fig. 11.1.

Any charged particle moving with a velocity $\vec{v}$ in the presence of a magnetic field $\vec{B}$ will experience a force, which is given by the expression

$$\vec{F_L} = q\left(\vec{v} \times \vec{B}\right). \tag{11.3}$$

This is known as the *Lorentz force.* In the present case, the force acts laterally across the sample, as shown. As a result, moving charges will be displaced to the side of the sample, resulting in an internal transverse electric field, E_H. The

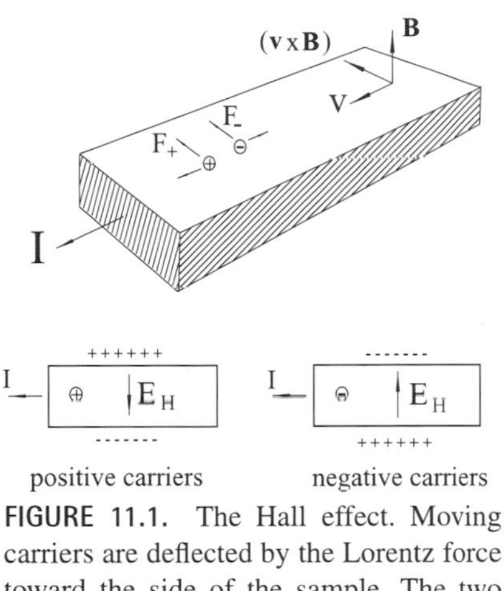

FIGURE 11.1. The Hall effect. Moving carriers are deflected by the Lorentz force toward the side of the sample. The two top views illustrate how the polarity of the charge carriers is reflected in the polarity of the transverse voltage.

polarity of this field is determined by the sign of the free carriers. The magnitude of E_H can be found as follows.

In equilibrium, a sufficient amount of transverse charge displacement has occurred so that the Lorentz force is just balanced by the electric force. That is,

$$q\,vB = qE_H, \tag{11.4}$$

so the Hall field is

$$E_H = vB. \tag{11.5}$$

The bias current in a sample of cross section $A = wt$, with w being the width and t the thickness (in the direction of B), can be expressed $I = qAvn$ for negative carriers (electrons) with drift velocity v, and $I = qAvp$ for positive carriers (holes). Using Eq. (11.5) and the definition of transverse Hall voltage $V_H = wE_H$,

$$V_H = \left[\frac{I}{qtn}\right]B \tag{11.6}$$

for negative carriers, and

$$V_H = \left[\frac{I}{qtp}\right]B \tag{11.7}$$

for positive carriers.

The Hall voltage measured across the sample is therefore a linear function of applied magnetic field and is the basis of a very useful sensor. The factors in square brackets represent the sensitivity of the probes, and clearly this is enhanced by choosing a relatively large bias current or a thin sensor element.

Example

Suppose a thin ribbon of copper is placed in a magnetic field of strength $B = 0.1$ T. The ribbon thickness is 0.2 mm. Because copper is a metal, the free carriers are electrons; at room temperature, the density of conduction electrons is $8.5 \times 10^{28}\ \mathrm{m}^{-3}$. For a bias current of one amp, the Hall voltage would then be

$$V_H = \frac{1.0}{1.6 \times 10^{-19} \times 2.0 \times 10^{-4} \times 8.5 \times 10^{28}} \times 0.1 = 37\ \mathrm{nV!}$$

The sensitivity in this case is thus 0.37 μV per tesla. This reveals the small size of the effect in metals for moderate fields, currents, and dimensions.

Because of the appearance of carrier concentrations (n or p) in the denominators of Eqs. (11.6) and (11.7), semiconductors will exhibit much larger Hall voltages than metals.

Commercial Hall probes are available in the form of small encapsulated integrated circuits that have on-chip amplifiers. Three-, four-, and eight-pin plastic packages are common. A three-pin package is shown in Fig. 11.2.

Active areas within which the magnetic field is actually sensed range around 0.1×0.1 inches or less. These ICs require power supply voltages of around 5–10 V and corresponding bias currents of only 1–10 mA. Their sensitivities tend to be ~1–10 volts per tesla.

Complete Hall effect gaussmeters that include electronics, digital readout, and suitable probes are also widely available. Resolutions to below 10 mG are possible with these systems.

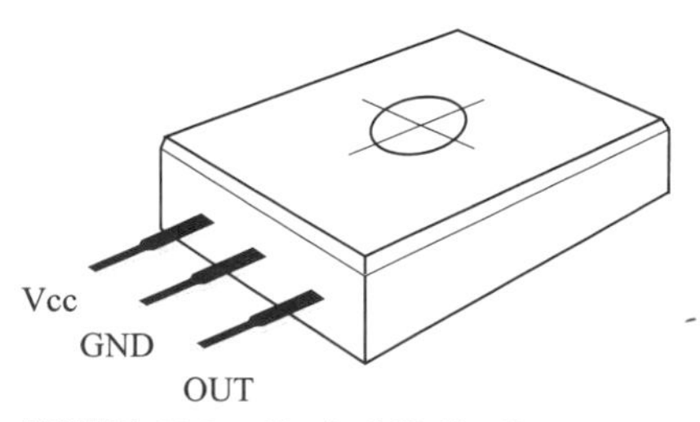

FIGURE 11.2. Typical Hall effect sensor packaged in a three-pin module. Integrated electronics including amplifiers simplify the connections so that only two power supply leads and a sensor output are required.

11.2 FLUXGATE MAGNETOMETERS

The central element in a fluxgate magnetometer is a coil with a ferromagnetic core. To explain the operation of this type of magnetometer, some fundamentals of solenoids must first be reviewed.

Fundamentals

For the particular case of a long, straight, thin wire coil wound at n turns per meter and carrying an energizing current of I amps, the internal magnetic field within the solenoid is given by

$$B_0 = \mu_0 n I. \tag{11.8}$$

If the coil is filled with a magnetic core (see Fig. 11.3), the internal field will increase:

$$B = B_0 + \mu_0 M, \tag{11.9}$$

where M is the *magnetization* (amp–turns per meter) of the core caused by the tendency of magnetic dipoles to align with the driving field. The net B field is thus intensified by the core.

Another quantity H known as the *magnetic intensity* is defined by

$$H = \frac{B_0}{\mu_0}. \tag{11.10}$$

Clearly, H is directly tied to the driving current; it has the same units as M, namely amp–turns per meter or amps/m. For the solenoid,

$$H = nI. \tag{11.11}$$

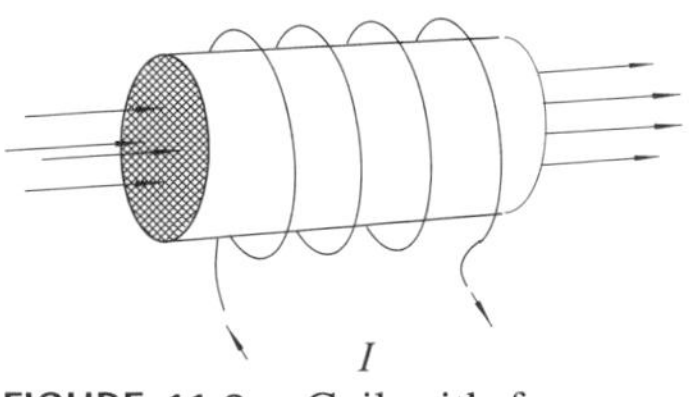

FIGURE 11.3. Coil with ferromagnetic core energized by a drive current I.

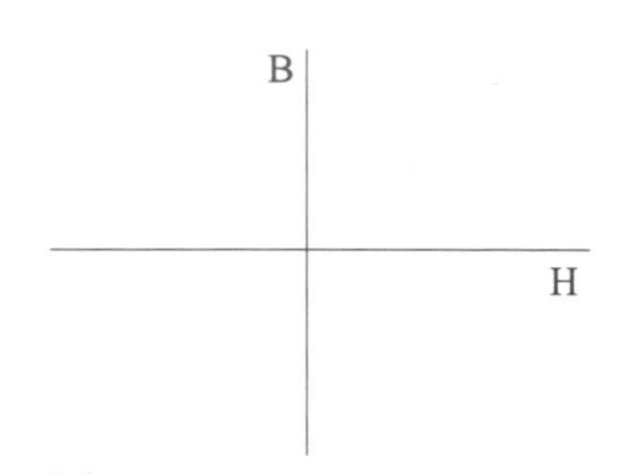

FIGURE 11.4. Hysteresis curve for a magnetic sample.

From Eqs. (11.9) and (11.10),

$$B = \mu_0 (H + M). \tag{11.12}$$

The magnetization is a response by the core to H, so

$$M = \chi H, \tag{11.13}$$

where χ is the susceptibility of the core material. Thus,

$$B = \mu_0 (1 + \chi) H = \mu H, \tag{11.14}$$

where μ is the permeability of the core material. Ferromagnetic materials are quite nonlinear, meaning that the permeability is not a simple constant. A plot of B versus H typically exhibits nonlinearity and hysteresis, as shown in Fig. 11.4.

Fluxgate Principle

The subject of fluxgate theory and design is complex and specialized. Good reviews are to be found in the literature [2, 3, 4, 5, 6, 7, 8]. A number of different geometries have emerged over the past five decades. The discussion here will focus on the so-called Vacquier parallel-gated flux sensor illustrated in Fig. 11.5.

Two parallel cores are wound in series from a single wire in such a way that the resulting coils are equal but opposing. This excitation coil is then driven by a suitable time-varying current I_{ex}. The pickup coil surrounds both excitation coils and is used to detect any net time-varying flux. According to Faraday's law of induction, the voltage that will appear across the pickup coil will be

$$V_{pickup} = n \frac{d\Phi}{dt}, \tag{11.15}$$

where n is the number of turns in the coil and Φ is the net magnetic flux.

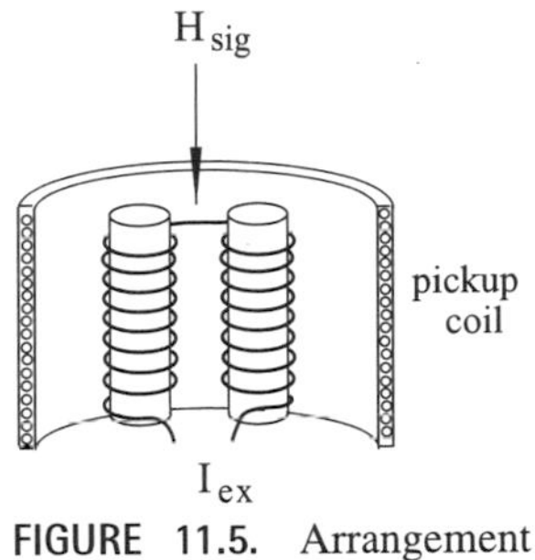

FIGURE 11.5. Arrangement of coils and cores for a Vacquier-type fluxgate magnetometer. The pickup coil is shown cut away to reveal the inner components.

In the absence of any external field ($H_{sig} = 0$), the energizing current I_{ex} will produce oppositely directed induced fields in the coils. Hence, the net flux within the pickup coil is zero, and no output voltage will be generated.

To see what happens when a constant external field is present, consider Fig. 11.6. To keep a focus on basics, the hysteresis loops have been collapsed and simplified to piecewise-linear magnetization curves (this argument follows [2]). In one core (coil #1), the linear segment is $H = H_{ex} + H_{sig}$, while in the other

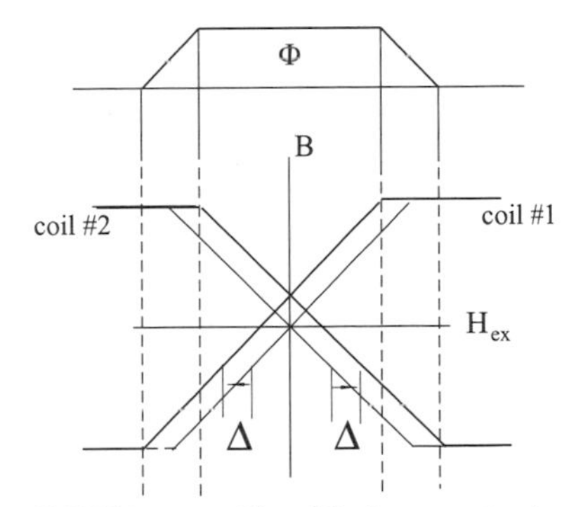

FIGURE 11.6. Simplified magnetization curves for the two coils in a Vacquier-type fluxgate sensor. When H_{sig} is present, the curve for coil #1 is shifted to the left by $\Delta \approx H_{sig}$ and the curve for coil #2 is shifted to the right by the same amount. The upper curve shows the net resulting flux as a function of the excitation H_{ex}.

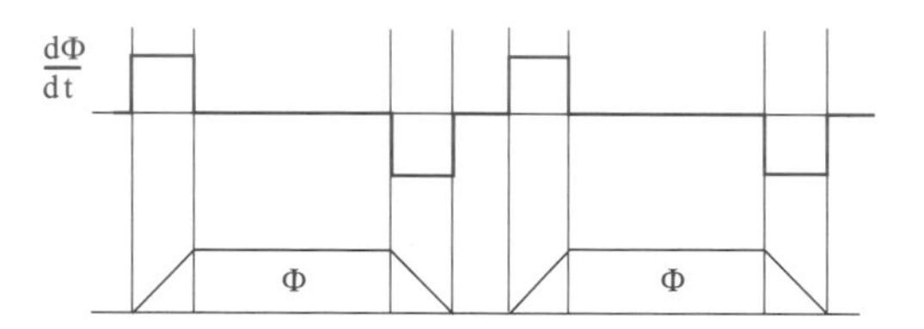

FIGURE 11.7. Time-varying flux and flux derivative resulting from a periodic excitation field H_{ex}.

core (coil #2) it is $H = -H_{ex} + H_{sig}$, the minus sign arising from the winding reversal. (More precisely, the H_{sig} terms need to be reduced somewhat because of a geometrical effect, which is accounted for by so-called demagnetizing factors.) The linear segments for either core terminate in horizontal sections corresponding to constant B states arising from magnetic saturation of the cores. As shown in the figure, for sufficiently strong excitation H_{ex} to the left or right, the cores are both saturated and the B fields cancel, leaving a net flux $\Phi = 0$. When the excitation is strong enough to saturate one, but not both, of the cores, then the net flux ramps up or down linearly, as shown. For the inner range of excitation, which does not saturate either core, the net flux is constant.

A periodic excitation H_{ex} of sufficient amplitude to drive the cores into saturation will result in a periodic flux $\Phi(t)$ as illustrated in Fig. 11.7. Then, according to Eq. (11.15), the pickup coil will sense a sequence of positive and negative voltage pulses. Each of the Φ envelopes is generated in a half-cycle of the excitation, so the induced voltage pulses appear at twice the excitation frequency. Electronic circuits tuned to the second harmonic can then be employed for signal detection in these magnetometers.

11.3 COMPARATIVE PERFORMANCE

In addition to the two-core Vacquier design discussed in the previous section, fluxgate magnetometers have been made with double core/double pickup, single cores, and ring cores. Excitation frequencies are typically several kilohertz. The useful range for fluxgate magnetometers is approximately 100 nT (1 mG) to 0.2 mT (2 G), in contrast to Hall effect sensors, which can be used over the range 0.1 mT (1 G) to 10 T (100,000 G). Both fluxgate and Hall effect devices are directionally sensitive, so vector magnetometers can be devised. The resolution of fluxgates can go down to around a gamma (1 nT), which is about 100 times more sensitive than a typical Hall sensor.

The high sensitivity, ruggedness, and good stability of the fluxgate has made it a sensor of choice for aerospace work. Triaxial fluxgates were deployed on the moon as lunar surface magnetometers during the Apollo 12 mission.

BIBLIOGRAPHY

[1] Ramón Pallás-Areny and John G. Webster, *Sensors and Signal Conditioning* (Wiley-Interscience, New York, 1991), pp. 187–191.
[2] Daniel I. Gordon, "Recent advances in fluxgate magnetometry," IEEE Trans. Magn. 8, 76 (1972).
[3] Fritz Primdahl, "The fluxgate mechanism, Part I: The gating curves of parallel and orthogonal fluxgates," IEEE Trans. Magn. 6, 376 (1970).
[4] F. Primdahl, "The fluxgate magnetometer," J. Phys. E: Sci. Instrum. 241 (1979).
[5] Steven A. Macintyre, "Magnetic field sensor design," Sensor Rev. 11, 7 (1991).
[6] Donald C. Scouten, "Sensor noise in low-level flux-gate magnetometers," IEEE Trans. Magn. 8, 223 (1972).
[7] Jacob Fraden, *AIP Handbook of Modern Sensors* (AIP Press, New York, 1995), pp. 483–485.
[8] R.A. Dunlap, *Experimental Physics: Modern Methods* (Oxford University Press, Oxford, U.K., 1988), pp. 364–368.

CHAPTER 12

Strain

Forces applied to solid bodies that are anchored or fixed in position typically result in deformations. Depending on the stiffness of the sample, a given force might produce large or small deformations. This chapter deals with techniques for measuring the small deformations commonly experienced by structural materials such as metals. This would include applications such as the instrumental determination of stretching in the skin of aircraft wings, the bending of steel beams in bridges and buildings, twisting in shafts, and so forth.

12.1 STRAIN

Consider a sample of simple geometry as shown in Fig. 12.1. The indicated forces place the sample in tension (reversing the directions causes compression). The sample responds to this tension by stretching so that the original length L becomes $L + \Delta L$. Longitudinal strain is defined as the proportional length change

$$\epsilon_\ell = \frac{\Delta L}{L}. \qquad (12.1)$$

ϵ is dimensionless, but is commonly expressed as inches per inch, mm per mm, etc. Because strains are typically such small quantities for solid samples, a *microstrain*, $\mu\epsilon$ (which represents a relative deformation of 1 part per million), can be a convenient alternative unit. For example, a relative length change of 0.005 could be stated as $\epsilon = 0.005$ or equivalently as 5000 microstrains.

When applied forces are in line as in the previous example, pure extension or compression results. However, parallel forces, as shown in Fig. 12.2, produce

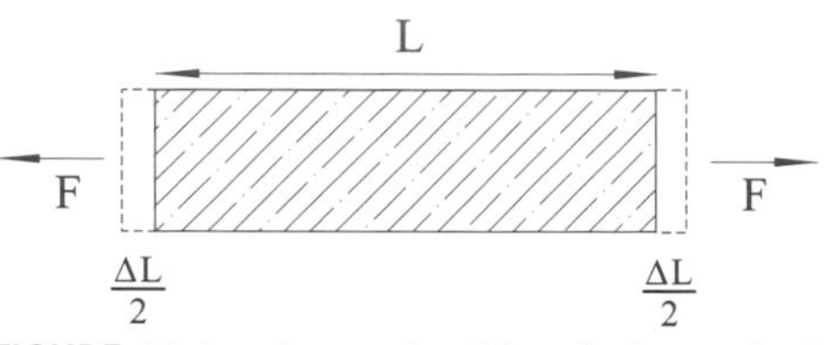

FIGURE 12.1. A sample of length L stretched by an amount ΔL because of the applied force F.

shearing effects. The horizontal displacement of the sample per unit transverse distance is $\tan\gamma$, and for small angles, $\tan\gamma \simeq \gamma$. The shearing strain is γ.

When a sample is stretched as in Fig. 12.1, it also narrows, as indicated in Fig. 12.3. Here, the transverse strain would be

$$\epsilon_t = -\frac{\Delta w}{w}. \tag{12.2}$$

Poisson's ratio ν is defined as

$$\nu = \frac{-\epsilon_t}{\epsilon_\ell}. \tag{12.3}$$

As an example, consider a solid cylindrical sample of length L and radius r. The volume is simply $V = \pi r^2 L$. Therefore,

$$\Delta V = \frac{\partial V}{\partial L}\Delta L + \frac{\partial V}{\partial r}\Delta r$$

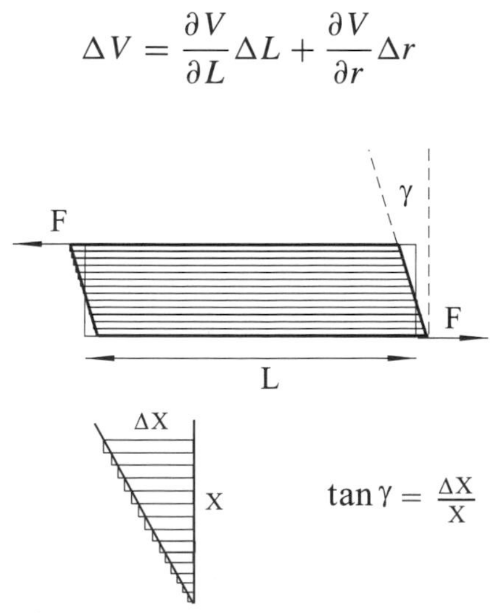

FIGURE 12.2. Shear strain in a sample with parallel applied forces. Shearing is visualized as lateral slippage of adjacent material planes.

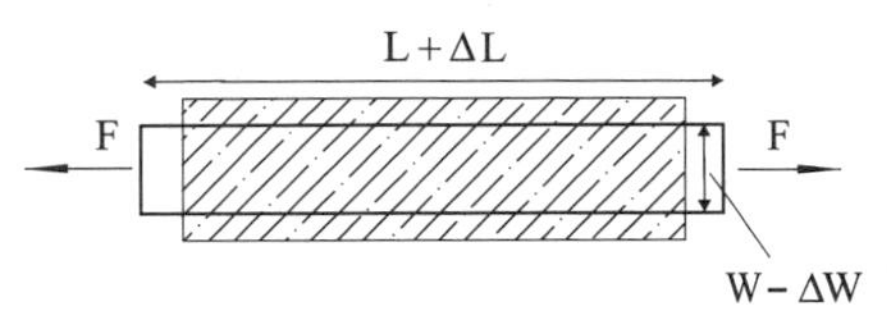

FIGURE 12.3. An elongated sample experiencing a concurrent reduction in its lateral dimension.

or

$$\Delta V = (\pi r^2)\Delta L + (2\pi r L)\,\Delta r.$$

The relative change in volume is thus

$$\frac{\Delta V}{V} = \frac{(\pi r^2)\Delta L + (2\pi r L)\,\Delta r}{\pi r^2 L}$$

or

$$\frac{\Delta V}{V} = \frac{\Delta L}{L} + 2\frac{\Delta r}{r},$$

and from previous definitions,

$$\frac{\Delta V}{V} = \epsilon_\ell + 2\epsilon_t. \tag{12.4}$$

Using Eq. (12.3),

$$\frac{\Delta V}{V} = \epsilon_\ell\,(1 - 2\nu)\,. \tag{12.5}$$

In the special case where the volume of the sample does not change under deformation, Eq. (12.5) yields a Poisson's ratio

$$\nu = \frac{1}{2}. \tag{12.6}$$

In real materials such as steel, copper, or aluminum ν is somewhat smaller, with typical values [2] in the range 0.25–0.40. It is evident from Eq. (12.5) that a value of Poisson's ratio less than 0.50 indicates that the volume is not invariant but actually increases when a sample is stretched ($\epsilon_\ell > 0$) and decreases when a sample is compressed ($\epsilon_\ell < 0$).

Just as strains were defined as proportional deformations, forces may be expressed in appropriately scaled ways. Thus, if a force F is applied uniformly over an area A, then the force per unit area is known as the applied stress σ.

12.2 RESISTIVE STRAIN GAGES

A sample of cross sectional area A and length L that is composed of a material with resistivity ρ will have a net resistance $R = \rho L/A$. The sensitivity of this resistance to variations in the parameters is then

$$\delta R = \frac{\partial R}{\partial L}\delta L + \frac{\partial R}{\partial A}\delta A + \frac{\partial R}{\partial \rho}\delta\rho. \tag{12.7}$$

Hence,

$$\delta R = \frac{\rho}{A}\delta L - \frac{\rho L}{A^2}\delta A + \frac{L}{A}\delta\rho,$$

so the relative change in resistance is

$$\frac{\delta R}{R} = \frac{\delta L}{L} - \frac{\delta A}{A} + \frac{\delta\rho}{\rho}. \tag{12.8}$$

In the case of a circular cross section, $A = \pi r^2$ and $\delta A = \frac{\partial A}{\partial r}\delta r = 2\pi r\delta r$, in which case $\frac{\delta A}{A} = 2\frac{\delta r}{r}$. But $\frac{\delta L}{L}$ is the longitudinal strain, and $\frac{\delta r}{r}$ is the transverse strain. Consequently,

$$\frac{\delta R}{R} = \epsilon_\ell - 2\epsilon_t + \frac{\delta\rho}{\rho}. \tag{12.9}$$

This result is valid also for noncircular cross sections because area is essentially quadratic in linear dimensions. As we have seen, Poisson's ratio describes the interrelation between longitudinal and transverse strains. Hence,

$$\frac{\delta R}{R} = \epsilon_\ell + 2\epsilon_\ell\nu + \frac{\delta\rho}{\rho}$$

or, finally,

$$\frac{\delta R}{R} = \epsilon_\ell(1 + 2\nu) + \frac{\delta\rho}{\rho}. \tag{12.10}$$

An important sensor parameter is the so-called *Gage Factor* (G), which is defined as

$$G = \frac{\delta R/R}{\delta L/L}. \tag{12.11}$$

The gage factor is thus the ratio of relative change in resistance to relative change in length. Using Eq. (12.10),

$$\boxed{G = (1 + 2\nu) + \frac{1}{\epsilon_\ell}\frac{\delta\rho}{\rho}}. \tag{12.12}$$

Bonded Metal Films

If the resistivity does not change, then the first term on the right-hand side of Eq. (12.10) implies that the relative change in sample resistance is directly proportional to the longitudinal strain. This behavior is the basis of metal-film strain gages. In these sensors, a thin film of metallic alloy (typical materials are constantan, nickel–chromium, or iron–chromium–aluminum) is photoetched to a desired shape and backed by a plastic, electrically insulating layer (typical materials are polyimide, epoxy, or epoxy-phenolic). The device is then glued directly on the object whose strain is to be monitored so that the deformations of the test object are matched by the sensor. Special adhesives are available for this purpose. Because the combination of foil and backing is fairly flexible, these gages can be attached satisfactorily to surfaces that have some degree of curvature.

Temperature changes can lead to errors because of possible differences in the thermal expansion properties of the object being monitored and the gage attached to it. Expansion coefficients for metals such as steel and aluminum are 10–20 ppm/°C. Manufacturers can process the gage alloys so that the temperature characteristics of available sensors will be matched to a given specimen material such as ferric steel, aluminum, or titanium, for example.

The gage pattern is usually chosen to produce a large equivalent length by folding, as indicated in Fig. 12.4. The sensitive axis is along the dimension L, so the gage must be cemented to the test object with this in mind. The two large pads serve as solder tabs to which external wiring is connected. A considerable range of patterns and sizes is available, with lengths of about 10 mm and widths of 5 mm being representative. Standard sensor resistance values are 120 Ω and 350 Ω. Gages are rated up to maximum elongations of typically 50,000 $\mu\epsilon$ which is equivalent to a strain of 5%. For a 10 mm sensor length, this is an elongation of 0.5 mm. Working strain ranges of 1 or 2% are usual. Poisson's ratio is typically 0.3, and the gage factor for metal-film devices is generally taken to be $G \simeq 2.0$. The exact value for each particular sensor is supplied by the manufacturer.

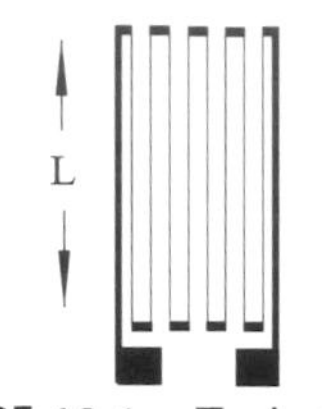

FIGURE 12.4. Typical layout of a metal-film strain gage. In this example, the equivalent length is $8L$.

Single Gage, Quarter Bridge

From Eq. (12.11), $\delta R = R\,G\,\delta L/L$. This would, for example, give $\delta R = 2.4 \times 10^{-3}\ \Omega$, assuming $R = 120\ \Omega$, $G = 2.0$, and $\epsilon_\ell = 10^{-5}$ (ten parts per million). The standard technique for resolving such small resistance changes involves the use of a Wheatstone bridge. Figure 12.5 illustrates the use of a single strain gage in a bridge—the so-called quarter-bridge configuration. It was shown in Chapter 4 that the bridge has optimum sensitivity when all four resistors are equal. Thus, for a strain gage of nominal resistance R_g, one should choose $R_1 = R_2 = R_3 = R_g$. This means that $V_{\text{out}} = 0$ in the unstrained state. Then, if under strain $R_g \to R_g + \Delta R_g$, the off-balance output voltage would become [refer to Eq. (4.1)]

$$V_{\text{out}} = V_{\text{bias}}\left[\frac{R_g}{R_g + R_g + \Delta R_g} - \frac{R_g}{R_g + R_g}\right]. \tag{12.13}$$

Then,

$$\frac{V_{\text{out}}}{V_{\text{bias}}} = \frac{-\frac{\Delta R_g}{R_g}}{4 + 2\frac{\Delta R_g}{R_g}}. \tag{12.14}$$

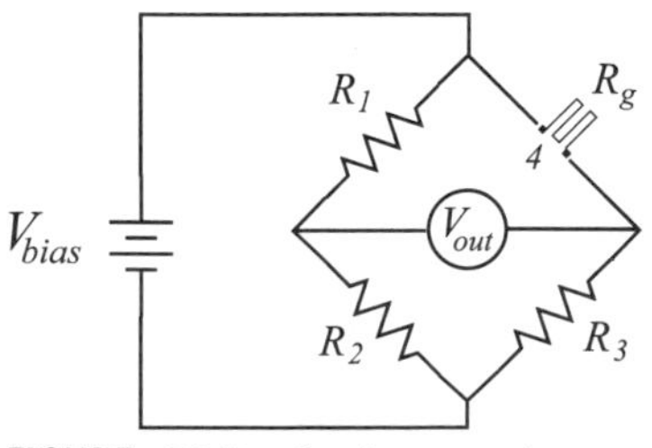

FIGURE 12.5. Strain gage in one arm of a Wheatstone bridge—the quarter-bridge configuration.

Note that for positive strains (elongations) this voltage will be negative, the reason being the position of the gage in the bridge and the fact that the gage resistance is increasing. Using Eq. (12.11),

$$\frac{\Delta R_g}{R_g} = G\frac{\Delta L}{L} = G\epsilon. \tag{12.15}$$

Thus, it follows that

$$\frac{V_{\text{out}}}{V_{\text{bias}}} = \frac{-G\epsilon}{4 + 2G\epsilon}. \tag{12.16}$$

This may be rearranged to give an explicit expression for the strain in terms of the measured relative output voltage $V^* = V_{\text{out}}/V_{\text{bias}}$.

$$\boxed{\epsilon = \frac{-4V^*}{G(1 + 2V^*)}}. \tag{12.17}$$

As an example, suppose $G = 2.0$, $V_{\text{bias}} = 10.0$ V, and the measured output is $-50\ \mu$V. Then,

$$\epsilon = \frac{4 \times 5 \times 10^{-6}}{2(1 - 2 \times 5 \times 10^{-6})} = 10 \text{ parts per million.}$$

Two Gages, Half Bridge

It is possible to make strain measurements using two gages instead of one, as previously. One possible configuration, shown in Fig. 12.6, with both sensors attached to the test sample in such a way as to subject each to the same extension

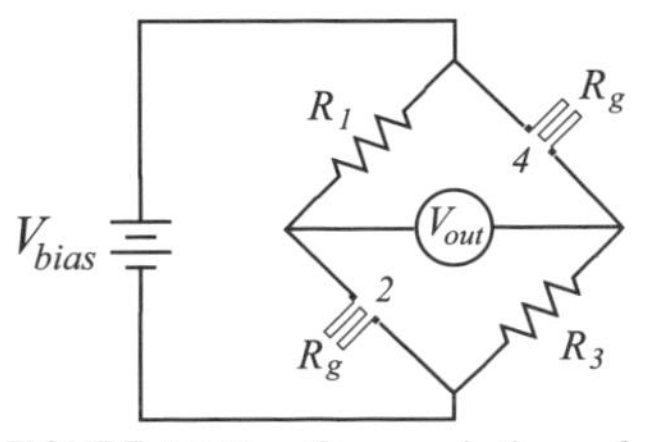

FIGURE 12.6. One variation of the half-bridge configuration.

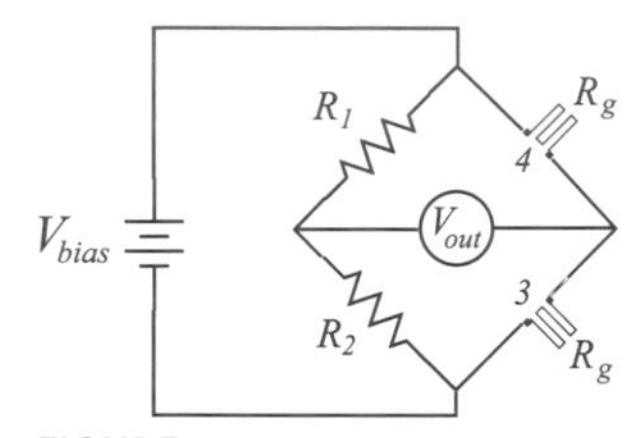

FIGURE 12.7. Half-bridge arrangement combining a gage operating in the extension mode (#4) with one in the compression mode (#3).

strain ϵ, results in a bridge voltage

$$\frac{V_{\text{out}}}{V_{\text{bias}}} = \frac{R_g}{R_g + R_g + \Delta R_g} - \frac{R_g + \Delta R_g}{R_g + R_g + \Delta R_g} \tag{12.18}$$

from which

$$\frac{V_{\text{out}}}{V_{\text{bias}}} = \frac{-G\epsilon}{2 + G\epsilon}, \tag{12.19}$$

which is twice the output of the quarter bridge—Eq. (12.16). The strain is thus

$$\boxed{\epsilon = \frac{-2V^*}{G(1 + V^*)}}. \tag{12.20}$$

Another configuration involving two gages is shown in Fig. 12.7. In this case, one device (#4) must be stretched while the other (#3) must be compressed; otherwise, the bridge output is trivially zero. The output voltage is

$$\frac{V_{\text{out}}}{V_{\text{bias}}} = \frac{R_g - \Delta R_g}{R_g + \Delta R_g + R_g - \Delta R_g} - \frac{1}{2} \tag{12.21}$$

or

$$\frac{V_{\text{out}}}{V_{\text{bias}}} = \frac{-G\epsilon}{2}, \tag{12.22}$$

which is a slightly larger output than for the other half-bridge geometry—Eq. (12.19). The strain is thus

$$\boxed{\epsilon = \frac{-2V^*}{G}}. \tag{12.23}$$

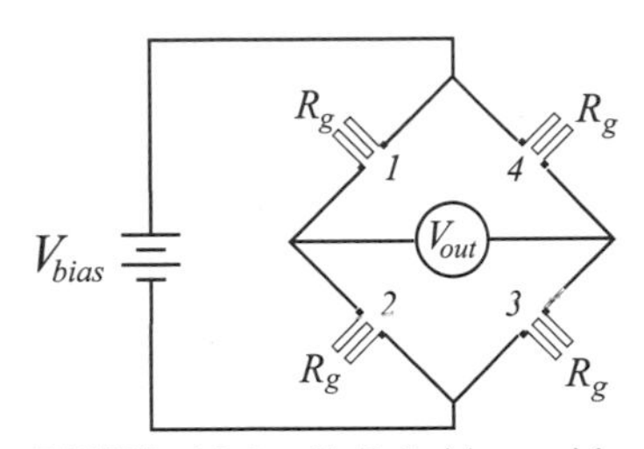

FIGURE 12.8. Full bridge with four gages. In this example, #1 and #3 are compressed, while #2 and #4 are stretched by the strained test object.

Four Gages, Full Bridge

Finally, consider four gages in a bridge as shown in Fig. 12.8. Two (#1 and #3) are in compressive mode, and the two others (#2 and #4) are in extension mode. The output voltage is

$$\frac{V_{\text{out}}}{V_{\text{bias}}} = \frac{R_g - \Delta R_g}{R_g - \Delta R_g + R_g + \Delta R_g} - \frac{R_g + \Delta R_g}{R_g + \Delta R_g + R_g - \Delta R_g} \tag{12.24}$$

or

$$\frac{V_{\text{out}}}{V_{\text{bias}}} = -G\epsilon. \tag{12.25}$$

Hence,

$$\boxed{\epsilon = \frac{-V^*}{G}}. \tag{12.26}$$

Piezoresistive Devices

Consider the final term in Eq. (12.10):

$$\frac{\delta R}{R} = \epsilon_\ell (1 + 2\nu) + \frac{\delta \rho}{\rho}.$$

For semiconductors such as silicon, this factor $\delta\rho/\rho$ is itself a strain-dependent quantity. This is because deformations of the atomic lattice resulting from applied forces make it possible for current to flow more easily along some internal directions than in others. Thus, the resistivity will depend on both the magnitude and the direction (with respect to the principal axes of the crystal structure) of the

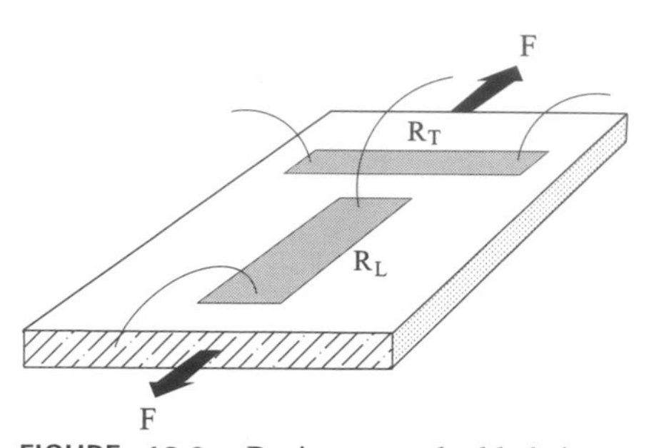

FIGURE 12.9. Resistors embedded in a semiconducting matrix subject to an applied strain σ. One resistor is oriented along the direction of the strain, while the other is transverse. Current leads are indicated for each resistor.

applied stress as well as the direction of the intended current. This is in contrast to the earlier situation for metal films, where simple macroscopic deformations produced changes to sample geometry and hence to the overall resistance. For this reason, piezoresistive sensors must be made from suitably aligned, single-crystal blanks. With proper lattice orientation, an applied stress directed along one axis will lead to a change in resistivity in that direction as well as in a direction orthogonal to it.

This last remark is illustrated in Fig. 12.9. A complete description of piezoresistivity requires quantum mechanics, crystallography, knowledge of stress-tensor representations, and so forth. Such detail is beyond the scope of this book. However, a somewhat simplified analysis can be given, as follows [3].

In a thin planar sample, stress components which are parallel (σ_l) and/or transverse (σ_t) to a diffused surface resistor lead to a change in resistance according to [8, 7]

$$\frac{\delta R}{R} = \pi_l \sigma_l + \pi_t \sigma_t, \tag{12.27}$$

where π_l and π_t are piezoresistive coefficients. For p-type resistors diffused into an n-type single-crystal wafer with (100) surface orientation, $\pi_l = -\pi_t = \pi_{44}/2$; for lightly doped silicon, the coefficient π_{44} has the numerical value [8] 1.38×10^{-5} cm^2/N. Note that, for this orientation, the transverse effect is equal in magnitude and opposite in polarity to the longitudinal effect.

The applied force shown in Fig. 12.9 would produce a longitudinal stress on silicon resistor R_L resulting in the change

$$\frac{\delta R_L}{R_L} = \pi_l \sigma \tag{12.28}$$

and a transverse stress on R_T resulting in

$$\frac{\delta R_T}{R_T} = \pi_t \sigma. \tag{12.29}$$

For materials within their elastic limits, stress and strain are related through the classic equation

$$\epsilon = \frac{\sigma}{Y}, \tag{12.30}$$

where Y is Young's modulus. Thus, since $\frac{\delta R}{R} = G\epsilon$, the gage factor for these semiconductor resistors is

$$G = \pi Y. \tag{12.31}$$

Young's modulus for silicon is 1.7×10^7 N/cm^2 and hence $G \approx 120$, a value that is much larger than typical metal-foil strain gages provide.

Fabrication of these strain gages employs techniques common to integrated circuit manufacture. The pattern defining the piezoresistor geometry is masked onto an oriented single-crystal wafer. Ion implantation through the mask openings onto the semiconductor surface and subsequent thermal diffusion into the material forms the required devices. By choosing n-type base material and p-type resistors (created by doping with boron to a hole concentration $p \approx 3 \times 10^{17}$ cm^{-3}), a result is achieved in which the resistor-sensors are isolated from the base semiconductor by reverse-biased p-n junctions. Thus, a number of piezoresistors in any desired geometry can be embedded in the host n-type silicon.

Semiconductor piezoresistive sensors have very large gage factors so they are sensitive, but they also have some limiting attributes. The variation of resistance with strain is actually nonlinear and for silicon may be approximated as [1]

$$\frac{\delta R}{R} = 120\,\epsilon + 4\,\epsilon^2 : \text{p-type},$$

$$\frac{\delta R}{R} = -110\,\epsilon + 10\,\epsilon^2 : \text{n-type}.$$

The maximum permitted strain for these gages, typically about 3000 $\mu\epsilon$, is smaller by a factor of more than 10 compared to the corresponding numbers for metal-foil devices. Semiconductor sensors also exhibit strong temperature dependences which require some form of compensation.

PROBLEMS

Problem 12.1. A strain gage is placed in one arm of a bridge, as shown in Fig. 12.10. If the strain is $\epsilon = 0.005$, the bias supply is 10 V, and $R_1 = R_2 = R_3 = R_g$ (where R_g is the unstrained sensor value of 350 Ω), determine the bridge output voltage. [Ans. 0.0236 V].

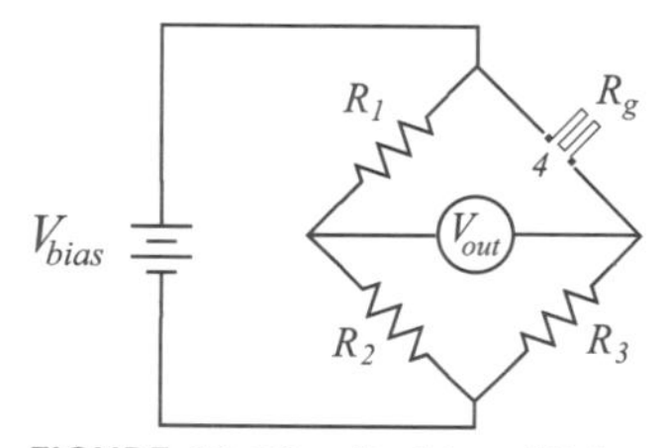

FIGURE 12.10. Problem 12.1.

Problem 12.2. A strain gage has an associated gage factor of 1.90. Assuming the resistivity does not change, what is the value of Poisson's ratio? [Ans. $\nu = 0.45$].

BIBLIOGRAPHY

[1] Application Note 290-1, "Practical strain gage measurements," (Hewlett-Packard Co., Palo Alto, CA, 1981); reprinted in *The Pressure, Strain and Force Handbook* (Omega Engineering, Inc., Stamford, CT).

[2] F.H. Newman and V.H.L. Searle, *The General Properties of Matter* (Edward Arnold Publishers, London, 1957), p. 108.

[3] Anthony J. Wheeler and Ahmad R. Ganji, *Introduction to Engineering Experimentation* (Prentice-Hall, Englewood Cliffs, NJ, 1996) pp. 189–199.

[4] John P. Bentley, *Principles of Measurement Systems*, third edition (Longman Group, London, 1995), pp. 139–142.

[5] Stanley Wolf and Richard F.M. Smith, *Student Reference Manual for Electronic Instrumentation Laboratories* (Prentice-Hall, Englewood Cliffs, NJ, 1990), pp. 375–382.

[6] H.R. Winteler and G.H. Gautschi, *Piezoresistive Pressure Transducers* (Kistler Instrumente AG, Winterthur, Switzerland, 1989).

[7] Jacob Fraden, *AIP Handbook of Modern Sensors* (AIP Press, New York, 1993), pp. 377, 378.

[8] O.N. Tufte, P.W. Chapman, and Donald Long, "Silicon diffused-element piezoresistive diaphragms," J. Appl. Phys. 33, 3322 (1962); see Table I and Eq.(6).

[9] H. Tanigawa, T. Ishihara, M. Hirata, and K. Suzuki, "MOS integrated silicon pressure sensor," IEEE Trans. Electron Devices 32, 1191 (1985).

[10] Ramón Pallás-Areny and John G. Webster, *Sensors and Signal Conditioning* (Wiley-Interscience, New York, 1991), p. 47.

CHAPTER 13

Pressure

It is not difficult to think of a long list of instances in which pressure is an important parameter: pressure in a hydraulic fluid, atmospheric pressure, gas pressure in jet engines, pressure distributions over the surface of aircraft wings, pressure in engine cylinders and manifolds, blood pressure, etc. Pressure is by definition force per unit area. Hence, the usual units of pressure are pounds per square inch (psi) or newtons per square meter (N/m^2), which is also called a *pascal* (Pa). Mercury-filled barometers gave rise to pressure units of mm of Hg and inches of Hg. In meteorology, atmospheric pressure is expressed in bars or millibars. Conversion between these units is simplified with the following relationships.

$$
\begin{aligned}
1 \text{ psi} &= 6895 \text{ Pa}, \\
1 \text{ psi} &= 51.72 \text{ mm Hg}, \\
1 \text{ psi} &= 2.036 \text{ inches Hg}, \\
1 \text{ psi} &= 0.0689 \text{ bar} = 68.95 \text{ millibar}, \\
1 \text{ Pa} &= 1.4503 \times 10^{-4} \text{ psi}, \\
1 \text{ Pa} &= 7.502 \times 10^{-3} \text{ mm Hg}, \\
1 \text{ Pa} &= 2.953 \times 10^{-4} \text{ inches Hg}, \\
1 \text{ Pa} &= 10^{-5} \text{ bar} = 10^{-2} \text{ millibar}, \\
1 \text{ atmosphere} &= 14.696 \text{ psi} = 101325 \text{ Pa} = 760 \text{ mm Hg}.
\end{aligned}
$$

To measure pressure, a modern solid-state sensor will generally employ a membrane that is acted upon from both sides and deforms as a result of the net pressure difference. By the use of strain gages or some other technique, this deformation can be converted to a signal that reflects the existing pressure difference. It should be noted, however, that a long and interesting history exists concerning the theory and development of pressure measuring instruments. Prior

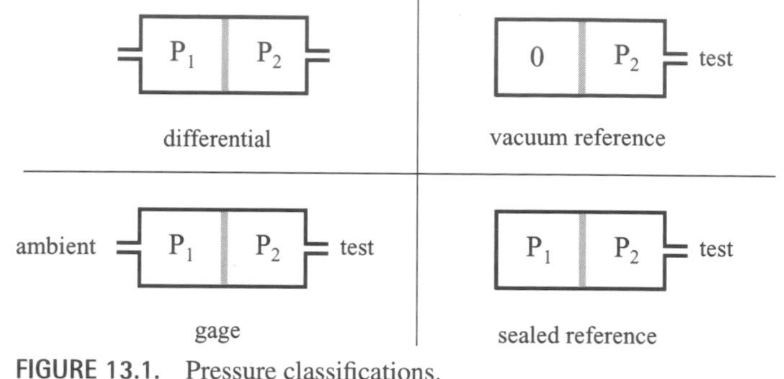

FIGURE 13.1. Pressure classifications.

to the emergence of solid-state electronics and more recently the integrated circuit, pressure was determined with mercury and oil manometers, Bourdon-tube gages, and mechanically coupled bellows diaphragms. Detailed treatments of many such transducers, including flow meters and viscosity meters, may be found in [1].

Several possible arrangements of measurement cells are indicated in Fig. 13.1. In each situation, one or two input ports admit the test environment(s) (gas or liquid) into sensor cavities, which are separated by a sensing diaphragm. The differential mode responds to the pressure difference $P_2 - P_1$. For gage pressure, only magnitudes above or below ambient or atmospheric pressure are of interest. This might apply, for example, when tire or balloon pressures are measured; the degree of inflation depends on the excess pressure above atmospheric. Absolute pressure can be obtained with the reference side pumped out to a vacuum and permanently sealed. Therefore, absolute pressure = gage pressure + atmospheric pressure. In some instances, a sealed pressure other than zero could be desirable.

13.1 PIEZORESISTIVE GAGES

Piezoresistive strain gages were discussed in the previous chapter. Two important properties of these sensors are: (1) high sensitivity; (2) the ease with which they can be fabricated on a host slice of single-crystal silicon. Silicon is itself well suited as a diaphragm material because of its excellent elastic properties. This combination of attributes means that cylindrical or rectangular holes can be chemically etched almost through small single-crystal silicon blocks leaving an integral diaphragm capping the resulting cavity. Piezoresistors can then be diffused onto the surface of the membrane. Solid-state monolithic pressure sensors are thus constructed as suggested by Fig. 13.2. As the diaphragm flexes

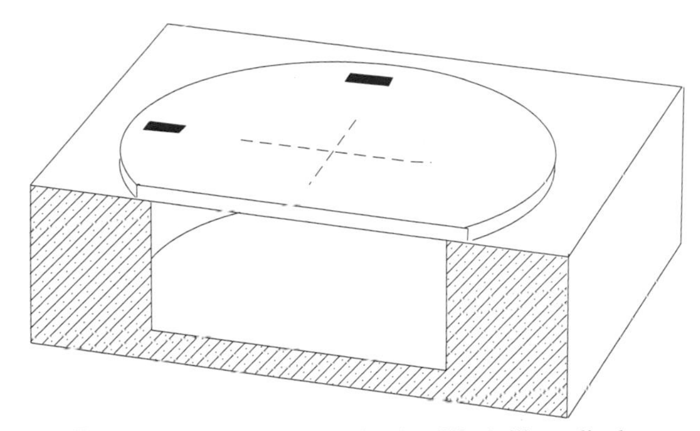

FIGURE 13.2. Cutaway view of a simplified silicon diaphragm pressure sensor. For absolute pressure readings, the cavity is a sealed vacuum. Otherwise, the cavity would be provided with an access port. Two diffused piezoresistors are indicated.

inward or outward under the action of a differential pressure, the piezoresistors are strained and their resistance changes.

A discussion of the stresses and strains experienced by diaphragms may be found in [3]. A primary reference for such problems is [4]. For a circular plate with clamped edges, the stress components, which are directed either along a radial line or tangent to a radial line, are (see Eqs. (63) and (64) of [4]):

$$\sigma_r(r) = \frac{3}{8}\frac{\Delta p}{t^2}[R^2(1+\nu) - r^2(3+\nu)], \tag{13.1}$$

$$\sigma_t(r) = \frac{3}{8}\frac{\Delta p}{t^2}[R^2(1+\nu) - r^2(1+3\nu)], \tag{13.2}$$

where Δp is the differential pressure across the plate, ν is Poisson's ratio, t is the plate thickness, and R is the plate radius. These two stress functions are plotted in Fig. 13.3. Notice that the upper surface is in compression around the center of the diaphragm but in tension near the circumference. Consequently, particular radii exist at which either the tangential or radial stresses are zero.

Gage Sensitivity

These facts suggest that diffused piezoresistors placed near the outside edge, where $\sigma_t = 0$, would experience only radial tensile stress. According to Eq. (13.2), this happens at

$$\frac{r^2}{R^2} = \frac{1+\nu}{1+3\nu}. \tag{13.3}$$

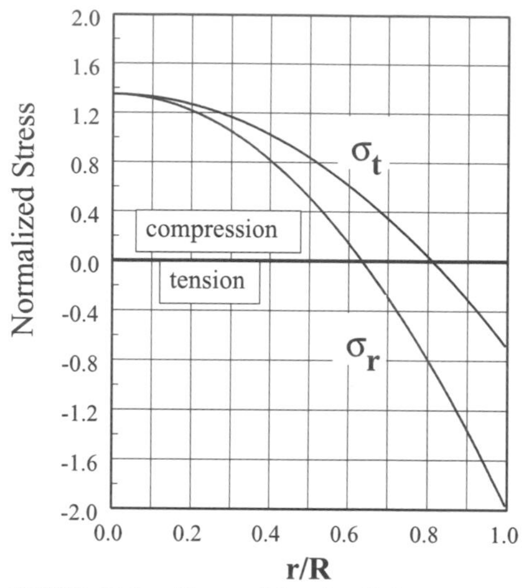

FIGURE 13.3. Tangential and radial stresses at the upper surface of a disk of radius R. The stresses are normalized by the factor $\frac{3}{8}(\frac{R}{t})^2\Delta p$, where t is the thickness of the disk and Δp is the differential pressure across it. Poisson's ratio was taken as $\upsilon = 0.35$ for this example.

For a resistor oriented toward the center of the diaphragm (see Fig. 13.2), this would produce a longitudinal effect as described in Chapter 12,

$$\frac{\delta R}{R} = \frac{\pi_{44}}{2}\sigma_r, \tag{13.4}$$

and for the resistor orthogonal to the radial line,

$$\frac{\delta R}{R} = -\frac{\pi_{44}}{2}\sigma_r. \tag{13.5}$$

The two resistors thus experience equal but opposite proportional changes. These variable resistors can be wired into the arms of a Wheatstone bridge. If four diffused resistors are positioned around the periphery, two of each orientation, then a full bridge can be employed, as indicated in Fig. 13.4. The bridge output in this case is

$$\frac{V_{\text{out}}}{V_{\text{bias}}} = \frac{\delta R}{R} = \frac{\pi_{44}}{2}\sigma_r \tag{13.6}$$

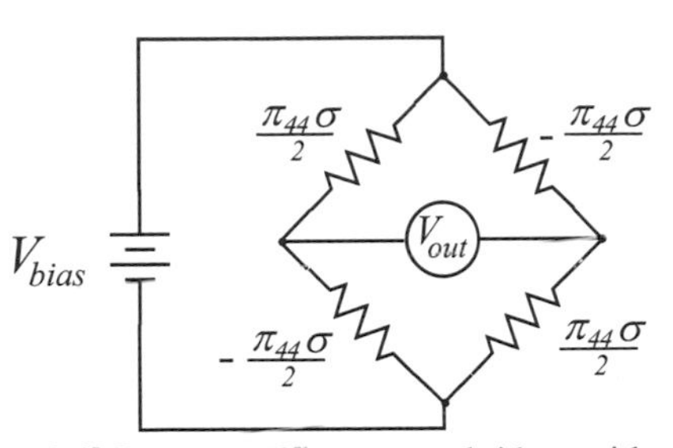

FIGURE 13.4. Wheatstone bridge with four piezoresistors, two under longitudinal stress and two under transverse stress.

and, using Eq. (13.1) with Eq. (13.3) for the radius,

$$\frac{V_{\text{out}}}{V_{\text{bias}}} = \left[\frac{3}{8}\frac{R^2}{t^2}\left(\frac{1-\nu^2}{1+3\nu}\right)\pi_{44}\right]\Delta p. \tag{13.7}$$

This demonstrates that the output voltage is a linear function of the differential pressure across the diaphragm.

The sensitivity of this arrangement is then

$$\frac{\Delta V_{\text{out}}}{\Delta p} = V_{\text{bias}}\left[\frac{3}{8}\frac{R^2}{t^2}\left(\frac{1-\nu^2}{1+3\nu}\right)\pi_{44}\right]. \tag{13.8}$$

Choosing for purposes of illustration $V_{\text{bias}} = 10$ V, $R = 0.1$ cm, $t = 30$ micron, $\nu = 0.30$, and $\pi_{44} = 1.38 \times 10^{-5}$ cm^2/N, the sensitivity is found to be 275 V/Pa, or equivalently 39.9 mV/psi. This is in reasonable agreement with the rated sensitivities of real pressure sensors.

Diaphragm Deflection

An expression is given on page 55 of [4] for the maximum deflection at the center of the circular plate,

$$d_{\max} = \frac{3R^4(1-\nu^2)}{16\,Yt^3}\Delta p, \tag{13.9}$$

where Y is Young's modulus (stiffness).

As an example, consider a silicon diaphragm that is 0.5 cm in radius and 0.1 mm in thickness. Using $\nu = 0.3$, $Y = 1.7 \times 10^7$ N/cm^2, and a differential pressure of, say, 5 psi, this center deflection is found to be 0.022 mm, which is only about 20% of the diaphragm thickness. In other words, the disk actually bends very little.

Resonant Frequency

According to [5, 6], the fundamental frequency of vibration of a circular disk that is rigidly supported at its periphery is given by

$$f_1 = \frac{t}{R^2}\sqrt{\frac{\pi^2 Y}{48(1-\nu^2)\rho}}, \tag{13.10}$$

where the symbols are as previously defined and ρ is the material density. With the same numerical values used above and the density of silicon $\rho = 2.33\ \text{g/cm}^3$, the fundamental frequency in this example turns out to be $f_1 = 1624$ Hz.

From Eqs. (13.1) and (13.2) it is apparent that the stress in the slightly deformed diaphragm is proportional to the square of the ratio of radius to thickness,

$$\sigma \propto \frac{R^2}{t^2}. \tag{13.11}$$

Thus, if both the thickness and radius of the membrane are reduced in such a way that their ratio remains fixed, the internal stress will stay constant while the natural frequency will rise. Generally speaking, it is desirable to push this mechanical resonance above any intended working frequency for the sensor, so small, thin diaphragms are the usual choice.

Sensor Specifications

It should be noted that practical pressure sensors, while following the basic principles just outlined, nevertheless employ individual technical variations with respect to diaphragm geometry, piezoresistor placement, and so forth. A representative design is described in detail by Tanigawa et al. [7]. Two piezoresistors 100 μm long and 20 μm wide were diffused into the surface of a 1 mm^2 rectangular diaphragm that was etched to a thickness of only 30 μm. The resistors were integrated into an on-chip, half-bridge circuit with an NMOS operational amplifier for output conditioning. The measured sensitivity was given as 350 V/Pa (51 mV/psi) before amplification and 7×10^3 V/Pa (1.02 V/psi) after amplification.

Commercial sensors are available covering pressure ranges of $0 \rightarrow 1$ psi to about $0 \rightarrow 100$ psi. These are four-terminal devices with contacts corresponding to the four points of a Wheatstone bridge, as in Fig. 13.4; in other words, two terminals for external dc excitation and two output terminals. Typical rated sensitivities are 5 to 30 mV/psi—values quite consistent with the earlier theoretical estimate for a full bridge. Transient response times of better than 1 msec are common, as are resonant frequencies in excess of 100 kHz.

13.2 PIEZOELECTRIC GAGES

Piezoelectricity was discovered by Pierre and Jacques Curie in 1880. This phenomenon arises in certain crystals, notably quartz (SiO_2), which because of the geometric configurations of their atoms exhibit an interdependence between mechanical deformation and electrical polarization. When such a crystal is strained by an applied force, the distortion of the lattice results in charge appearing at the surfaces of the sample. Conversely, the application of an electric field across a crystal leads to dimensional changes. The first effect is the basis of old-fashioned phonograph pickups and modern solid-state barbecue igniters; the second is fundamental to the operation of quartz clocks and watches as well as ultrasonic generators.

Depending on the precise orientation at which a sample slab is cut from a starting single-crystal bar of quartz, three different forms of piezoelectric effect can be observed. As illustrated in Fig. 13.5, these are longitudinal, shear, and transverse. The relations between applied force and resulting charge for these

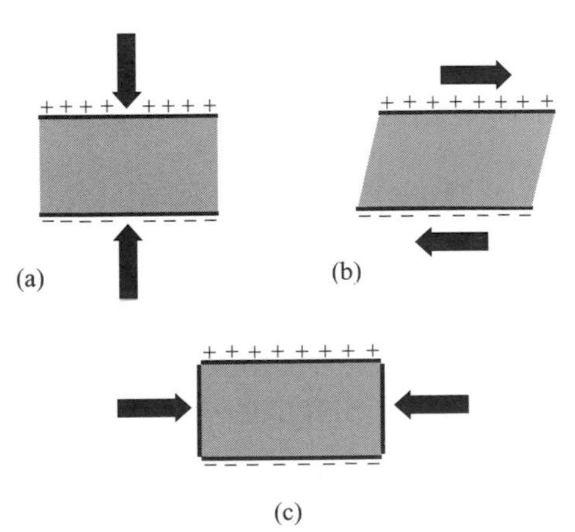

FIGURE 13.5. Piezoelectric effect. (a) longitudinal mode; (b) shear mode; (c) transverse mode.

elements have the form [8, 1, 10]

$$\begin{aligned} Q &= d_{11}F \quad \text{longitudinal,} \\ Q &= kd_{11}F \quad \text{transverse,} \\ Q &= 2d_{11}F \quad \text{shear,} \end{aligned} \tag{13.12}$$

where the piezoelectric charge coefficient for quartz is $d_{11} = 2.3 \times 10^{-12}$ C/N and k is a geometrical factor not very different from unity. This would mean, for example, that a compressive force of 1 N applied to a 1 cm × 1 cm element (1.45 psi) would generate 2.3 pC (picocoulomb) over this same area. For comparison purposes, a 1000 μF capacitor charged to a potential difference of 10 V would hold a total charge of $Q = CV = 0.01$ C. Thus, the charges produced by the piezoelectric effect, even for pressures of thousands of psi, are small.

Quartz has a high equivalent spring stiffness and a Young's modulus $Y = 80 \times 10^9$ N/m^2. Its resistivity is very large: $\rho = 10^{14}$ Ω–cm. This finite device resistance means that the charge induced by a steady applied force will inevitably decay back to zero with a time constant given by the product of the net leakage resistance times the effective capacitance. Care must be exercised to prevent poor cable insulation from degrading this situation and producing faster decays.

Thus, a constant force (or pressure) does not result in a constant charge. Consequently, a sensor based on monitoring the piezocharge given in Eq. (13.12) as a means of deducing pressure cannot extend to dc, and so must be restricted to applications involving time varying or, at best, quasi-static forces and pressures. This would include, say, measurements of impact dynamics in ballistics or cylinder pressures in combustion engines.

The stiffness of quartz mentioned earlier implies reasonably high resonant frequencies for these sensors; values of 50 → 100 kHz are common. The equation describing the relative amplitude response of a forced second-order linear system such as a driven pendulum (at small angles) or an excited quartz crystal is

$$a(f) = \frac{1}{\sqrt{\left[1 - \left(\frac{f}{f_0}\right)^2\right]^2 + \frac{1}{Q^2}\left(\frac{f}{f_0}\right)^2}}, \tag{13.13}$$

where f_0 is the resonant frequency and the parameter Q characterizes the damping (high Q corresponds to low dissipation). This equation is plotted in Fig. 13.6. The sharp peak at the resonant frequency is quite evident. It is desirable to operate a sensor below this peak region, where the frequency response is nearly flat. Hence, for piezoelectric transducers, there is a working range of frequencies, the lower end of which is set by leakage effects and the upper end by mechanical resonance.

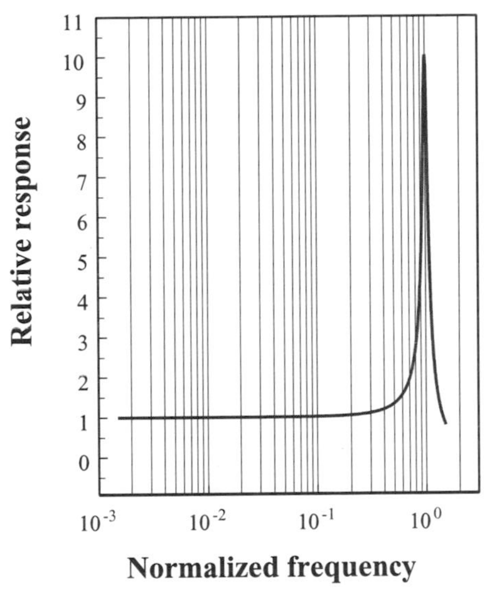

FIGURE 13.6. Typical frequency response of a system which has a resonance. Here the frequencies are given in units of the resonant frequency, and the output is normalized to the low frequency value. The resonant Q was chosen to be 10 for this example.

Charge Amplifier

A pressure meter must convert the induced charge Q into a usable electrical signal. The technical difficulty of this requirement is that the charge may be small but the impedance of the crystal large. A solution to this problem is provided by the so-called *charge amplifier* circuit [3, 1, 10]. The basic idea is conveyed by the PSpice schematic in Fig. 13.7.

The feedback path of an FET input op-amp contains a capacitor. The high input impedance to the op-amp is essential for minimizing charge leakage. The current from the source then flows through capacitor C_1, and because of the virtual ground at the inverting input,

$$I_1 = I_C = -\frac{d}{dt}(C_1 V_{\text{out}}) \tag{13.14}$$

or

$$V_{\text{out}} = -\frac{1}{C_1}\int I_1\, dt = -\left[\frac{I_1}{C_1}\right] t. \tag{13.15}$$

Either of these expressions states that the output voltage is a linear function of time when the drive current is constant, as it is in this example.

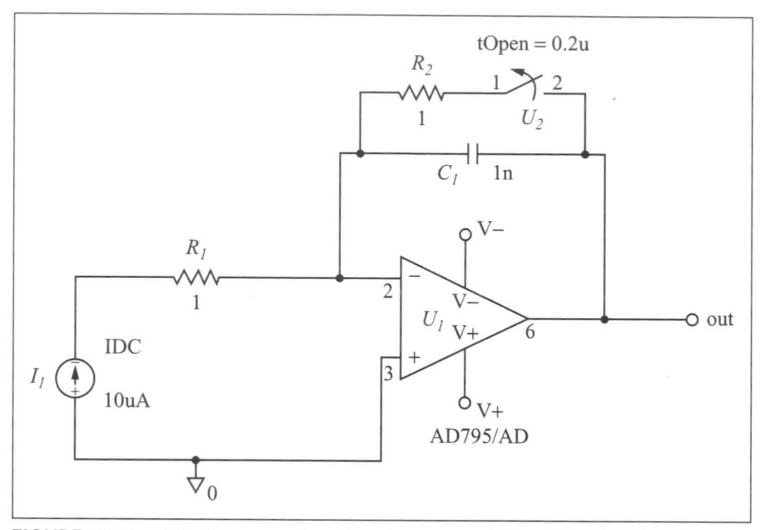

FIGURE 13.7. PSpice example of a charge amplifier. The timed switch on resistor R_2 prevents drift at startup. Resistor R_1 does not affect the circuit behavior but is required by PSpice for initial bias calculations.

PSpice simulation results are shown in Fig. 13.8. The slope of the output voltage gives a rate of change of 10^4 V/sec, which agrees with the prediction of Eq. (13.15): $\frac{I_1}{C_1} = \frac{10\times10^{-6}}{10^{-9}}$.

More generally, Eq. (13.14) can be expressed

$$\frac{dQ}{dt} = -C_1\frac{dV_{\text{out}}}{dt}$$

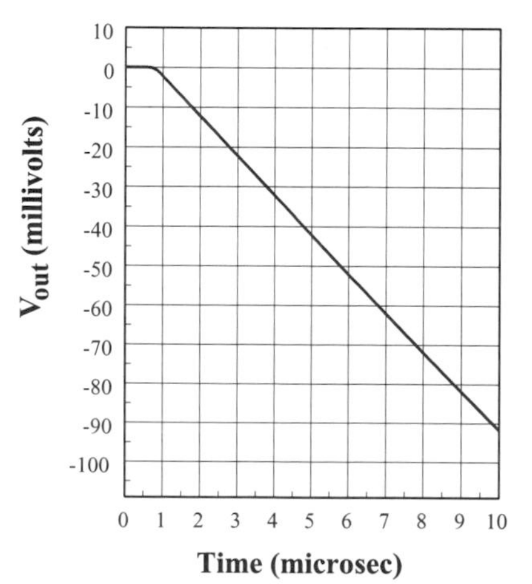

FIGURE 13.8. Output from the current-driven charge amplifier.

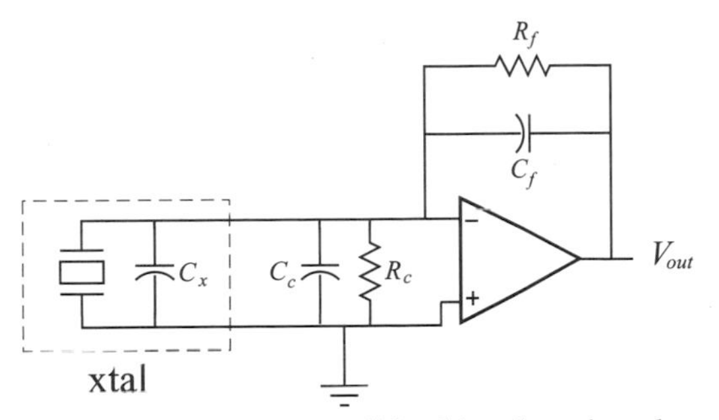

FIGURE 13.9. Charge amplifier driven by a piezoelectric crystal with associated capacitance C_X. The feedback (range) capacitor is C_f; the resistor R_f sets the time constant. C_C and R_C are possible cable parameters.

or

$$V_{\text{out}} = -\frac{1}{C_1} Q(t). \tag{13.16}$$

This equation suggests the name "charge amplifier," but more correctly one can see that this circuit is a charge-to-voltage converter.

When a quartz piezoelectric sensor is to be used, the arrangement of Fig. 13.9 is employed. R_C represents cable insulation resistance, which should be large ($>10^{13}$ Ω) to prevent undesired charge leakage. Assuming for the moment that the resistors are large enough to be ignored, the equation applicable to this circuit is, as before,

$$V_{\text{out}} = -\frac{1}{C_f} \int I \, dt = -\frac{Q}{C_f}. \tag{13.17}$$

Thus, the piezocharge has been converted to an output voltage. Moreover, the op-amp will have a low output impedance, which permits driving auxiliary electronics such as digitizing circuits and meter displays.

The piezoelectric charge coefficient for quartz is $d_{11} = 2.3 \times 10^{-12}$ C/N, so a force of 1 N will generate 2.3 pC. A feedback capacitor of, say, 10 pF would then yield an output of 0.23 V.

This charge amplifier has a time constant

$$\tau = R_f C_f, \tag{13.18}$$

which characterizes the exponential rate at which the initial charge Q decays away (even in the case when the force applied to the quartz crystal is constant). If R_f is associated with the insulation resistance of the feedback capacitor, then values for R_f may be presumed to be on the order of 10^{14} Ω, in which case $C_f \approx$ 10 pF implies $\tau \approx 1000$ sec. Choosing larger values for the feedback capacitor can effectively lengthen this time constant, with 10^4 pF giving $\tau \approx 10^6$ sec. These are representative values for commercial charge amplifiers. Large time constants are desirable for quasi-static situations.

It should be noted also that nonideal properties of the op-amp imply that the inverting input is at a small but nonzero offset potential $V_- = V_{\text{off}} \neq 0$; in other words, the virtual ground is not exactly true. A drift current $I_d = \frac{V_{\text{off}}}{R_C}$ then flows through the cable insulation, and an output drift voltage is developed according to Eq. (13.15) or Eq. (13.17).

$$V_d = \frac{1}{C_f} \int I_d \, dt = \left[\frac{V_{\text{off}}}{R_C C_f} \right] t. \tag{13.19}$$

Thus, a linearly increasing or decreasing (depending on the sign of the offset) output voltage arises. It will ultimately saturate at the positive or negative power supply level. Clearly, amplifiers with low offset and cables with very good insulation are remedies for drift.

Sensor Specifications

Commercial piezoelectric force and pressure sensors are available in conventional high-impedance form (charge output) as well as with integrated charge amplifiers, which are referred to as low-impedance devices (voltage output). Direct-force sensors are available with measuring ranges from a sensitive ± 5 lb to maximum values in excess of 100,000 lb. Pressure ranges of $0 \rightarrow 30$ psi to over $0 \rightarrow 100{,}000$ psi are available with resonant frequencies as high as 500 kHz. Diaphragm diameters for pressure sensors are typically less than 0.5 inches. Transient response is characterized by rise times of the order of a few microseconds.

BIBLIOGRAPHY

[1] Patrick J. O'Higgins, *Basic Instrumentation: Industrial Measurement* (McGraw-Hill, New York, 1966).

[2] Jacob Fraden, *AIP Handbook of Modern Sensors* (AIP Press, New York, 1993), Chapter 9.

[3] H.R. Winteler and G.H. Gautschi, *Piezoresistive Pressure Transducers* (Kistler Instrumente AG, Winterthur, Switzerland, 1989).

[4] S. Timoshenko and S. Woinowsky-Krieger, *Theory of Plates and Shells*, second edition (McGraw-Hill, New York, 1959), pp. 54–56.

[5] Eugen Skudrzyk, *Simple and Complex Vibratory Systems* (The Pennsylvania State University Press, State College, PA, 1968), pp. 247–249 and Eq. (8.40).

[6] Mary Désirée Waller, *Chladni Figures: A Study in Symmetry* (G. Bell and Sons, London, 1961), p. 122, Eq. (5).

[7] H. Tanigawa, T. Ishihara, M. Hirata, and K. Suzuki, "MOS integrated silicon pressure sensor," IEEE Trans. Electron Devices, 32, 1191 (1985).

[8] R. Kail and W. Mahr, *Piezoelectric Measuring Instruments and their Applications* (Kistler Instrumente AG, Winterthur, Switzerland, 1984).

[9] Ramón Pallás-Areny and John G. Webster, *Sensors and Signal Conditioning* (Wiley-Interscience, New York, 1991), pp. 247–257 and p. 285.

[10] John P. Bentley, *Principles of Measurement Systems*, third edition (Longmans, London, 1995), pp. 163–168.

CHAPTER 14

Displacement and Rotation

In this chapter, we are concerned with the measurement of the position and motion of an object. Because so many techniques for accomplishing these tasks are possible, some of which are rather specialized and dependent on particular applications, only a few of the most commonly available and widely used sensors will be discussed here. The focus is on direct readings of position, but of course velocity can be computed (numerical differentiation) from a series of position coordinates taken at known times. For other types of sensors that respond instead to velocity, the output signal must be integrated to yield position. In the case of acceleration, sensors can provide precise readings without numerical processing.

14.1 DISPLACEMENT

The sensor which is most often the choice for monitoring displacements as large as 25 cm or more is the linear variable differential transformer (LVDT). An LVDT is, fundamentally, a mutual inductance device.

Mutual Inductance

Inductance was introduced in Chapter 3 through the expression

$$V_L = L\frac{dI_L}{dt},$$

which relates the reverse voltage that appears across a coil when a time-varying current flows through it. More correctly, this should be called self-inductance, since the effect is localized within a single coil.

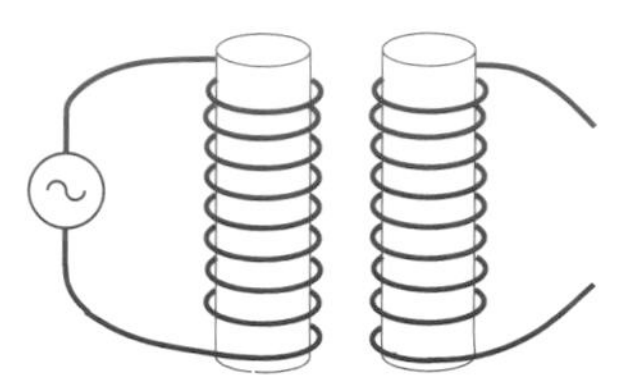

FIGURE 14.1. Mutual inductance: a coil with time-varying current (the primary) induces a signal in an adjacent coil (the secondary).

It is also true that a changing current in one circuit containing a coil can induce a response in a second circuit containing another coil if the magnetic fields generated by the first link to the second. With reference to Fig. 14.1, this can be expressed as

$$V_S = M_{ps}\frac{dI_P}{dt}, \tag{14.1}$$

where the secondary voltage V_S is seen to be a linear function of the time rate of change of the primary current I_P. The polarity of V_S depends upon the direction of the windings in the coils.

The proportionality constant is the coefficient of *mutual inductance* between primary and secondary, M_{ps}. Units are henrys, the same as for self-inductance. It is a fact that Eq. (14.1) remains valid under role reversal. In other words, it does not matter which coil is chosen to be the primary and which the secondary. Therefore, $M_{ps} = M_{sp}$, and hence it is only necessary to speak of the mutual inductance M between any given pair of coils.

The concept of complex impedance was introduced in Chapter 3. For an inductor, this was expressed as $Z = j(\omega L)$. In a situation such as illustrated in Fig. 14.1, mutual inductance between the coils would lead to

$$V_S = \pm j(\omega M)\,I_P, \tag{14.2}$$

where the correct polarity $\pm$ is determined by the way in which the two coils are wound with respect to each other. V_S and I_P are complex quantities in this equation.

LVDT

A linear variable differential transformer [1] consists of two pickup coils located on either side of an excitation coil, as illustrated in Fig. 14.2. The pickup coils are wound in opposition so their emf's subtract. All three coils are placed on

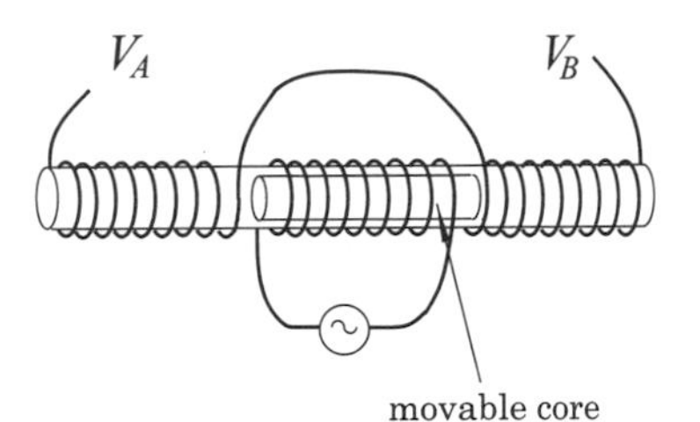

FIGURE 14.2. Linear variable differential transformer (LVDT). Secondary coils A and B are wound in opposition. The movable core can be displaced along the axis of the sensor, thereby altering thc coupling between the driven primary and each of the two secondaries.

a cylindrical form so the LVDT has the appearance of a thin tube. Within the tube is a ferromagnetic core, which can move freely along the system axis. The object whose linear motion is to be detected must be physically connected to this movable core via a suitable nonmagnetic pushrod (not shown), which extends out of one end of the LVDT assembly. Thus, the motion of the object is matched by motion of the ferromagnetic core.

An equivalent representation of the LVDT is given in Fig. 14.3. Each secondary is magnetically linked to the primary via a mutual inductance. The momentary position of the core affects the two mutual inductances M_1 and M_2. The sensor is constructed so that the center position of the core yields equal secondary voltages, which then cancel. Hence, the output has a null at the midpoint position. If the core were to move from this center so as to increase M_1, then M_2

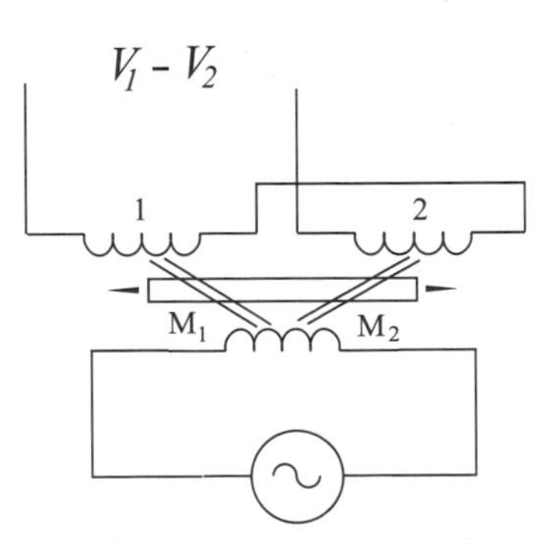

FIGURE 14.3. Schematic of an LVDT. M_1 and M_2 are mutual inductances between the drive coil and the secondaries.

would decrease. As a result, the signal from coil 1 would increase but the signal from coil 2 would diminish. Thus, the net output $V = V_1 - V_2$ would increase as a positive number. The converse to these events also applies. Opposite displacement of the core from the zero position would lead to increasing negative values for $V = V_1 - V_2$. In either direction, the resulting output voltage amplitude is a nearly linear function of displacement, provided the core shift is not too large.

In complex notation,

$$V(t) = j\omega(M_1 - M_2)\, I_P \tag{14.3}$$

and

$$I_P = \frac{V_e}{Z_P}, \tag{14.4}$$

with V_e being the primary excitation. As a guide, excitation frequencies are chosen to be about 10 times the expected vibrational characteristics of the system being monitored. Thus, mechanical oscillations at 100 Hz would suggest a drive at 1 kHz. Excitation frequencies of a few kilohertz are common with LVDTs.

Information about the instantaneous core position is contained in the ac signal $V(t)$. A standard technique for converting this to a useful dc voltage is by means of a phase synchronous demodulator (see p. 217 of [1], and [3]), which filters the absolute value of $V(t)$ to determine the magnitude of the displacement, and measures the phase shift of $V(t)$ relative to the excitation voltage as a means of knowing the direction of displacement (e.g., left or right).

The precision achievable with the method of synchronous demodulation depends on a number of factors, including the stability of both the amplitude and frequency of the excitation source. A method has been found to reduce this dependence; it is described in a paper by S.C. Saxena and S.B.L. Seksena [4]. The voltage generated by a secondary is some function of core permeability μ, primary excitation current I_P, primary excitation frequency f_P, and the core displacement x relative to the zero position. Therefore, for a symmetrically constructed LVDT,

$$\begin{aligned} V_1 &= k\, F_1(\mu, I_P, f_P, x), \\ V_2 &= k\, F_2(\mu, I_P, f_P, -x), \end{aligned} \tag{14.5}$$

where $V_{1,2}$ refers to the voltages across each of the secondaries. The functions $F_{1,2}$ can further be decomposed to

$$\begin{aligned} F_1 &= H\,(\mu, I_P, f_P)\, F(x), \\ F_2 &= H\,(\mu, I_P, f_P)\, F(-x), \end{aligned} \tag{14.6}$$

The sensor described in [4] utilized a dual set of secondary windings, one pair generating the difference $V_1 - V_2$, and the other pair wound so that a sum is produced, $V_1 + V_2$. From these, a ratiometric output was created:

$$V_{out} = \frac{V_1 - V_2}{V_1 + V_2}. \tag{14.7}$$

Using Eqs. (14.5) and (14.6), this becomes

$$V_{out} = \frac{kH(F(x) - F(-x))}{kH(F(x) + F(-x))}$$

or

$$V_{out} = \frac{F(x) - F(-x)}{F(x) + F(-x)}. \tag{14.8}$$

Notice that the output voltage in this scheme does not depend on the excitation amplitude or frequency, nor on the core permeability, which can be temperature-sensitive. Furthermore, the sum term in the denominator, $F(x) + F(-x)$, remains constant because, as we saw, as one mutual inductance goes up with displacement the other goes down.

At least one commercial special-purpose integrated circuit operates on this same principle but without the necessity of dual secondaries. A simplified block diagram of the Analog Devices AD598 [5] is shown in Fig. 14.4. This chip combines a triangle waveform oscillator, wave shaper, and amplifier to generate a sinusoidal drive for the LVDT primary coil. The output frequency can be set within the range 20 Hz to 20 kHz and the output amplitude from 2 V rms

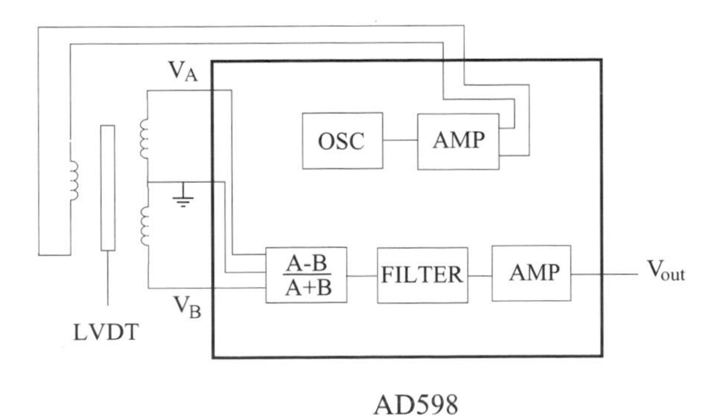

FIGURE 14.4. LVDT interface using Analog Devices AD598 integrated circuit.

to 24 V rms. The coil signals V_A and V_B are processed by an internal analog computing loop to obtain the required ratio, which is then filtered and amplified. The final output is an adjustable scaled version of $(A - B)/(A + B)$.

Sensor Specifications

LVDTs are precision sensors, but they are also rugged. The cores are not in physical contact with the coils, so there is little wear, and with delrin bearings long operating lifetimes are assured. LVDTs are available [6] with strokes ranging from ± 1 mm to over ± 300 mm, with 5 to 10 mm being common. Core masses vary, but a few grams is typical. Representative overall dimensions might be 50 mm in length by 10 or 20 mm in diameter. Sensitivities of commercial units are commonly 50 to 250 mV/V/mm (or equivalently about 2.5 mV/V/mil, where 1 mil = 0.001 inch). Excitation frequencies are usually in the range 1 to 10 kHz.

14.2 ROTATION

Many examples can be found of systems in which rotation about an axis occurs and for which the angular coordinate is of physical significance. A very simple example is the measurement of the direction in which a satellite dish (or weather vane) points. An important practical application requiring angular sensing is the monitoring of mechanical joint motions in robotics.

Two different sensors for measuring shaft angles are now discussed—the angular resolver and the optical shaft-encoder.

Resolver

Physically, angular resolvers resemble miniature motors. They are cylindrical in shape, with a steel shaft emerging from a bearing in one end. The resolver is fixed in place, while its shaft must be mechanically coupled to the shaft of the test object. Angular displacements of the test object are thus transferred to a rotor within the sensor. Good alignment of the two rotational axes is necessary to prevent binding.

Three coils are arranged within the resolver. Two are internally mounted to the sensor body and are thus immovable. These stator windings are configured as a pair of orthogonal coils. This situation is suggested by the two coil planes in the diagram of Fig. 14.5. A third coil is wound on the rotor, which is able to turn freely on low-friction axial bearings. Figure 14.6 shows the simplified schematic usually chosen to represent a resolver.

When the rotor coil is energized with an external drive signal $I_0 \sin \omega t$, an oscillating magnetic field $B_0 \sin \omega t$ is created in a direction that is perpendicular

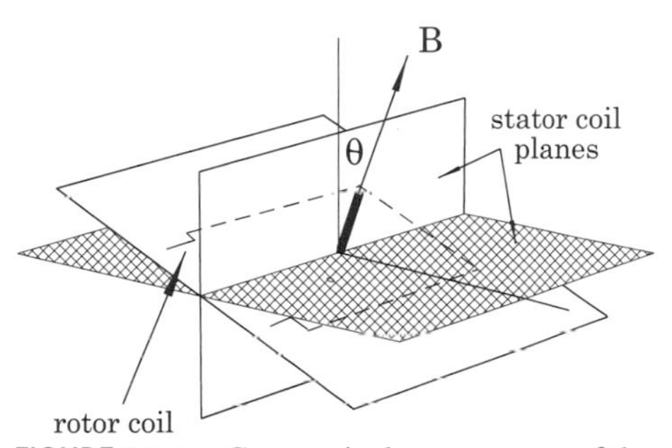

FIGURE 14.5. Geometrical arrangement of the coil planes in an angular resolver. For a rotor angle θ, the stator coils will sense components of the excitation field proportional to $\sin\theta$ and $\cos\theta$.

to the plane of the rotor coil. The component of this magnetic field which is then perpendicular to either stator coil plane will, by Faraday's law of induction, induce an emf in that coil. If the rotor is oriented at an angle θ as shown in the figure, then the stator signals will be proportional to $\sin\omega t \sin\theta$ and $\sin\omega t \cos\theta$.

A technique for extracting the desired information, namely the shaft angle θ, from this sensor is illustrated in Fig. 14.7. This block diagram represents the on-chip components of decoders made by Analog Devices [5] for this purpose. An up–down counter holds a value ϕ, which is the current approximation to θ. Analog functions generate $\cos\phi$ and $\sin\phi$, then multiply these by the signals from the stator coils, as indicated. The difference amplifier output is thus set by the equality

$$\sin\theta\cos\phi - \sin\phi\cos\theta = \sin(\theta - \phi)\,.$$

The combination of phase-sensitive detector, integrator, and voltage-controlled oscillator comprises a feedback control loop whose purpose is to null the error signal $\sin(\theta - \phi)$; in other words, to achieve the condition $\phi = \theta$. The digital output is therefore the required shaft angle.

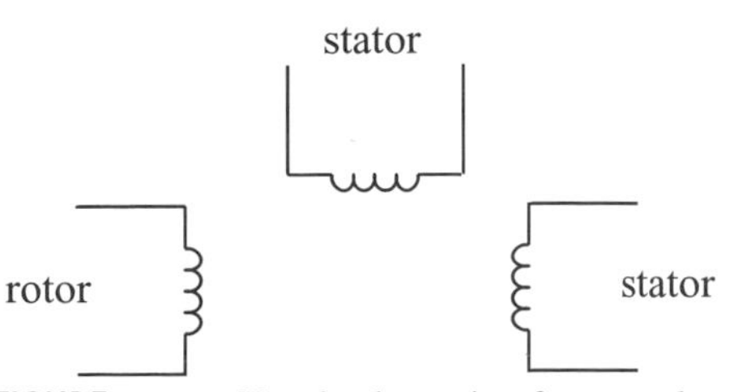

FIGURE 14.6. Usual schematic of an angular resolver indicating the rotor (excitation coil) and two separate stator secondaries.

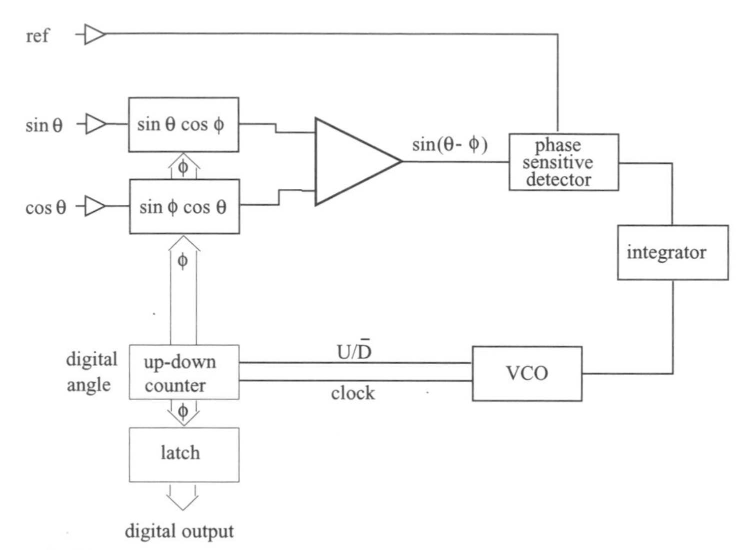

FIGURE 14.7. Block diagram of a tracking resolver to digital converter.

Sensor Specifications

Combinations of resolvers and tracking decoders are available to cover a variety of performance requirements. Resolutions of 10 to 16 bits are common, with tracking rates of at least 10 to 50 revolutions per second. The following table shows the correspondence between bit position and angular resolution in the decoder output.

Bit Number	Degrees
1	180.00
2	90.000
3	45.000
4	22.500
5	11.250
6	5.6250
7	2.8125
8	1.4063
9	0.7031
10	0.3516
11	0.1758
12	0.0879
13	0.0440
14	0.0220
15	0.0110
16	0.0055

Excitation signals for the rotor are usually 5 to 10 V at 2 to 5 kHz. Just as the cores of LVDTs add mass to the system under test, the rotor of a resolver adds moment of inertia. A representative value for rotor moment of inertia would be 3.5 $gm-cm^2$.

Incremental Optical Shaft Encoders

A second important and widely used sensor of angular displacement is based on the chopping of a light beam by a thin metal disk that is pierced by a number of equally spaced slots, as illustrated in Fig. 14.8. By simply counting pulses generated by the interruptions of the beam, the angular coordinate can, in principle, be ascertained. It is necessary to know whether the wheel is turning clockwise or counterclockwise, since this determines whether the count should be incremented or decremented at the next optically generated pulse.

Integrated components are available [7], which incorporate light-emitting diodes, collimating lenses, multidiode photodetector arrays, and analog processing elements. The configuration of the photodetector arrays is such that when one group "sees" light coming through a slot, the other group is blocked. The detector circuitry then generates two digital channels whose signals are in quadrature, meaning that they are out of phase with each other by 90 degrees. This quadrature scheme also permits the determination of the direction of rotation

FIGURE 14.8. Basic principle of an incremental optical shaft encoder. Light from the emitter passes through slots in the wheel to reach the photodetector. Pulse counting determines the shaft rotation angle.

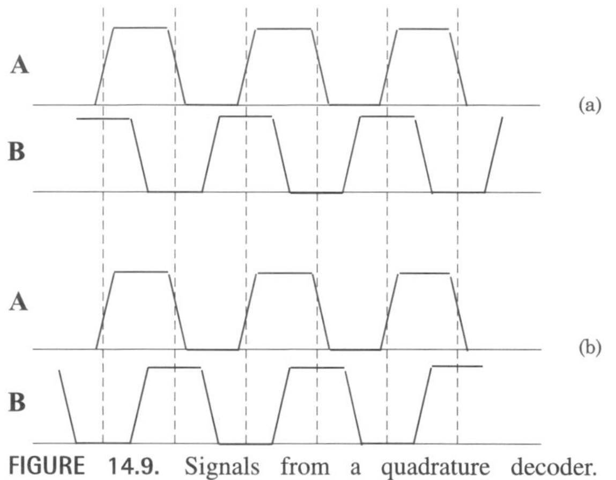

FIGURE 14.9. Signals from a quadrature decoder. (a) Channel A lags Channel B by 90 degrees. (b) Channel A leads Channel B by 90 degrees.

by noting which is the leading channel, as illustrated in Fig. 14.9. For example, if case (a) corresponded to clockwise rotation, then case (b) would represent counterclockwise motion.

Note that, with quadrature signals, four subintervals occur in the time between the passage of one slot and the next one [8]. They can be specified as: (1) A high, B high; (2) A high, B low; (3) A low, B low; (4) A low, B high. The sequence $1 \rightarrow 2 \rightarrow 3 \rightarrow 4 \rightarrow 1 \rightarrow$ is case (a), while case (b) has the order reversed: $2 \rightarrow 1 \rightarrow 4 \rightarrow 3 \rightarrow 2 \rightarrow$.

Sensor Specifications

Codewheels are typically made from thin foil and have diameters of 1 or 2 inches. The number of slots (cycles) available ranges from 500 to 2048. Because of the quadrature decoding, the four subintervals explained earlier then would produce up to 8192 counts per revolution (13 bits). As complete commercial units [9], these encoders have an appearance similar to the resolvers described earlier—they resemble small motors. They are capable of monitoring rapidly spinning shafts with speeds to above 10,000 rpm. Moments of inertia typically range over $0.5 \rightarrow 50$ gm$-$cm^2.

BIBLIOGRAPHY

[1] Ramón Pallás-Areny and John G. Webster, *Sensors and Signal Conditioning* (Wiley-Interscience, New York, 1991), pp. 159–166.

[2] Anthony J. Wheeler and Ahmad R. Ganji, *Introduction to Engineering Experimentation* (Prentice-Hall, Englewood Cliffs, NJ, 1996), pp. 200–204.

[3] Jacob Fraden, *AIP Handbook of Modern Sensors* (AIP Press, New York, 1995), p. 278.

[4] S.C. Saxena and S.B.L. Seksena, "A self-compensated smart LVDT transducer," IEEE Trans. Instrum. Meas., 38, 748 (1989).

[5] AD598 LVDT Signal Conditioner (Analog Devices, One Technology Way, Norwood, MA).

[6] *The Pressure, Strain and Force Handbook* (Omega Engineering, Inc., Stamford, CT, 1995). See, for example, data on series LD200 sensors.

[7] For example, HEDS-9000 Two Channel Optical Incremental Encoder Module and HEDS-6100 Codewheel from Hewlett-Packard, Palo Alto, CA.

[8] For example, HCTL-2000 Quadrature Decoder/Counter Interface IC from Hewlett-Packard, Palo Alto, CA.

[9] For example, optical incremental shaft encoders from Encoder Division, Litton Systems, Inc., Chatsworth, CA.

CHAPTER 15

Acceleration

Acceleration is the instantaneous rate of change of velocity:

$$\vec{a} = \frac{d\vec{v}}{dt}.$$

Units are, commonly, cm/sec^2, m/sec^2, and ft/sec^2. The acceleration imparted to a freely falling object due to the earth's gravity is referred to as 1 g, where

$$1\ g = 9.80665\ \text{m/sec}^2 = 32.174\ \text{ft/sec}^2.$$

For example, a sports car that can reach 100 kilometers per hour in 6 sec has an acceleration of 0.472 g.

Note that because velocity is a vector quantity defined by both magnitude (speed) and direction, so too is acceleration. Sensors for acceleration (accelerometers) must therefore respond in a directional fashion; that is, they must have a sensitive axis.

The key to accelerometer operation lies in Newton's second law:

$$\vec{F} = m\vec{a},$$

which says that force equals mass times acceleration. In other words, if a mass is to accelerate, then a force must act on it. Hence, by measuring a force, one may determine an acceleration

$$\vec{a} = \frac{\vec{F}}{m}. \tag{15.1}$$

Suppose the acceleration of some object (robot arm, car, rocket, shake table, impact hammer, etc.) is to be determined. A sensor having its own internal test

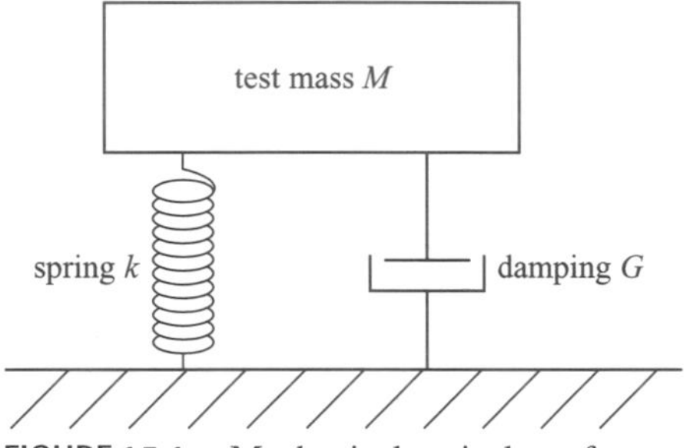

FIGURE 15.1. Mechanical equivalent of an accelerometer attached to object under test.

mass m together with associated components must be rigidly attached to the object under study, with the presumption that the addition of the sensor does not alter the behavior of the system. Generally speaking, small-mass sensors are best.

A general equivalent mechanical model of an accelerometer is shown in Fig. 15.1. Here, the sensitive axis is up/down in the diagram. With the vertical coordinate denoted by y and velocity-dependent damping, the equation of motion of the mass M will be

$$M \frac{d^2 y_m}{dt^2} = -k\,(y_m - y_b - L) - G\,\frac{d}{dt}\,(y_m - y_b), \tag{15.2}$$

where y_m locates the test mass and y_b corresponds to the base (sample under test). The natural length of the spring is L. Defining a relative elongation $\epsilon = (y_m - y_b - L)$, and with $\omega_0^2 = \frac{k}{M}$ and $\beta = \frac{1}{2}\frac{G}{M}$, this equation becomes

$$\frac{d^2\epsilon}{dt^2} + 2\beta\frac{d\epsilon}{dt} + \omega_0^2\epsilon = -a_b(t), \tag{15.3}$$

where a_b denotes the acceleration of the base—the quantity to be determined—which could be constant or, more generally, time-dependent.

To aid the discussion at this point, a stylized accelerometer is depicted in Fig. 15.2. A sensor test mass is located at the end of a flexible beam; the sensitive axis is thus up/down in the figure. The body of the device is attached to the object under test. The deformation of the elastic beam caused by acceleration can then be used to generate a sensor signal.

A variation of this approach is illustrated in Fig. 15.3. Now, the cantilever is placed between two conducting plates. As the cantilever deflects, the capacitance changes. This can be detected with appropriate auxiliary circuitry, and hence an electronic signal proportional to acceleration can be generated.

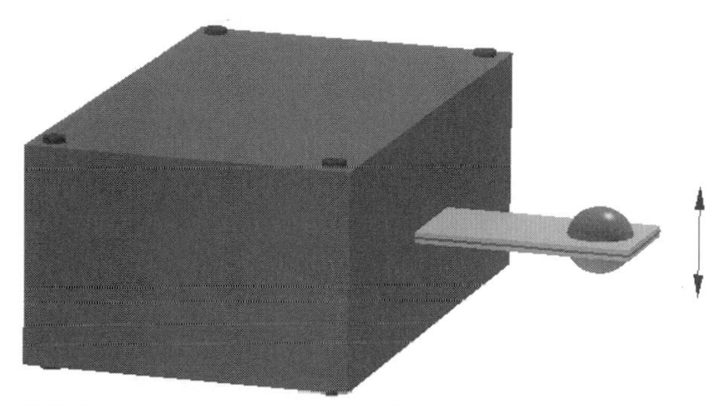

FIGURE 15.2. Simplified depiction of a flex-beam accelerometer. The body of the sensor is attached to the object under test so that acceleration up or down leads to deformation of the beam.

In both of these hypothetical accelerometers, the sensor mass and its elastic supporting structure (the beam) are modeled by Eq. (15.3).

The differential equation (15.3) is second-order, linear, and inhomogeneous. Its complete solution is the sum of the general solution of the reduced equation (when $a_b = 0$) together with any particular solution to the full equation [1]:

$$\epsilon = \left[A \exp\left(\sqrt{\beta^2 - \omega_0^2}\right) t + B \exp\left(-\sqrt{\beta^2 - \omega_0^2}\right) t \right] e^{-\beta t} + \epsilon_p. \quad (15.4)$$

Noticc that the exponential factor to the right of the square bracket will force

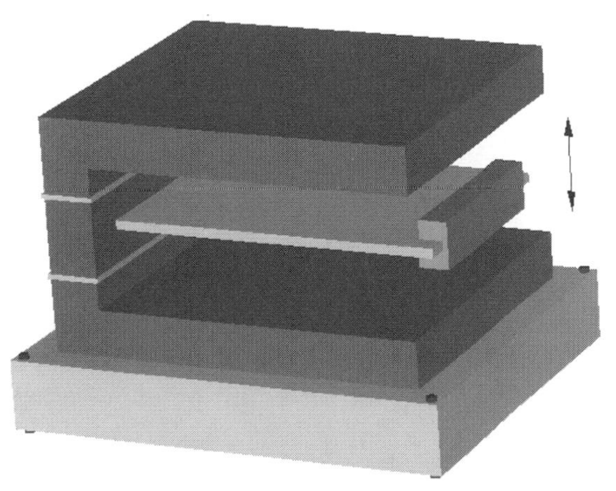

FIGURE 15.3. Stylized accelerometer with capacitive sensing.

a decay of that term, which is therefore a transient, ultimately leaving only ϵ_p. Critical damping occurs when $\beta = \omega_0$.

Constant Acceleration

For the special case of constant base acceleration $a_b = a$, Eq. (15.3) has the equilibrium solution

$$\epsilon_p = -\left[\omega_0^2\right]^{-1} a. \tag{15.5}$$

As expected, positive upward acceleration of the base results in a steady-state compression of the spring. In some types of accelerometer, this linear relationship between deformation and acceleration plays a key role in sensor operation.

Harmonic Acceleration

For the case of harmonic motion of the object under test (i.e., with the base vibrating up and down) as, for example, $y_b = Y \cos(\omega t)$, then the base acceleration would be $a_b = -Y\omega^2 \cos(\omega t)$ and

$$\frac{d^2\epsilon}{dt^2} + 2\beta\frac{d\epsilon}{dt} + \omega_0^2\epsilon = Y\omega^2 \cos(\omega t). \tag{15.6}$$

The transient portion of the general solution remains as before, but now

$$\epsilon_p = \frac{Y\omega^2}{\sqrt{\left(\omega_0^2 - \omega^2\right)^2 + 4\beta^2\omega^2}} \cos(\omega t - \delta) \tag{15.7}$$

with

$$\delta = \tan^{-1}\left[\frac{2\beta\omega}{\omega_0^2 - \omega^2}\right]. \tag{15.8}$$

As Eqs. (15.7) and (15.8) reveal, both the amplitude and phase shift of the resulting spring extension ϵ_p exhibit frequency dependencies. A useful change of variables involves

$$Q = \frac{\omega_0}{2\beta}, \tag{15.9}$$

in which case

$$\epsilon_p = Y \left[\frac{\Omega^2}{\sqrt{(1-\Omega^2)^2 + \Omega^2/Q^2}} \right] \cos(\omega t - \delta) \tag{15.10}$$

and

$$\delta = \tan^{-1}\left[\frac{\Omega/Q}{1-\Omega^2}\right], \tag{15.11}$$

where $\Omega = \omega/\omega_0$ is the normalized excitation frequency.

The square-bracketed factor in Eq. (15.10) is plotted in Fig. 15.4. This illustrates the fact that there is a peak at, or close to, the characteristic frequency ω_0, and that the peak becomes progressively sharper as the parameter Q is increased. An amplitude of unity in this plot corresponds to the situation where the test mass oscillates according to

$$\epsilon_p = Y \cos(\omega t - \delta);$$

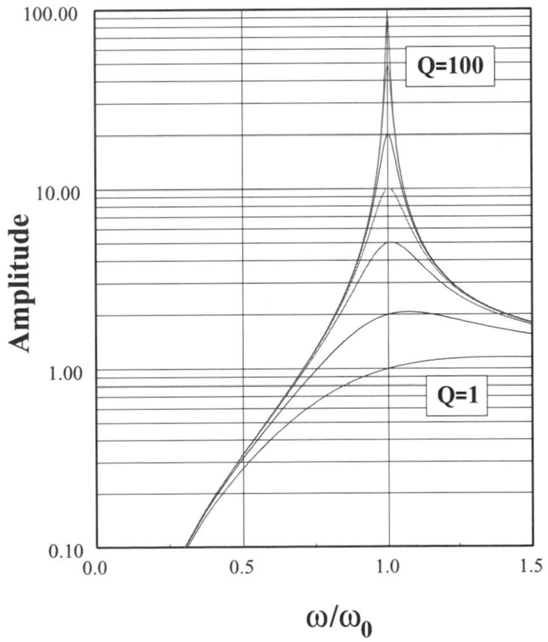

FIGURE 15.4. Amplitude response of test mass. Curves are plotted for Q values 1, 2, 5, 10, 20, 50, 100.

in other words, the amplitude of the spring extension happens to equal the amplitude of the base oscillation.

By taking the derivative of expression (15.10) with respect to Ω, it can be shown that the peak of the response is at

$$\Omega_R = \frac{1}{\sqrt{\left(1 - \frac{1}{2Q^2}\right)}}. \tag{15.12}$$

For large Q, the resonant peak lies approximately at $\omega = \omega_0$, but as Q decreases the peak shifts to higher frequencies. This effect can be seen in Fig. 15.4.

It should be noted that this behavior is slightly different from the more commonly discussed case of damped motion with harmonic forcing [1]. In that situation, the applied *force* is known, whereas here the *motion* of the base (object under test) is specified. As a consequence, the force on the test mass is created by the stretching or compression of the elastic spring. Thus, for example, the resonant frequency now shifts up rather than down.

The frequency dependence of the phase shift δ is plotted in Fig. 15.5. At low frequencies, the test mass moves in phase with the base, but for high frequencies, it oscillates completely out of phase with the base ($\delta = \pi$).

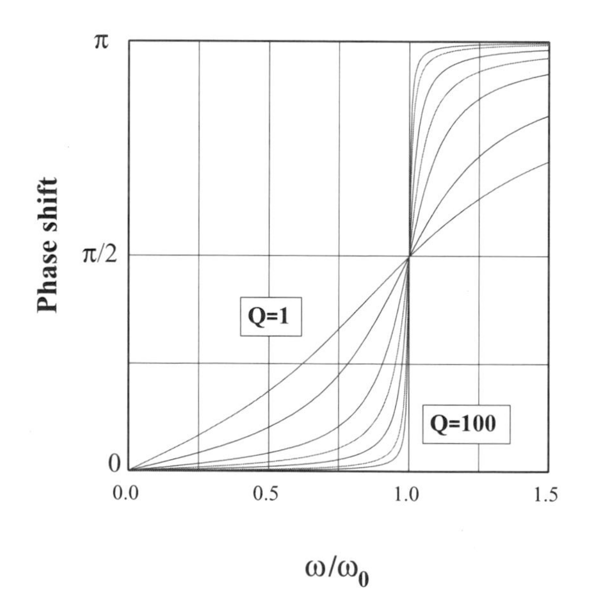

FIGURE 15.5. Phase shift (δ) between the motion of the test mass and the motion of the mounting base. Curves are plotted for Q values of 1, 2, 5, 10, 20, 50, 100.

15.1 MICROMACHINED SENSORS

As the previous discussion makes clear, an accelerometer can be constructed by attaching a test mass to an elastic support and then measuring displacements from equilibrium. Extremely small devices with feature sizes of only a few microns ($1\ \mu = 10^{-6}$ m) can now be manufactured with the techniques of micromachining [2], which utilize the repertoire of integrated circuit production processes, including diffusion, implantation, deposition, patterning, and etching. These fabrication methods have been developed for silicon in its role as the standard semiconductor for devices. Fortuitously, silicon also possesses nearly ideal mechanical properties, including a modulus of elasticity like steel.

The term micro-electro-mechanical systems (MEMS) is sometimes applied to these structures. Many intriguing MEMS implementations are currently in development; these include micromotors, micromirror arrays, microrelays, microresonators, and so forth.

The two principal categories of processing are *bulk micromachining*, which is commonly employed to create pressure transducers, and *surface micromachining*, which is used to fabricate devices such as accelerometers. In bulk micromachining, structures are chemically carved down into a base slab of silicon, whereas in surface micromachining, the desired mechanical structure is assembled above the surface of a silicon base.

A crucial step in surface micromachining is the process of laying down and later removing a sacrificial layer. Thus, it is possible to form something like a beam or plate suspended above the silicon surface by following a sequence of steps: (1) deposit a sacrificial layer onto the surface of the base silicon; (2) deposit the beam or plate material in the desired geometry directly onto the sacrificial layer; (3) etch away the underlying sacrificial layer. Except for the necessary support posts, the beam or plate has now been cut free to float above the base. An illustration of the results that can be achieved is shown in Fig. 15.6.

Using this processing technology, it is possible to fabricate miniature sensors of the general type depicted schematically in Fig. 15.3; The K-BEAM series of accelerometers from Kistler Instrument Corp. [3] are examples of commercial products based on cantilevered seismic masses with capacitive readout. Measurement ranges from $\pm 2\ g$ to $\pm 20\ g$ are available, with respective sensitivities from 2.5 V/g to 0.25 V/g. The frequency response for these accelerometers extends from dc to approximately 1 kHz, with nominal resonant frequencies of around 2.5 kHz. The devices are quite compact and have masses of less than 3 grams.

Another approach, depicted in Fig. 15.7, consists of a micromachined central beam attached to supporting posts (anchors) by box shaped "springs," sometimes

FIGURE 15.6. Scanning electron microscope (SEM) photo of micromachined structure. (Courtesy of Analog Devices, Wilmington, MA.)

called tethers. This is the basic structure of an accelerometer produced by Analog Devices [4]. Also represented in the figure are tabs, protruding from the beam, which serve as movable intermediate plates in capacitors with fixed outer plates.

When accelerated along the axis of the beam, one spring will compress while the other extends, as suggested in Fig. 15.8. The shift in beam-tab location

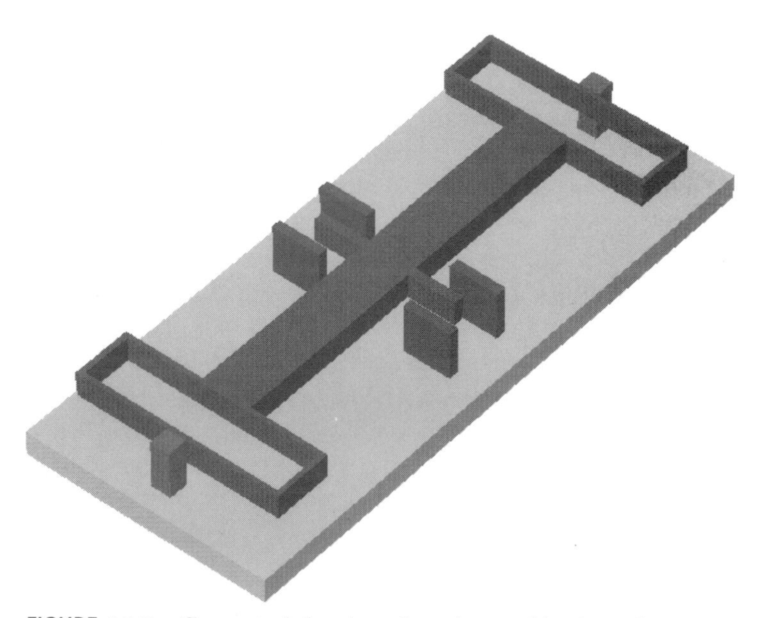

FIGURE 15.7. Conceptual drawing of a micromachined accelerometer. The test mass consists of the central beam which is terminated at either end by box springs and support posts. The sensitive axis is along the beam.

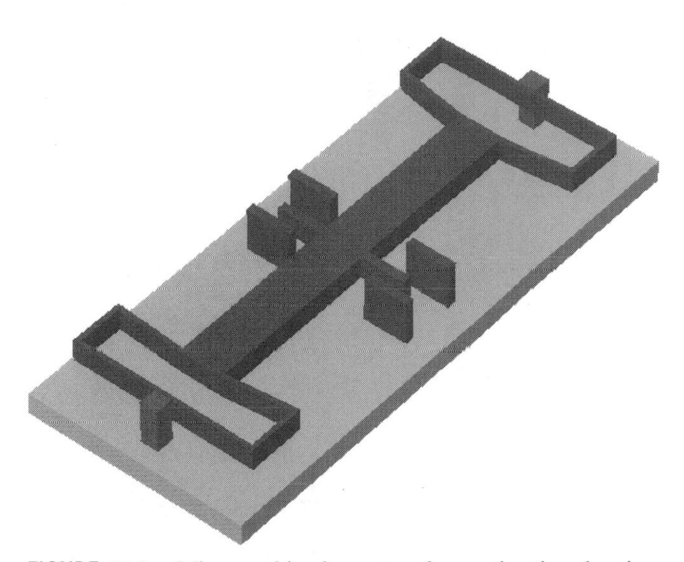

FIGURE 15.8. Micromachined sensor under acceleration showing the spring deformations and changes in the capacitors.

between the fixed plates results in an altered capacitance, which is sensed by on-chip electronics. Because this design utilizes capacitance sensing, dc (static) accelerations can be measured; it is thus suitable for applications requiring tilt monitoring.

In the actual ADXL05 accelerometer, a total of 46 tabs are employed. The complete sensor is contained in a small 10-pin metal package with a mass of only 5 grams. Typical measurement ranges are $\pm 2\ g$, and the onboard electronics generates an output voltage which is of the order of 0.1–1 V per g. Frequency response is dc to about 4 kHz, while the resonant frequency of the elastic structure is 12 kHz.

Two-axis accelerometers are also available. An SEM view of the micromachined chip for a sensor of this type [5] is shown in Fig. 15.9, and a magnified portion is presented in Fig. 15.10. Very fine interdigitated capacitors are formed along all four edges of the central plate. For this accelerometer, the range is $\pm 2\ g$, and in most other respects the performance is similar to the single-axis ADXL05.

Another example of a complex micromachined device [6] is shown in Fig. 15.11.

The large polysilicon butterfly-shaped central structure is an inertial rotor. Suppose x and y define the plane of the chip, with the z axis being vertical (out of the picture). The rotor is tethered at its center in such a way that it can both rotate by small amounts clockwise or counterclockwise about the z

FIGURE 15.9. Scanning electron microscope view of a complete two-axis accelerometer chip. (Courtesy Analog Devices, Wilmington, MA.)

FIGURE 15.10. Enlarged SEM view of the two-axis accelerometer illustrating the intricate patterning possible with surface micromachining. (Courtesy Analog Devices, Wilmington, MA.)

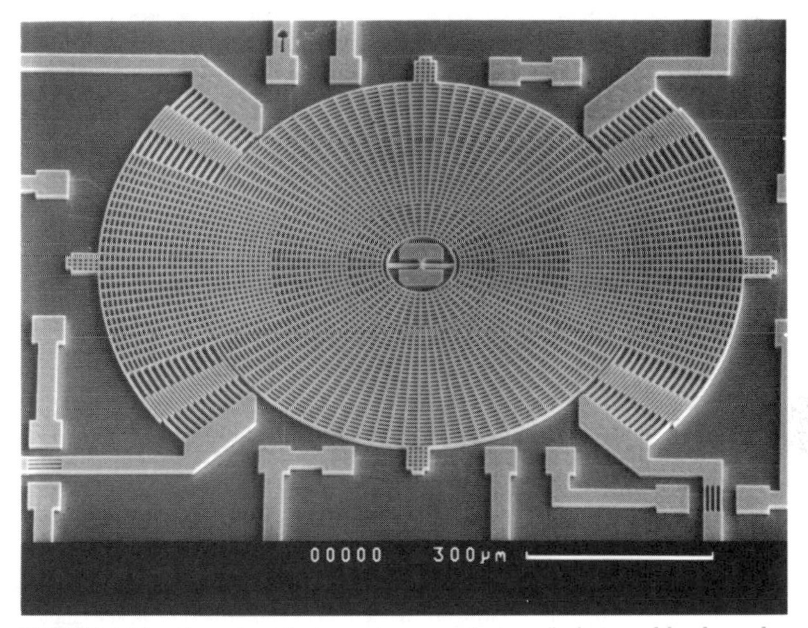

FIGURE 15.11. Scanning electron microscope image of micromachined angular rate sensor. Note the scale at the bottom of the figure. (Courtesy of Robert Bosch GmbH, Stuttgart.)

direction, and also wobble (tilt) on that axis. In operation, the rotor is maintained in oscillations in the $x - y$ plane at a frequency of approximately 1.5 kHz by electrical excitation of the four comb structures visible in the photo. A close-up view of a comb is shown in Fig. 15.12.

Rotation of the sensor chip about either of the two in-plane axes (x or y) causes the rotor to wobble, an effect that is then capacitively detected.

15.2 PIEZOELECTRIC SENSORS

The piezoelectric effect was discussed in Chapter 13 in connection with pressure sensors. In piezo materials, mechanical deformations result in the appearance of surface charges. Several different types of accelerometer make use of this effect. It should be noted that because of charge leakage effects, these sensors cannot be used at dc. In contrast to the capacitively based micromachined accelerometers discussed in the previous section, the frequency response of piezoelectric accelerometers will have a lower limit as well as an upper cutoff.

FIGURE 15.12. Magnified view of comb structure in the angular rate sensor. (Courtesy of Robert Bosch GmbH, Stuttgart.)

Bimorph Beam

The conceptual drawing of Fig. 15.2 shows a cantilever beam at the end of which a seismic mass is situated. Acceleration causes flexing of the beam.

In bimorph sensors, the beam is fabricated as a ceramic sandwich with a metallic center layer. As the beam bends, one of the outer layers goes into compression while the other is in extension. The resulting piezoelectric charges must then be converted to a usable output signal. This is typically done with a charge amplifier, as described in Chapter 13.

One interesting application of a hybrid bimorph accelerometer was in an x–y plotter [7]. Real-time sensing of the pen acceleration and subsequent integration of the waveform was used to optimize writing speed. For this particular device, the five-millimeter-long bimorph beam was made from lead zirconate titanate. An acceleration of 1 g deflected the beam tip by only 0.02 μ, but with the associated electronics an output sensitivity of 1 V/g was achieved. The resonant frequency was 7.8 kHz and the overall frequency response was approximately 5 Hz to a few kHz.

Bimorph accelerometers are commercially available [8]. Typical specifications include measurement ranges from ± 5 g to ± 50 g, sensitivities of around 1 V/g, resonant frequencies of 10–20 kHz, and frequency responses from 0.5 Hz to 2 kHz or more. Sensor masses run around 5 grams.

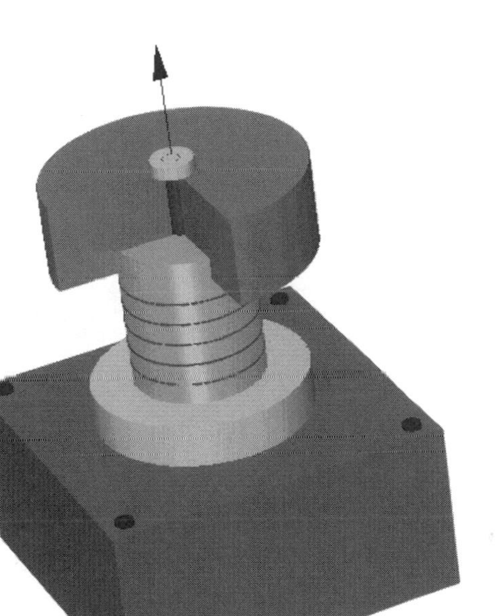

FIGURE 15.13. Conceptual representation of compression-mode piezoelectric accelerometer. The seismic mass is shown cut away to reveal the stack of piezo material fixed by a preload bolt.

Compression Mode

A conceptual representation of a compression-mode piezoelectric accelerometer is shown in Fig. 15.13. A preload bolt clamps the seismic mass to a stack of piezoelectric elements. When acceleration occurs along the axis of the device, the state of compression of the crystals changes, thus inducing surface charges. Again, because of the large but finite resistivity of the piezo materials, these charges will ultimately decay away. Therefore, this type of sensor cannot measure constant acceleration, but it is well suited to applications involving shock and vibration. Charge amplifiers must be employed.

Shear Mode

In shear-mode sensors, the piezo crystals and seismic mass are positioned in such a way that acceleration along the sensitive axis leads to shearing deformations; see Fig. 15.14. The shear charge sensitivity is twice that of the longitudinal mode [see Eq. (13.12)]; for this and other reasons, shear-mode devices are a very common choice [9, 10, 11].

FIGURE 15.14. Conceptual representation of a shear mode piezoelectric accelerometer. The annular seismic mass will cause shear forces to act on the piezo material.

Many different performance ratings can be found for accelerometers of either type. Typical frequency responses extend from around 1 Hz to 10 kHz or higher. Measurement ranges can exceed 1000 g, making some models suitable for shock and impact studies. Sensitivities are of the order of 10–100 mV/g. Sensor masses are 5–20 grams.

BIBLIOGRAPHY

[1] Tai L. Chow, *Classical Mechanics* (Wiley, New York, 1995), pp. 245–256, 261–278.
[2] See article by J. Byyzek, K. Petersen, and W. McCulley in IEEE Spectrum, pp. 20–31 (May 1994).
[3] Kistler Instrument Corporation, Amherst, New York.
[4] ADXL05, manufactured by Analog Devices, Norwood, MA.
[5] ADXL202, manufactured by Analog Devices, Norwood, MA.
[6] See article by D. Teegarden, G. Lorenz, and R. Neul in IEEE Spectrum, pp. 67–75 (July 1998).

[7] Model 7225A plotter manufactured by Hewlett-Packard Company, Palo Alto, CA; see Philip P. Maiorca and Norman H. MacNeil, "A closed-loop system for smoothing and matching step motor responses," Hewlett-Packard *J.* 18–23 (February 1979).

[8] For example, PiezoBEAM sensors from Kistler Instrument Corporation, Amherst, New York.

[9] Endevco Corporation, San Juan Capistrano, CA.

[10] K-SHEAR accelerometers from Kistler Instrument Corporation, Amherst, New York.

[11] *The Pressure Strain and Force Handbook* (Omega Engineering, Inc., Stamford, CT).

PART III

Measurements

CHAPTER 16

DC Measurements

16.1 INTRODUCTION

A fundamental task which almost any instrumentation system must be capable of performing is the measurement of voltages or currents. Generally speaking, electrical signals are time-varying quantities. The speed with which currents or voltages change ultimately determines the choice of measurement circuit.

In this chapter, various techniques are discussed for acquiring either constant (dc) or slowly varying signals. Voltage generation and detection is covered first; current sensing is left as a final topic. Time-dependent waveforms (ac) are discussed in the next chapter.

Analog meters have had a long and distinguished record as indispensable tools throughout the history of science and engineering. In these devices, a multi-turn coil is mounted on precision bearings and suspended within the field of a permanent magnet. Any current flowing through the coil causes it to rotate against a restoring torque produced by a fine spiral spring. A pointer attached to the coil indicates this deflection on a curved scale. When calibrated, the amount of deflection of the needle corresponds to the actual value of input voltage or current.

There are, however, drawbacks. Analog meters have delicate mechanical parts which can be damaged by shocks and rough handling. The physical inertia of the meter movement makes them comparatively slow. Readout is by visual inspection of pointer and scale, a procedure which consequently limits precision and renders these instruments unsuitable for applications that require automatic data logging. For these reasons, analog meters are now used mainly in applications requiring simple front-panel annunciators (for example, volume-level displays in recording studios and radio stations).

In the context of contemporary instrumentation systems, a *digital* meter measures a voltage or current by performing an *analog-to-digital* (A/D) conversion. A/D converters produce a numerical output from a voltage or current input, and they do so electronically. Essentially, there are no moving parts. A/D converters are rugged, fast, and accurate. The fact that readings are available in digital form makes this form of instrument suitable for data acquisition.

Various A/D converter designs are available; no single approach is "best." It is necessary to make tradeoffs such as simplicity versus complexity or speed versus precision.

16.2 DIGITAL-TO-ANALOG CONVERSION

Before discussing A/D converters, it is useful to first consider the inverse process: *digital-to-analog* conversion (D/A).

R-$2R$ Ladder

The circuit shown in Fig. 16.1 is a standard D/A converter based on an R-$2R$ ladder, named for the structure of the resistor chain. A reference voltage V_{ref} is applied to the input end of the ladder. To deduce what currents flow in various branches, it is instructive to consider a slight rearrangement of the ladder, as illustrated in Fig. 16.2. The switches in Fig. 16.1 terminate the $2R$ resistors either in a true ground or in a virtual ground (the op-amp inverting input). Thus, Fig. 16.2 is correct for all possible switch settings.

Notice that the ladder has the property that each nested subcircuit has exactly the same equivalent: two resistors $2R$ in parallel. In other words, every rectangle is replaceable with just R. This also applies to the complete ladder, independent of how many stages it has. The input, V_{ref}, "sees" an effective load of R, and

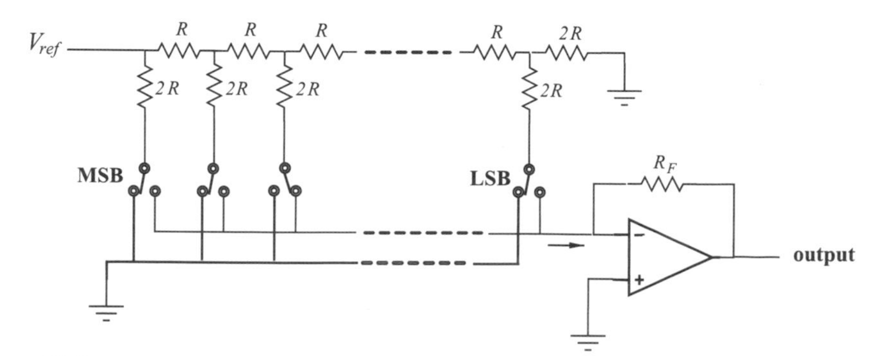

FIGURE 16.1. Digital-to-analog converter.

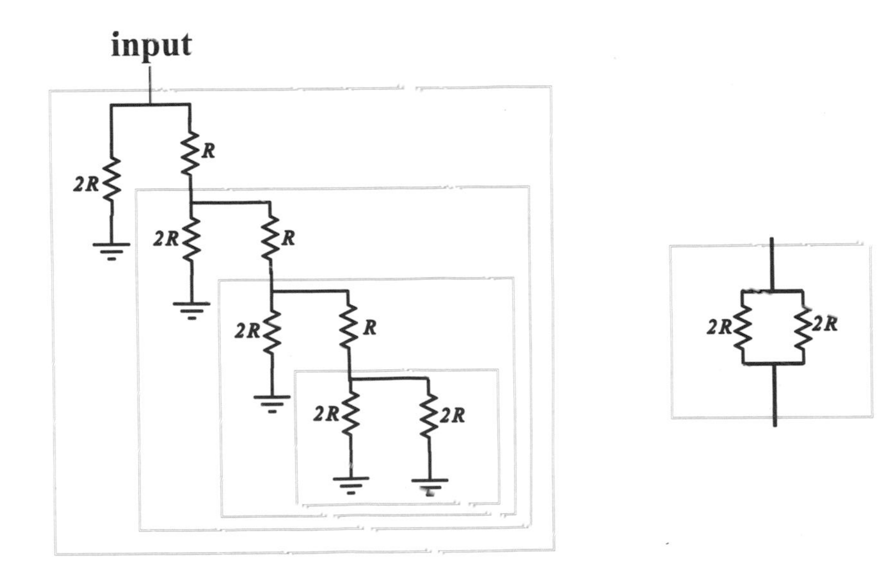

FIGURE 16.2. Rearrangement of R-$2R$ ladder. Every nested block is equivalent to the simple parallel combination on the right.

hence the total current flowing into the ladder must be

$$I = \frac{V_{\text{ref}}}{R}. \tag{16.1}$$

This current splits into two equal parts at the first ladder node, one half entering the switch marked MSB via a resistor $2R$, and the other half moving to the next ladder node, where it splits again, and so forth. The switch-controlled currents are thus $\frac{I}{2}, \frac{I}{4}, \frac{I}{8}, \frac{I}{16}, \ldots$.

The state of each switch decides whether its share of the input current is passed directly to ground or is routed to the op-amp. If the first switch (nearest the op-amp) is designated S_1 and the leftmost switch in Fig. 16.1 is S_N, then the total current summed at the inverting input is

$$I_{\text{total}} = \left(S_N \cdot 1 + S_{N-1} \cdot \frac{1}{2} + \cdots + S_1 \cdot \frac{1}{2^{N-1}} \right) \frac{V_{\text{ref}}}{2R}, \tag{16.2}$$

where $S_j = 0, 1$, depending on the jth switch setting. The bracket contains terms which decrease by powers of two and can be written

$$I_{\text{total}} = (b_N b_{N-1} b_{N-2} \cdots b_2 b_1) \left[\frac{V_{\text{ref}}}{2^N R} \right], \tag{16.3}$$

where $(b_N b_{N-1} b_{N-2} \cdots b_2 b_1)$ is a binary number whose N bits are set by the switches. The output voltage is just $V_{\text{out}} = -I_{\text{total}} R_F$, the negative sign resultir from the op-amp inversion. Hence, this circuit is a binary input D/A converter with a least significant bit (LSB) corresponding to a voltage increment

$$1\ \text{LSB} = \frac{R_F}{2^N R} V_{\text{ref}} \tag{16.4}$$

and maximum output voltage

$$V_{\max} = -\left[\frac{2^N - 1}{2^N}\right] \frac{R_F}{R} V_{\text{ref}}. \tag{16.5}$$

Current Sources

Another important form of D/A converter is shown in Fig. 16.3. Note that the op-amp is now connected to the left-hand side of the R-$2R$ ladder.

In this design, identical current sources are assigned to each node of the R-$2R$ ladder. The switches send current either into the node (bit = 1) or to ground (bit = 0). As a consequence of this arrangement, speed is improved and thermal transients are minimized because the switches (which are in fact transistors) handle the same current whether the bit is 0 or 1.

The switch settings in the schematic correspond to the binary word 0001 $\cdots$. Analysis of the various branchings reveals that the current injected by the fourth source splits such that a fraction $11/32$ is transferred to the left while the remaining $21/32$ passes down the ladder to the right. The current flowing to the left further subdivides at each of the next two nodes, some following the $2R$ paths to ground. The portions continuing towards the op-amp are, as indicated, reduced to $6/32$ and finally $4/32$. Thus, $1/8$ of the injected current at the fourth node actually reaches the op-amp input.

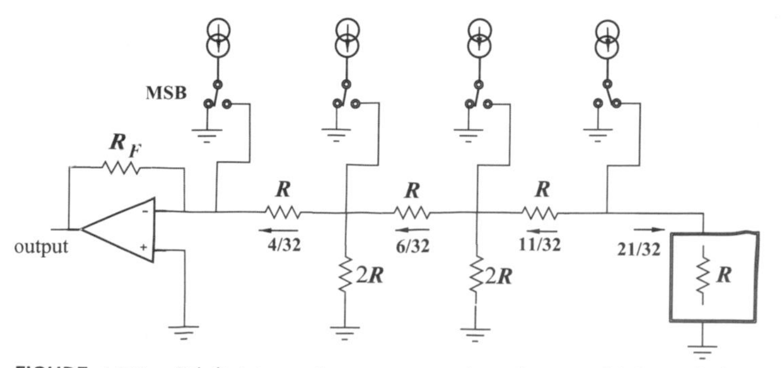

FIGURE 16.3. Digital-to-analog converter based on multiple switched current sources and R-$2R$ ladder. The first 4 bits are shown. The box represents the equivalent of the remaining portion of the ladder.

Similar analysis shows that the general rule for the proportions of injected current reaching the inverting input are: 1/1 for the first source, 1/2 for the second source, 1/4 for the third source, 1/8 for the fourth source, and so forth. In other words, the total current is

$$I_{\text{total}} = (b_N b_{N-1} b_{N-2} \cdots b_2 b_1) \left[\frac{I_S}{2^{N-1}} \right], \tag{16.6}$$

where I_S is the value of the (identical) current sources. Incidentally, in real devices a typical value for the source I_S is 1 mA. The D/A voltage output is then $V_{\text{out}} = -I_{\text{total}} R_F$.

Input Codes

As we have seen, the output currents or voltages of D/A converters are programmed by an input binary number. In the designs presented so far, the outputs are *unipolar*. Hence, for example, a D/A might deliver a voltage ranging from 0 to a maximum of, say, +10 V. Many applications are better served with *bipolar* capabilities; that is, ranges such as ±5 V. This could be accomplished with a modification of a unipolar circuit, which introduces a dc offset (−5 V in this example). In that particular case, a comparison between unipolar (binary) and bipolar (offset binary) versions is straightforward, as shown in the following table (where for all cases in this sample comparison, one LSB has been assigned the value 10/15 V).

Binary Input	Unipolar	Offset	Sign/Magnitude
1111	10.00	4.66	−4.66
1110	9.33	4.00	−4.00
1101	8.66	3.33	−3.33
1100	8.00	2.66	−2.66
1011	7.33	2.00	−2.00
1010	6.66	1.33	−1.33
1001	6.00	0.66	−0.66
1000	5.33	0.00	−0.00
0111	4.66	−0.66	+4.66
0110	4.00	−1.33	+4.00
0101	3.33	−2.00	+3.33
0100	2.66	−2.66	+2.66
0011	2.00	−3.33	+2.00
0010	1.33	−4.00	+1.33
0001	0.66	−4.66	+0.66
0000	0.00	−5.33	+0.00

The offset binary code adopts the convention that the first occurrence of 1 in the MSB is the zero location.

Also shown in the table is the sign+magnitude representation in which the apparent MSB is reserved as a sign bit. Thus, a 4-bit input word actually has 3-bit resolution.

Performance

The preceding discussions have assumed perfect components, such as ideal matching of resistors in the R-$2R$ ladder together with an absolutely known reference voltage V_{ref} or current source I_0. Imperfections are unavoidable and hence errors in the D/A conversion will occur. Nevertheless, advanced manufacturing techniques including final laser trimming of the thin-film resistors in the ladders now produce very-high-performance devices.

Monolithic D/A converters typically have internal references, although in some models an external source may be required. Few additional external components are usually required.

A fairly common output range is 10 V, and the number of binary bits on input can be anywhere from 10 or 12 to 16 or more for precision tasks. Settling times are typically a few μ sec or less.

Example

Consider an 8-switch D/A converter that employs a reference voltage of 5.0 V. Suppose the unit of ladder resistance is $R = 5\ \text{k}\Omega$ and the feedback resistor is 15 kΩ. Then, $V_{\max} \approx 14.9414$ V, and the value of an LSB is approximately 0.0586 V.

If the switch settings correspond to the binary number 01101011, then the D/A output voltage would be

$$(0 \cdot 128 + 1 \cdot 64 + 1 \cdot 32 + 0 \cdot 16 + 1 \cdot 8 + 0 \cdot 4 + 1 \cdot 2 + 1) \cdot \text{LSB},$$

or 6.2695 V.

16.3 VOLTAGE MEASUREMENT

Successive Approximation Converters

Consider the circuit illustrated in Fig. 16.4 and the logic flow chart depicted in Fig. 16.5. This arrangement is known as a *successive approximation* analog-to-digital converter. It operates as follows.

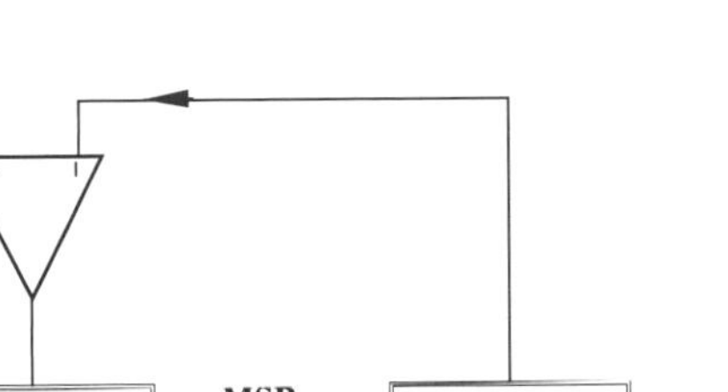

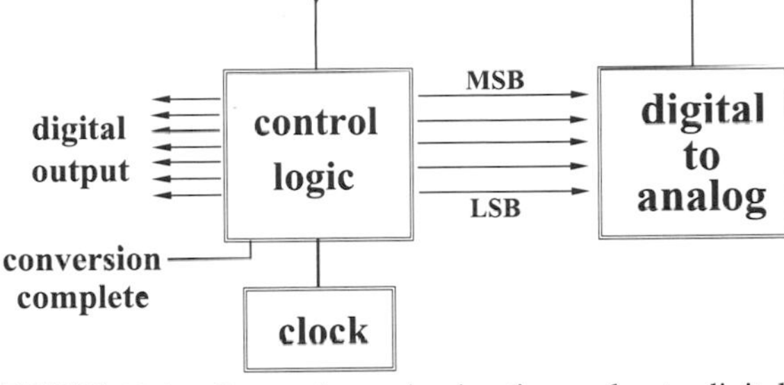

FIGURE 16.4. Successive approximation analog-to-digital converter.

Initially, the internal register is reset to all zeros, after which the most significant bit is set to 1. The D/A converter then generates an output which is compared with the input V_{in} (i.e., the voltage to be measured). If the D/A output still falls below V_{in}, then the next most significant bit is set to 1 and the test is repeated. As shown in Fig. 16.6, once a new bit causes the D/A output to exceed V_{in}, then that bit is reset. Essentially, the process consists of iteratively seeking the largest binary input for which the D/A output is $\leq V_{in}$. When the least significant bit has been tested, the conversion process terminates. The "correct" answer is revealed one bit at a time, rippling down from the MSB.

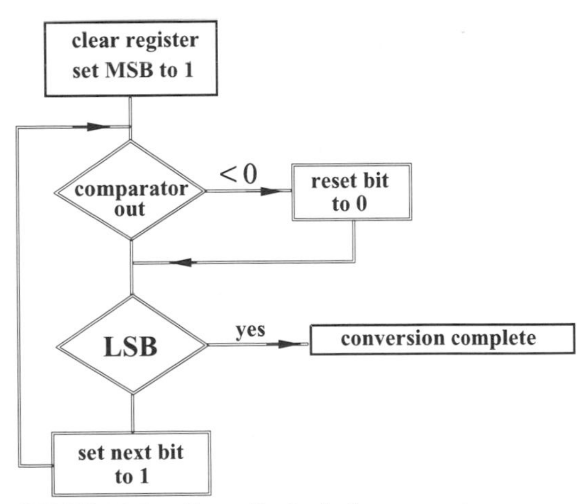

FIGURE 16.5. Controller logic for successive approximation A/D converter.

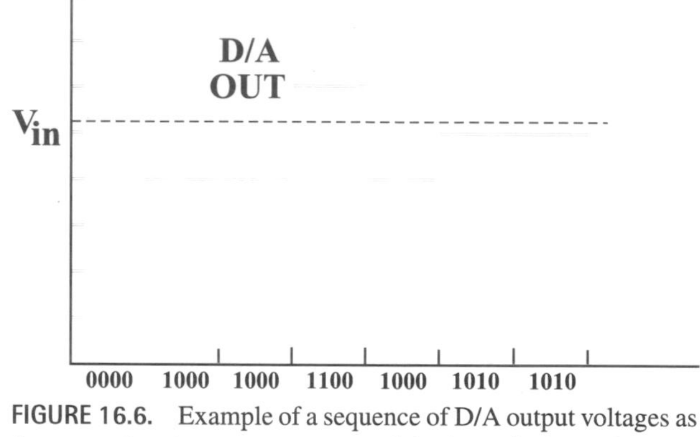

FIGURE 16.6. Example of a sequence of D/A output voltages as the approximation converges toward the input (unknown).

Each step in the iteration is made up of two substeps: a trial bit-setting followed by an interval for comparison and possible resetting. Thus, two clock periods are required to advance the iteration by one bit. For an N-bit conversion, a total of $2N$ clock cycles are required.

Output Codes

The conversion process just described can be visualized with reference to Fig. 16.7. Of particular importance is the fact that while the input V_{in} is generally a continuous function with infinitely fine gradations, the output from the D/A is discrete and has a granularity imposed by the total number of bits. As suggested in Fig. 16.7, the input voltage will certainly lie between some pair of possible D/A output levels, one produced by the binary word b and the other

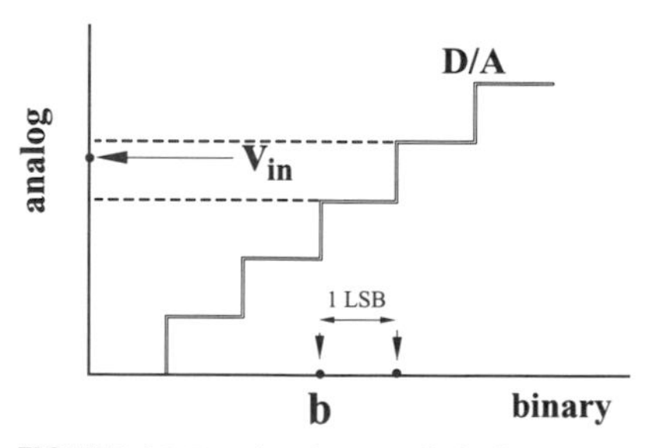

FIGURE 16.7. Analog-to-digital conversion based on comparing an input voltage with the output from a D/A converter. If the result from a successive approximation is the binary value b, then V_{in} must lie in the range indicated.)

by b incremented by 1 LSB. Successive approximation delivers the same digital result b for any voltage within this range.

A related matter is the effect that noise has on the conversion. Clearly, the issue is whether or not noise superimposed on the dc level carries the net voltage back and forth across zone boundaries. Sensitivity to noise will be largest for values of V_{in} that happen to coincide with step edges (dotted lines in the figure); in such a case, even infinitesimal noise will induce jitter in the least significant bit.

As explained earlier, the input word to digital-to-analog converters may follow one of a number of standard formats, including unipolar binary, offset binary, and sign/magnitude. The binary outputs of A/D converters also may be generated according to any of these formats. In the particular case of offset binary codes, it should be observed that for inputs just above or just below zero volts, the output will be either $011111\cdots$ or $100000\cdots$. In other words, *all* bits change if the input crosses zero. Such a *major carry* can sometimes result in transient glitches at the digital output terminals.

Performance

Successive approximation is perhaps the most commonly used type of analog-to-digital converter. Models having 8, 10, or 12 bits are commonly available with conversion times under 10 μsec. Inputs up to approximately ± 10 V can be accommodated. These devices are essentially complete and typically include onboard clock generators.

Tracking Converters

A second type of A/D converter is shown in Fig. 16.8. In operation, the voltage to be measured is compared with the output from a D/A, which is in turn fed from an up/down counter. If, for example, the count is too "high," then the D/A output will exceed V_{in} and the comparator output will be high. This then enables the gate to the down count input and also disables the up count input. Clock pulses sequentially lower the counter value until the D/A output falls below V_{in}, at which point the comparator output goes "low" and the logic gating switches the system to up-counting.

When a reading is initiated, the contents of the counter will creep up from zero, one clock pulse at a time, until it just exceeds the equivalent of V_{in}. After that, if the input voltage is noise-free, a down count will occur, followed by an up count, another down count, and so on. This process is illustrated in Fig. 16.9. From the description just given, it is apparent that even in the case of not quite constant (slowly varying with respect to the system clock frequency) V_{in}, the digital output from the counter can *track* the analog input.

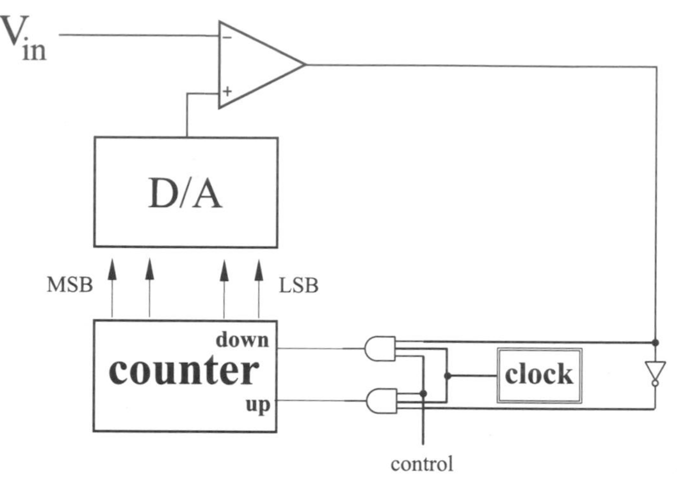

FIGURE 16.8. Schematic for a tracking analog-to-digital converter.

If the input is close to full scale, then almost all bits would have to reach 1. For an 8-bit system, this would require as many as 255 clock pulses. By comparison, an 8-bit successive approximation converter needs at most $2 \times 8 = 16$ clock cycles. Obviously, the tracking converter will be slow, but it is particularly well suited in applications such as peak detectors, where signal maxima can be "remembered" simply by disconnecting the down counter and permanently enabling the up-count function.

Voltage-to-Frequency Converters

Another possible method of performing analog-to-digital conversion is by means of a circuit that oscillates at a frequency that is set by a control voltage. In such a situation, measurement of voltage is replaced by measurement of frequency,

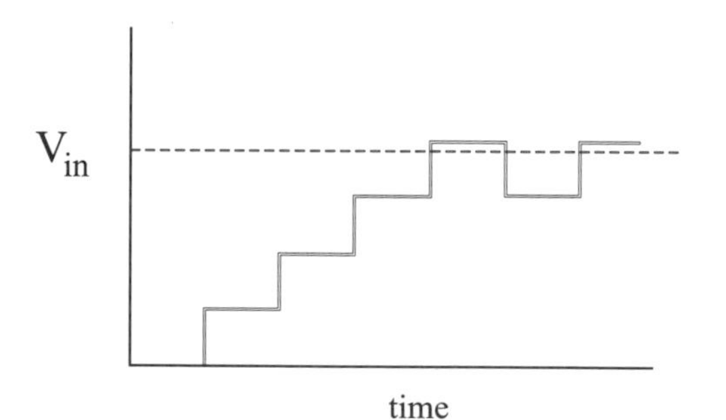

FIGURE 16.9. D/A output in a tracking analog-to-digital converter showing the initial rise to meet the input and subsequent dithering around the final state.

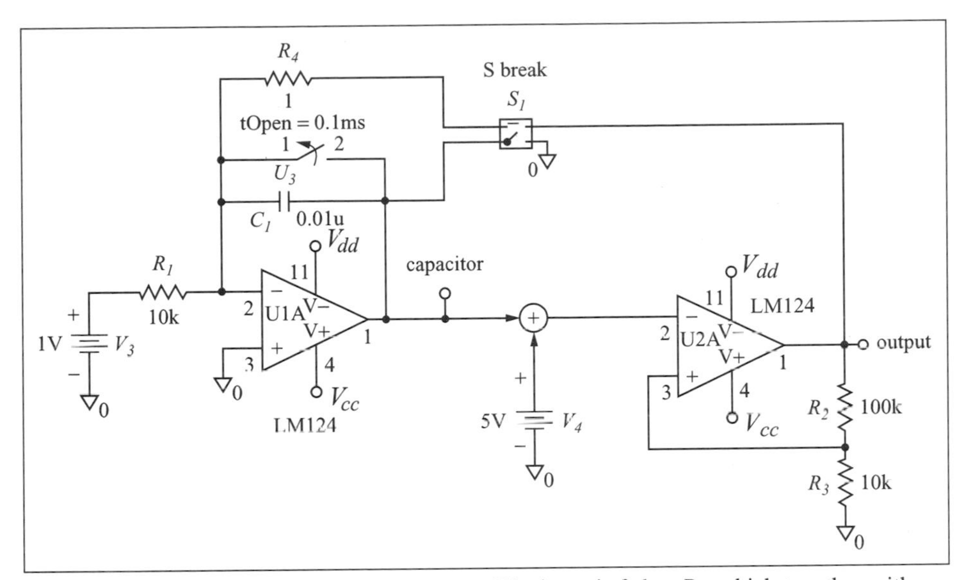

FIGURE 16.10. Voltage-to-frequency converter. The input is fed to R_1, which together with op-amp U_1 and capacitor C_1 forms an integrator. After level shifting, the ramp is passed to a Schmitt trigger, which in turn drives the capacitor resetting switch S_1.

which is essentially a counting operation. A PSpice schematic for a voltage-to-frequency (V/F) converter is given in Fig. 16.10. This example is not an optimum design; rather, it is presented to illustrate the basic operating principles of a V/F circuit.

The voltage input, here in the form of a 1 V source, is applied to resistor R_1. This resistor in combination with capacitor C_1 and the op-amp forms an integrator. For dc input, the output voltage from the op-amp (at the terminal labeled capacitor) is

$$V_{\text{capacitor}} = -\frac{V_{\text{in}}}{R_1 C_1} \int dt, \tag{16.7}$$

which is a negative-going ramp with slope proportional to V_{in}.

This ramp is then level shifted up by 5 V in this example and applied to the Schmitt trigger [2] (also see the discussion in Chapter 7 on the relaxation oscillator) formed by the second op-amp together with resistors R_2 and R_3. A Schmitt trigger possesses hysteresis—the output will be driven to positive saturation if the signal at the inverting input drops below

$$V_{T-} = -\frac{R_3}{R_2 + R_3} V_{\text{sat}} \tag{16.8}$$

and into negative saturation if the input rises above

$$V_{T+} = +\frac{R_3}{R_2 + R_3} V_{\text{sat}}. \tag{16.9}$$

With the components and power supplies used in this simulation, $V_T = \pm 1.09$ V.

When the Schmitt trigger output is high, the voltage controlled switch S_1 closes, thus discharging the capacitor. However, when the Schmitt trigger output goes negative, the switch opens, allowing a new integration ramp to be generated. The oscillations that result will have a period that is the time required for the ramp to fall from 0 (5 after level shifting) to −6.09 (−1.09 after level shifting) of

$$\tau = \frac{R_1 C_1}{V_{\text{in}}} 6.09, \tag{16.10}$$

so the frequency is

$$f = \frac{V_{\text{in}}}{R_1 C_1} 0.164. \tag{16.11}$$

For the components used in this PSpice simulation, $f = 1640$ Hz/V. Results of the PSpice simulation are shown in Fig. 16.11.

The periodic pulses that appear at V_{out} are in this case spaced at intervals of 0.679 msec. Repeating the simulation for a number of different input voltages yields the data plotted in Fig. 16.12. The observed voltage-to-frequency calibration differs slightly from the predicted value, possibly due to imperfect estimates of the op-amp saturation levels and dynamics, together with the response of the voltage-controlled switch.

Charge-Balance Converters

Most monolithic integrated circuit V/F converters operate on a principle known as *charge balance*. In simplified form, a representative schematic is shown in Fig. 16.13. The main components are an op-amp integrator, a comparator, a one-shot pulse generator, and a precision current source that can be controlled by the output of the one-shot.

The integrator output, for fixed input, is a negative ramp [see Eq. (6.22)]

$$V_C = -\frac{V_{\text{in}}}{RC} \int dt.$$

If started from zero, this ramp would proceed downward until the maximum negative op-amp output is achieved (saturation). However, before that can happen, the integrator output will reach a set point (for the purpose of this example, −0.6 V)

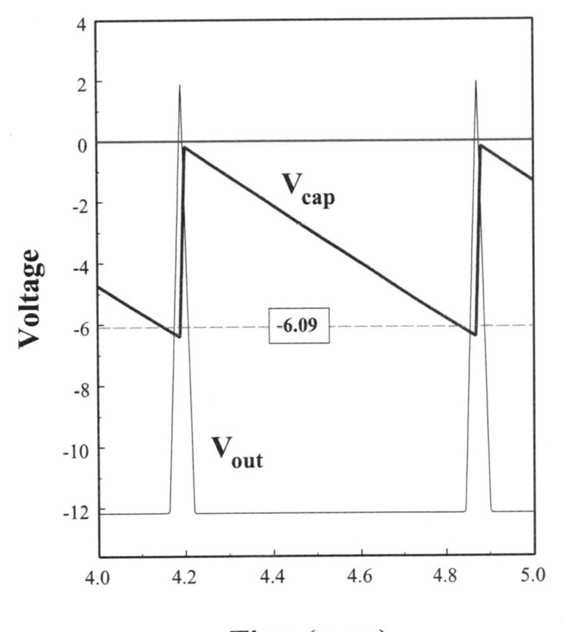

FIGURE 16.11. PSpice simulation waveforms for the V/F circuit with $V_{\text{in}} = 1.0$ V. At approximately -6.09 V, the Schmitt trigger fires and the capacitor is discharged. The negative ramp then repeats.

and the comparator will trigger the one-shot. This produces a fixed-duration output pulse, typically about $\Delta\tau \approx 2\,\mu$sec in width, which is used to enable the precision current source. A fairly standard design choice is $I_S = 1$ mA. While the source is enabled by the one-shot, charge is pumped off the capacitor in the amount

$$\Delta q = I_S \Delta\tau. \tag{16.12}$$

In one second, the input current to the converter delivers a charge

$$Q = I_{\text{in}} \tag{16.13}$$

to the capacitor.

If the one-shot is retriggered a total of N times in one second such that $N\Delta q = Q$, or

$$N = \frac{I_{\text{in}}}{I_S \Delta\tau}, \tag{16.14}$$

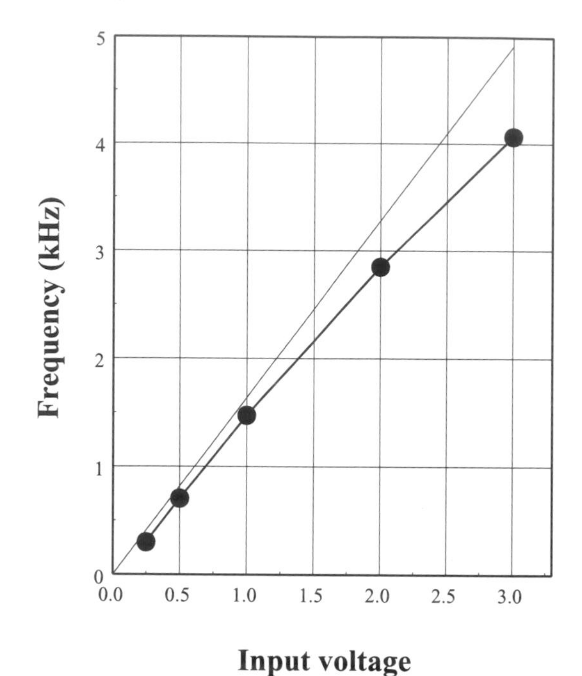

FIGURE 16.12. Calibration for the example V/F converter using data from PSpice simulation. The straight line is the predicted result: 1640 Hz/V.

then charge balance is achieved and the loss on the capacitor equals the gain. In this circumstance, the integrator output ramp will be maintained at the set point. The one-shot frequency is just N^{-1} and the input current is V_{in}/R. Hence,

$$f = \frac{V_{\text{in}}}{R I_S \Delta\tau}, \tag{16.15}$$

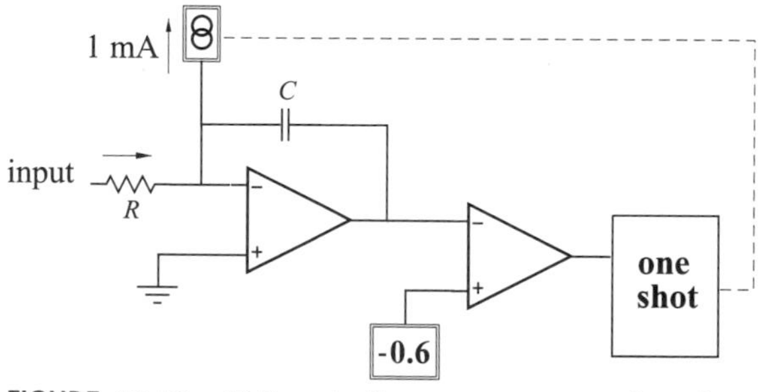

FIGURE 16.13. Voltage-to-frequency converter based on the principle of charge balance.

which is a linear V/F relationship with a proportionality constant determined by the external resistor and internal parameters I_S and $\Delta\tau$.

As an example, suppose $R = 20$ K. Then, each packet of charge is $\Delta q = 2$ nc and for 1 V input, the frequency is $f = 25$ kHz.

Performance

Monolithic V/F converters are available as integrated circuits. Typical specifications include settings for 10 kHz to 100 kHz at 10 V input, and linearity to better than 1/4%.

Dual-Slope Converters

The majority of voltmeters of the stand-alone benchtop variety are built around *dual-slope* analog-to-digital converters. A general layout of this type of A/D converter is shown in Fig. 16.14. An integrator is combined with a comparator, logic-controlled switch, and counter.

The conversion process takes place in two steps, as suggested in Fig. 16.15. For the first phase of the conversion, usually referred to as the run-up or integration, the control logic places the switch in the V_{in} position. The control logic then allows the resulting integration to proceed for a fixed time T_1 defined as exactly N cycles of the system clock

$$T_1 = N\tau, \tag{16.16}$$

with the clock period being τ.

At the end of run-up, the control logic flips the switch so that a precise reference voltage V_{ref} of opposite polarity to V_{in} is applied to the integrator. Now the second phase, referred to as the run-down or deintegration, proceeds until the comparator

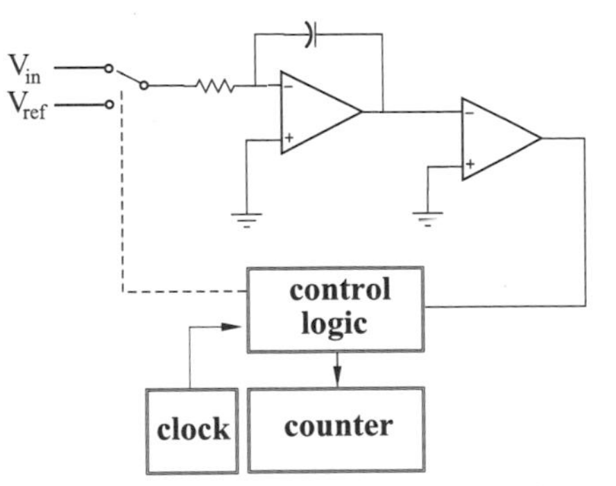

FIGURE 16.14. Schematic of a dual-slope analog-to-digital converter.

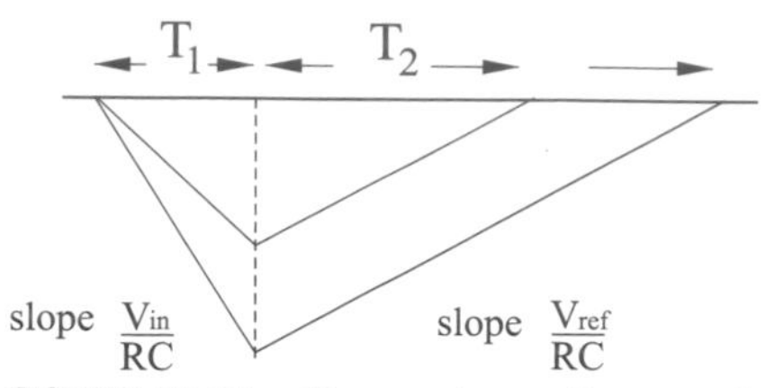

FIGURE 16.15. Run-up (away from zero) and rundown (back to zero) ramps for a dual-slope A/D converter. The waveforms that result from two different input voltages are suggested. Note that the run-up slopes change with V_{in} whereas the run-down slope remains fixed.

detects a zero crossing. The time required for the run-down phase, T_2, is defined by the number of clock cycles counted:

$$T_2 = n\tau. \tag{16.17}$$

From the geometry of Fig. 16.15, it is apparent that

$$T_1 \frac{V_{in}}{RC} = T_2 \frac{V_{ref}}{RC}$$

or

$$V_{in} = \frac{n}{N} V_{ref}. \tag{16.18}$$

Therefore, knowing the preset values of N and V_{ref} and measuring the run-down count n determines the unknown voltage V_{in}.

Resolution

The resolution of the dual-slope method can be analyzed as follows. Consider Fig. 16.16. The parallel lines represent the run-down ramps resulting from progressively larger input voltages starting with a value of V_{in}, for which the counter reads n. Because this is a clocked system, any action by the logic subcircuit will be synchronized with clock pulses. Therefore, the comparator output becomes "known" to the logic subsystem only at the discrete moments indicated.

The dashed lines cross zero between the $(n - 1)$st and nth polling moments, so the counter would stop at n for any of those voltages. The solid line has not quite crossed zero at the nth test moment, but does so by the time the next pulse occurs; hence, the counter stops at the value $(n + 1)$. According to Fig. 16.16, ΔV represents the maximum variation in voltage at the start of run-down that

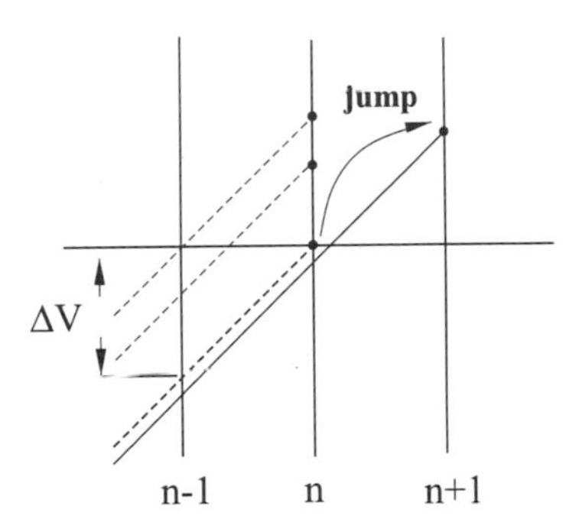

FIGURE 16.16. Zero-crossing portion of the run-down phase of a dual-slope A/D converter. The comparator output is sampled by the logic at discrete moments in time, leading to jumps in the zero-crossing clock count.

would not cause a change in counter value. From the figure,

$$\Delta V = \frac{V_{\text{ref}}}{RC}\tau, \tag{16.19}$$

where τ is the clock period.

But ΔV is related to a change in input voltage (refer to Fig. 16.15) by the expression

$$\Delta V = \frac{\Delta V_{\text{in}}}{RC}N\tau, \tag{16.20}$$

where N was defined previously as the preset number of clock cycles in the run-up phase. These two equations then yield

$$\Delta V_{\text{in}} = \left(\frac{1}{N}\right) V_{\text{ref}}. \tag{16.21}$$

This result shows that the largest possible change in V_{in} that could go undetected (i.e., fail to change the count value) is N^{-1} times the reference voltage. Not surprisingly, a large run-up count is desirable.

Performance

Monolithic integrated circuits are available which contain complete dual-slope A/D converters, often including the necessary input/output ports for interfacing to microprocessors. As a representative example, consider the ADC-800 manufactured by Datel [4]. This is a 15-bit converter. For this device, the run-up count

N is fixed at 16,384, and the maximum run-down count allowed is 32,768. From Eq. (16.18), it is apparent that the full-scale output will occur when $V_{in} = 2V_{ref}$. The recommended integrating capacitor value is 0.47 μF; the integrating resistor is selected from the criterion of a full-scale input current of 20 μA—hence R (megΩ) $= \frac{1}{20}$(full-scale voltage). For example, a required full-scale input of 4 V would dictate $R \approx 200\,\text{k}\Omega$.

A complete measurement cycle begins with a system zero phase (12,288 counts), followed by integrate (16,384 counts), deintegrate (32,768 counts max.), and finally an integrator zero phase (4096 counts). Thus, a total of 65,536 clock cycles are required for the entire conversion. Using a 2.4576 MHz external crystal, the on-chip clock generator will oscillate at 163.8 kHz. The total conversion time would thus be $65{,}536/(163.8 \times 10^3) = 0.4$ sec, or equivalently, the conversion rate would be 2.5 per second. These numbers underline the fact that dual slope is an accurate (and economical) but comparatively slow A/D protocol.

If, for example, the reference source is chosen to be $V_{ref} = 1.6384$ V, then 1 bit would be equivalent to [see Eq. (16.21)] $V_{ref}/\text{N} = 0.1$ mV. Consequently, apart from the position of the decimal point, the decimal equivalent of the binary output from the converter is the numerical value of the input voltage. Other choices of V_{ref} would require a scale conversion between binary meter reading and true voltage.

Multislope

The somewhat slow speed of the standard dual-slope design can be improved by the addition of multiple resistors to the input section, as suggested in Fig. 16.17. The run-up phase remains as before and involves the application of V_{in} to the integrator comprised of R_a and C for a preset number of clock cycles, N_{up}.

However, run-down now occurs in a distinct sequence in which resistors R_b, $R_b/100$ are alternated with $R_b/10$ and $R_b/1000$. These phases are illustrated in Fig. 16.18.

The development that led to Eq. (16.18) must now be modified to account for the multislope operation. To begin,

$$T_1 \frac{V_{in}}{R_a C} = T_2 \frac{V_{ref}}{R_b C}.$$

Thus,

$$\frac{N_{up} V_{in}}{R_a} = \frac{N V_{ref}}{R_b},$$

where the initial run-down crosses zero just prior to the Nth clock. Based on this

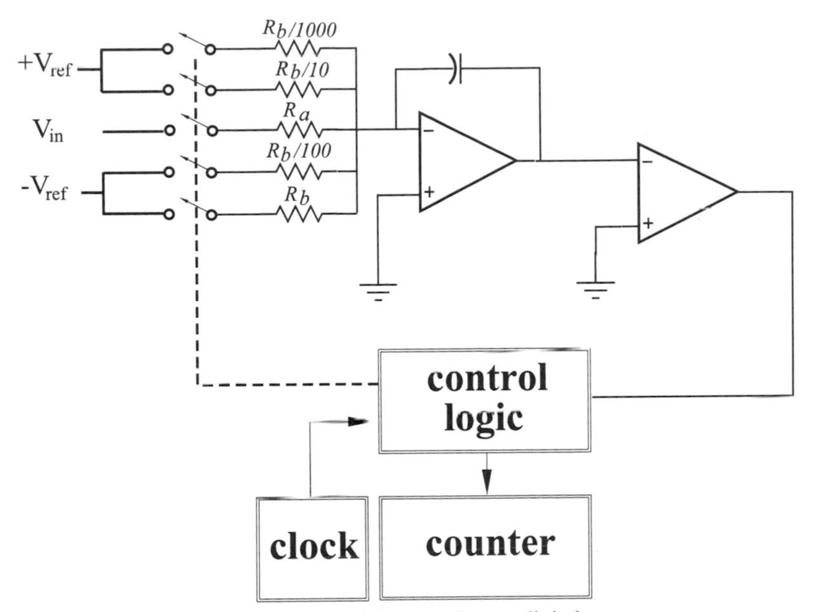

FIGURE 16.17. Schematic of multislope analog-to-digital converter.

expression, a first estimate of the input would be

$$V_{\text{in}} = \frac{N}{N_{\text{up}}}\frac{R_a}{R_b}V_{\text{ref}}. \tag{16.22}$$

This result actually is too big because T_2 is taken to be $N\tau$, whereas in fact $N\tau$ extends beyond the moment of zero crossing, so N is an overestimate.

Next, the logic-controlled switches substitute a new resistor and a reference of opposite polarity. The integration now proceeds in a reversed direction at

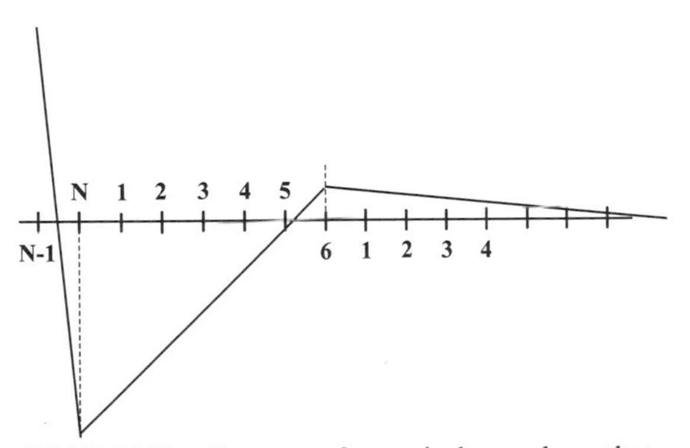

FIGURE 16.18. Sequence of ramps in the run-down phase of a multislope A/D converter. Each slope is one-tenth the preceding one.

one-tenth the rate of the preceding deintegration. This ramp will recross zero in a time N_1, which cannot exceed 10 additional clock cycles. In the example of Fig. 16.18, $N_1 = 6$. This provides a correction to the original estimate, which was too large: $N \Rightarrow N - \frac{N_1}{10}$. For the example, this is $N - 0.6$. But notice that this correction is itself too large.

Now resistor $R_b/100$ is switched in along with the original polarity for V_{ref} and another ramp back toward zero-crossing takes place. In a fashion similar to the above, the count N_2 gives a further correction to the previous correction: $N \Rightarrow N - \frac{N_1}{10} + \frac{N_2}{100}$.

Performance

Each correction stage lasts no more than 10 clock cycles, yet adds another factor of 10 in precision. This provides a tremendous speed/precision advantage over the original dual-slope conversion. For example, suppose the run-down count of a conventional converter happened to take 10,000 counts. With a multislope run-down, this same precision can be achieved in four iterations which would take at most 40 clock cycles—a speed gain of 250!

An example of a commercial product based on multislope run-down is the HP3458A digital multimeter [5]. This instrument also employs multislope run-up and achieves an acquisition rate of 100,000 measurements per second at a precision of $4\frac{1}{2}$ digits (14 bits).

Flash Converters

Of all the A/D converter types, flash converters are the speed champions. This performance is achieved by employing a parallel design strategy, as shown in Fig. 16.19.

The precision resistor ladder splits the reference voltage into equally spaced levels:

$$
\begin{aligned}
V_1 &= \left[\frac{V_{ref}}{256}\right], \\
V_2 &= 2\left[\frac{V_{ref}}{256}\right], \\
&\vdots \\
V_{254} &= 254\left[\frac{V_{ref}}{256}\right], \\
V_{255} &= 255\left[\frac{V_{ref}}{256}\right].
\end{aligned}
\tag{16.23}
$$

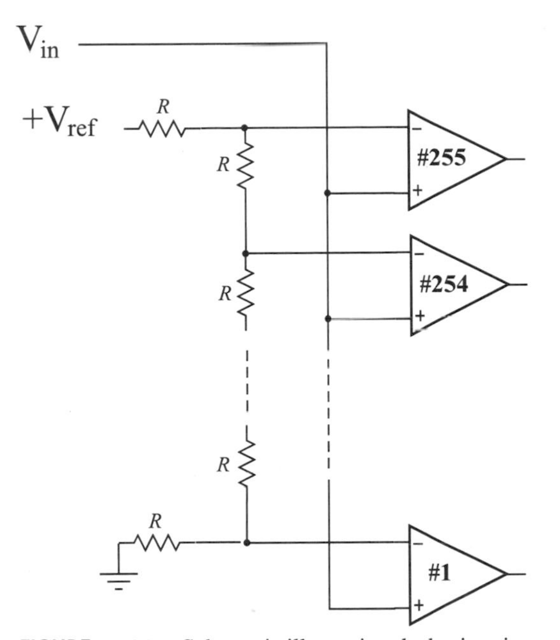

FIGURE 16.19. Schematic illustrating the basic principle of an 8-bit flash A/D encoder. For the 8-bit encoder illustrated, 255 comparators are required.

As the input voltage rises, additional comparators turn "on" one at a time, as illustrated in Fig. 16.20. Note that there are 256 states for this system, ranging from all comparators off and thus low, to all 255 outputs high.

The obvious speed advantage of this type of analog-to-digital converter is the outcome of the parallel structure. When the input voltage changes, all comparators can react at the same instant, in contrast to the methods described previously, which iterate toward a final result over many clock cycles. The final speed limit is set by the comparator settling times.

The principal disadvantage is the number of comparators which must be provided. Even 8 bits requires 255 on-chip op-amps. This constraint limits the number of bits of resolution that can be accommodated without real difficulty in terms of chip complexity and power dissipation. However, with combination serial/parallel architectures, it is possible to produce fast converters with fewer than $(2^n - 1)$ comparators. For example, the AD773A from Analog Devices [1] achieves 10-bit resolution with just 48 comparators instead of 1023, and is rated at 20 megasamples per second. The AD678, also from Analog Devices, employs a 4-bit flash core together with a recursive subranging algorithm to give 14-bit resolution at 200 thousand samples per second (kSPS).

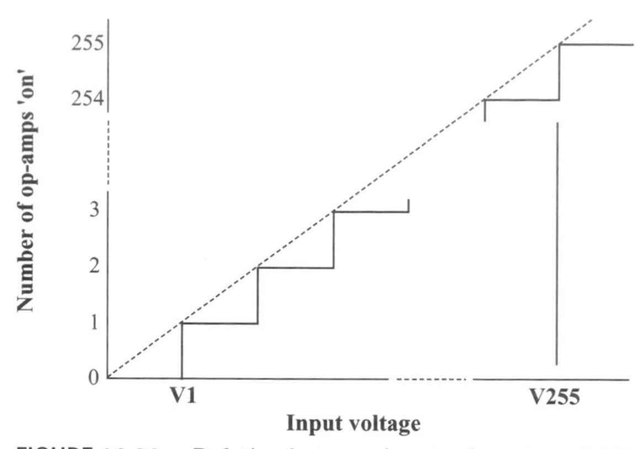

FIGURE 16.20. Relation between input voltage to an 8-bit flash encoder and the number of output amplifiers in the "on" state.

As already noted, the comparators turn on one at a time, creating (for 8 bits) a total of 256 possible states in response to the input voltage. Single-chip flash converters such as the Datel ADC-208 [4] include an additional internal 256 to 8 decoder so that true 8-bit binary outputs are provided. This device has a rated conversion time of just 50 nsec and conversion rate of 20 million samples per second (MSPS).

16.4 CURRENT MEASUREMENT

A dc current is most easily measured by routing the current through a precisely known resistance R_S and then determining the voltage that appears across that resistor. Hence, trivially,

$$I = \frac{V}{R_S}. \tag{16.24}$$

The difficulty inherent in this procedure is that a sensing resistor has, in effect, been made part of the circuit under test and will to some extent perturb that circuit (see Fig. 16.21). In other words, the measurement changes the thing being measured. The smaller the sensing resistor, the better. On the other hand, too small a value for R_S will lead to very small potential differences V, which may be difficult to measure with acceptable accuracy. A common choice for the sensing resistor in digital multimeters is 0.1 Ω.

The same sort of consideration applies to the various A/D converters discussed earlier since they will have non-ideal (finite) input impedance and consequently

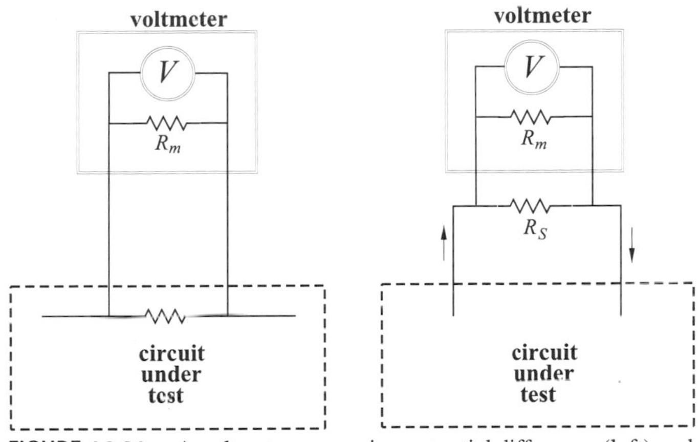

FIGURE 16.21. A voltmeter measuring potential difference (left) and current (right). R_m is the finite meter resistance; R_s is the precision series resistor needed for current sensing.

will drain away some small amount of current from a circuit under test. A perfect voltmeter has infinite input impedance. Practically speaking, the desirable situation is for the voltmeter impedance to greatly exceed the impedance across which it is placed.

16.5 RESISTANCE MEASUREMENT

Measuring resistance involves a direct application of Ohm's law

$$R = \frac{V}{I}. \tag{16.25}$$

In other words, R is the ratio of the voltage that appears across a resisitor divided by the current flowing through the resistor (which gives rise to that voltage). Two methods of determining resistance are illustrated in Fig. 16.22. The simpler method is a basic ohmmeter. A pair of test leads are attached to the unknown resistance, as shown. Within the ohmmeter, a precision current source delivers I through the test leads to the unknown. A voltmeter (A/D converter) measures the potential difference at the two output terminals. Clearly, $V = I(R + 2r)$, so

$$\frac{V}{I} = R + 2r. \tag{16.26}$$

Obviously, the ratio of the measured value V and the known value I is not exactly the sought-after value R. The resistance of the leads, $r + r$, which normally is

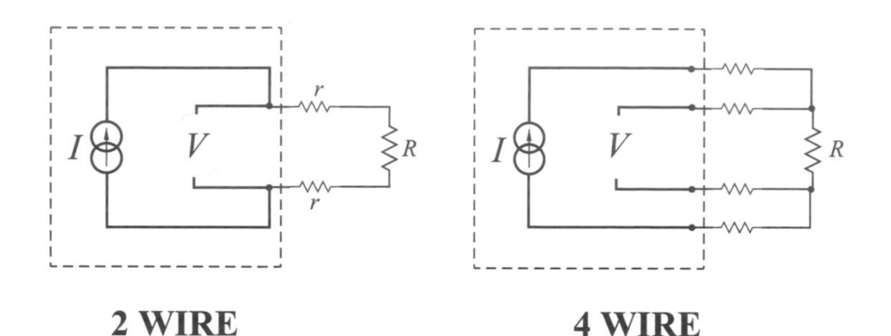

FIGURE 16.22. Circuit configuration for measuring resistance. Left: 2 wire; right: 4 wire. R is the unknown, and r is the resistance of the connecting leads.

not known, adds an error to the process. Of course, if the lead resistance is small enough, then

$$R \simeq \frac{V}{I}. \tag{16.27}$$

A second method is the 4-wire arrangement. Four test leads are employed as shown. The resistance of the current leads has no effect on the magnitude of I. Because the voltmeter has extremely high impedance, no current will flow in the voltage-sensing leads, so the measured voltage will be the true potential difference across R. Therefore, in this case,

$$R = \frac{V}{I}. \tag{16.28}$$

The result now is precise, in contrast to the 2-wire technique.

In digital multimeters, current sources ranging from 1 mA to 0.1 μA are commonly used, depending on the size of the unknown resistor; smaller currents are used for the larger resistance ranges.

PROBLEMS

Problem 16.1. Consider a dual-slope A/D converter.

1. If the desired resolution is 1 mV and the reference is 10.0 V, what is the required number of clock pulses N in the run-up phase? [Ans. 10,000].
2. How many clock cycles are required in the run-down phase if the "unknown" input is 6.0 V? [Ans. 6000].

Problem 16.2. Consider a 4-digit A/D converter based on the principle of multislope run-down.

1. Suppose the system clock has a frequency of 1 MHz and that a full-scale input produces a capacitor voltage of 9.999 V at the completion of the run-up phase. The output digits in this case would be 9999. If the reference in the circuit is 5.000 V, what value is appropriate for the first run-down time constant R_bC? [Ans. 5×10^{-6} sec].
2. The input "unknown" voltage is now changed to half of the full-scale value. Sketch the four run-down stages. What total conversion time is needed? [Ans. 10 μsec].

BIBLIOGRAPHY

[1] Daniel H. Sheingold, editor, *Analog-Digital Conversion Handbook* (Analog Devices, Inc., Norwood, MA, 1972).
[2] William D. Stanley, *Operational Amplifiers with Linear Integrated Circuits* (Merrill/Macmillan, New York, 1994); see pp. 316–324.
[3] Analog Devices, Norwood, MA.
[4] Datel, Inc., Mansfield, MA.
[5] See article in Hewlett-Packard Journal, April 1989, p. 8.

C H A P T E R 17

AC Measurements

The previous chapter covered a number of topics related to the measurement of static or quasi-static voltages or currents. But many, if not most, common signals are periodic in time. Periodic signals (voltage or current), whatever their shape, repeat exactly over some time interval τ; hence,

$$f(t+\tau) = f(t).$$

Because of their usefulness with respect to certain electrical properties, three particular ac quantities are of interest: peak, average-rectified, and root-mean-squared.

The *peak* of a periodic voltage or current waveform is simply the magnitude of the largest amplitude reached within any period τ.

Rectification converts a function that is at times positive and negative to one that is always positive, as suggested in Fig. 17.1. The *average-rectified* value is then the average of this waveform evaluated over precisely one period; that is,

$$|f|_{\mathrm{av}} = \frac{1}{\tau}\int_0^{\tau} |f(t)|\, dt, \tag{17.1}$$

where the double vertical bar denotes the absolute value.

Root-mean-squared (rms) is a numerical value computed from a periodic waveform by doing exactly what the name suggests—taking the square root of the average of the squared function; that is,

$$f_{\mathrm{rms}} = \sqrt{\frac{1}{\tau}\int_0^{\tau} [f(t)]^2\, dt}. \tag{17.2}$$

The importance of this quantity can be appreciated from the following argument.

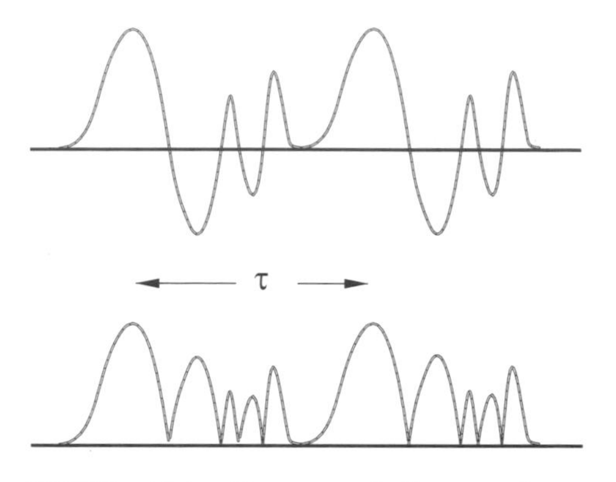

FIGURE 17.1. Upper: periodic waveform; lower: rectified version of upper waveform.

Suppose a periodic, but otherwise arbitrary, voltage waveform $v(t)$ appears across a resistor R. Then, the instantaneous power dissipation in the resistor is $P(t) = i(t) \times v(t)$, where $i(t)$ is the current flowing through the resistor. The average power dissipated during exactly one cycle is

$$\langle P \rangle = \frac{1}{\tau} \int_0^\tau iv\, dt = \frac{1}{R\tau} \int_0^\tau v^2\, dt = \frac{R}{\tau} \int_0^\tau i^2\, dt. \tag{17.3}$$

Now, consider a constant voltage V_{dc} (i.e., a battery) connected across an identical resistor, as shown in Fig. 17.2. In this case, the power dissipation is steady and is given by

$$P_{\text{dc}} = I_{\text{dc}} V_{\text{dc}} = \frac{V_{\text{dc}}^2}{R} = I_{\text{dc}}^2 R. \tag{17.4}$$

Equating the time-average dissipation in one case, $\langle P \rangle$, with the constant dissipation in the other, P_{dc},

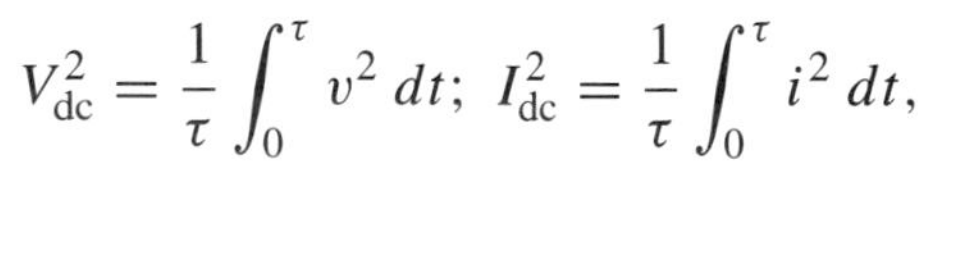

$$V_{\text{dc}}^2 = \frac{1}{\tau} \int_0^\tau v^2\, dt; \; I_{\text{dc}}^2 = \frac{1}{\tau} \int_0^\tau i^2\, dt,$$

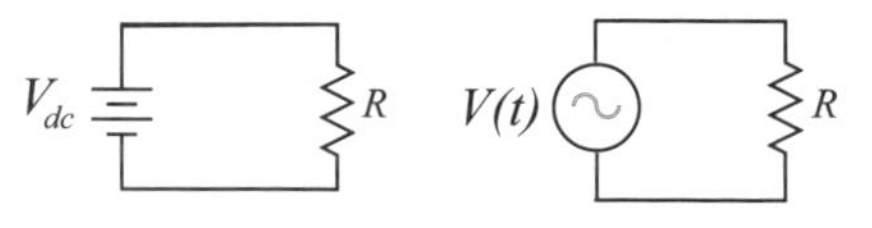

FIGURE 17.2. Resistive load driven by dc and time-dependent sources.

or,

$$V_{dc} = \sqrt{\frac{1}{\tau}\int_0^\tau v^2\,dt} = v_{rms} \tag{17.5}$$

and

$$I_{dc} = \sqrt{\frac{1}{\tau}\int_0^\tau i^2\,dt} = i_{rms}. \tag{17.6}$$

Therefore, the rms value associated with $v(t)$ is equal to the strength of a battery that would deliver the same Joule heating to the resistive load. The rms voltage and current thus play the role of "equivalent" or "effective" dc values in this context of power transfer.

Example: sine wave

Suppose a current or voltage waveform is given by a pure sine wave

$$f(t) = A\sin(\omega_0 t)\,.$$

The peak value is, trivially, the amplitude A. The average value of $f(t)$ is zero. The average-rectified value (see Fig. 17.3) can be calculated by integrating over just one-half a period; that is,

$$|f|_{av} = \frac{2}{\tau}\int_0^{\tau/2} A\sin(\omega_0 t)\,dt = \frac{2A}{\pi}. \tag{17.7}$$

The rms value of this waveform is

$$f_{rms} = \left[\frac{1}{\tau}\int_0^\tau A^2\sin^2(\omega_0 t)\,dt\right] = \frac{A}{\sqrt{2}}. \tag{17.8}$$

Example: square wave

Consider now the waveform shown in Fig. 17.4. The peak value is the amplitude A. The average-rectified value is, by inspection,

$$|f|_{av} = A. \tag{17.9}$$

The squared amplitude is simply a constant A^2, so the rms value is

$$f_{rms} = A. \tag{17.10}$$

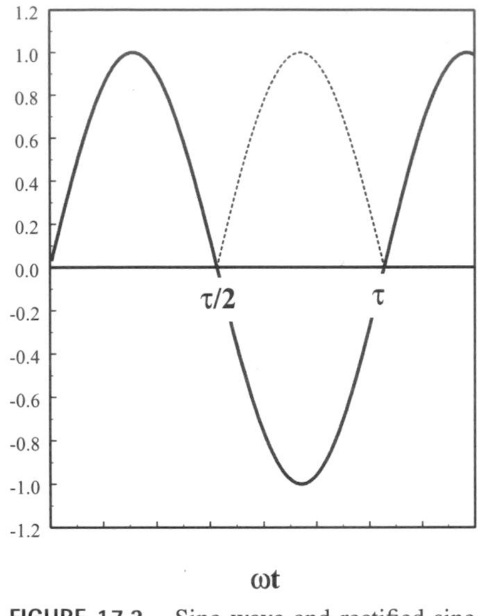

FIGURE 17.3. Sine wave and rectified sine wave.

Example: pulse train

A common periodic signal is illustrated in Fig. 17.5. The amplitude of the pulses is A, the period is τ, and the pulse width is w. A parameter called the duty cycle (η) is defined as the ratio of "on" time to the total; that is,

$$\eta = \frac{w}{\tau}.$$

This waveform is already rectified, so it is not difficult to see that

$$|f|_{\text{av}} = A\eta \tag{17.11}$$

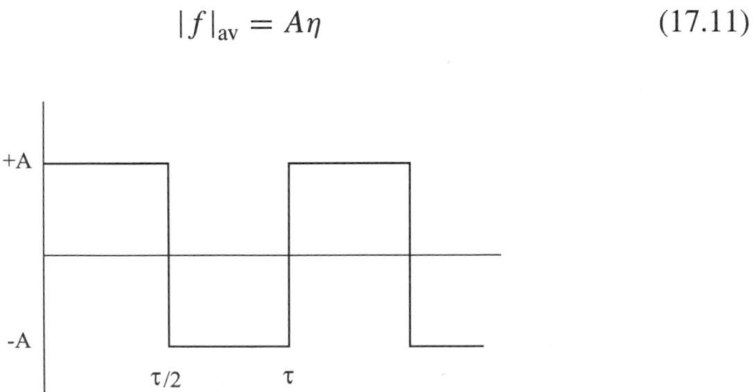

FIGURE 17.4. Symmetric square wave.

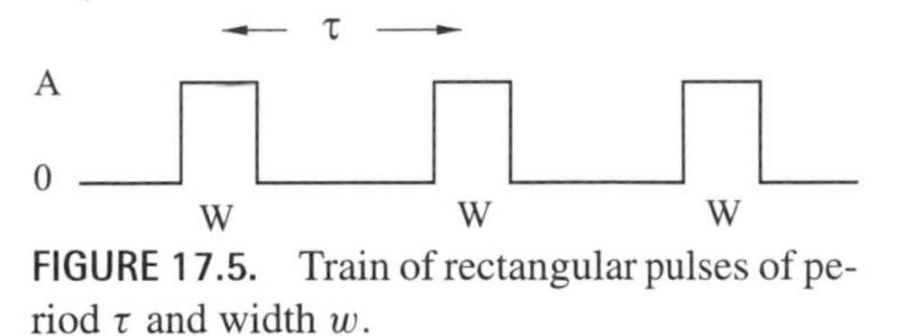

FIGURE 17.5. Train of rectangular pulses of period τ and width w.

and

$$f_{\text{rms}} = A\sqrt{\eta}. \tag{17.12}$$

17.1 ESTIMATED RMS

In making a measurement of a periodic waveform, the objective usually is to determine the rms value. However, the average-rectified voltage (or current) is more easily and economically obtained. This is because the input signal needs only to be rectified with diodes and then averaged with a low-pass filter before being measured with a conventional dc meter.

In the previous examples, it was seen that the shape of a specific waveform determines the relationship between the average-rectified and rms values. Thus, if the *form factor* is defined by the ratio

$$\text{FF} = \frac{f_{\text{rms}}}{|f|_{\text{av}}}, \tag{17.13}$$

then

	Av Rect	rms	FF
sine wave	$\frac{2A}{\pi}$	$\frac{A}{\sqrt{2}}$	$\frac{\pi}{2\sqrt{2}}$
square wave	A	A	1
pulse train	$A\eta$	$A\sqrt{\eta}$	$\frac{1}{\sqrt{\eta}}$

Many other waveforms, such as ramps and sawtooths, could be added to this list, each having its own unique connection between average-rectified and rms (see [1], pp. 108 and 391).

If the average-rectified value is *measured* with a meter, and the form factor for the particular waveform is *known*, then the root-mean-squared value can be deduced from Eq. (17.13).

By far the most common waveforms are sinusoidal, in which case the form factor is $\frac{\pi}{2\sqrt{2}} = 1.1107$. Less expensive ac meters (in contrast to true rms instruments, which will be discussed later) assume that all inputs will be sine waves, actually perform average-rectified readings, and then display "rms" results which are generated by simple multiplication by the numerical value of 1.111. If the signal is not a sine wave, the displayed result will be in error.

Example

Suppose such a meter reads 3.42 V "rms" with an input pulse train whose duty cycle is 20%. The value 3.42 represents 1.1107 times the average-rectified voltage (which was the actual measured quantity), but the form factor is really $1/\sqrt{0.2} = 2.2361$. Hence, the correct rms value would be

$$3.42 \times \frac{2.2361}{1.1107} = 6.89.$$

The meter reading is thus too small by about 100% !

Sine Wave + Harmonics

Complications also arise when the signal to be measured is composed not just of a fundamental sinusoid, but also of additional harmonics. Consider

$$f(t) = [\sin(\omega_0 t) + b \sin(n\omega_0 t + \phi)]\,. \tag{17.14}$$

This is made up of a fundamental at frequency ω_0 plus the nth harmonic of amplitude b. The harmonic is phase shifted with respect to the fundamental by an amount ϕ. An expression such as Eq. (17.14) might be used to represent a distorted signal.

The rms value of $f(t)$ can be shown to be

$$f_{\text{rms}} = \sqrt{\frac{1 + b^2}{2}}. \tag{17.15}$$

but the average-rectified value is a complex function of both b and ϕ. Therefore, the form factor will also be dependent on these parameters. The two most likely possibilities involve first or second harmonics.

n = 2: second-harmonic distortion

The average-rectified value of $f(t)$ as defined by Eq. (17.14) may be evaluated numerically. These results are plotted in Fig. 17.6.

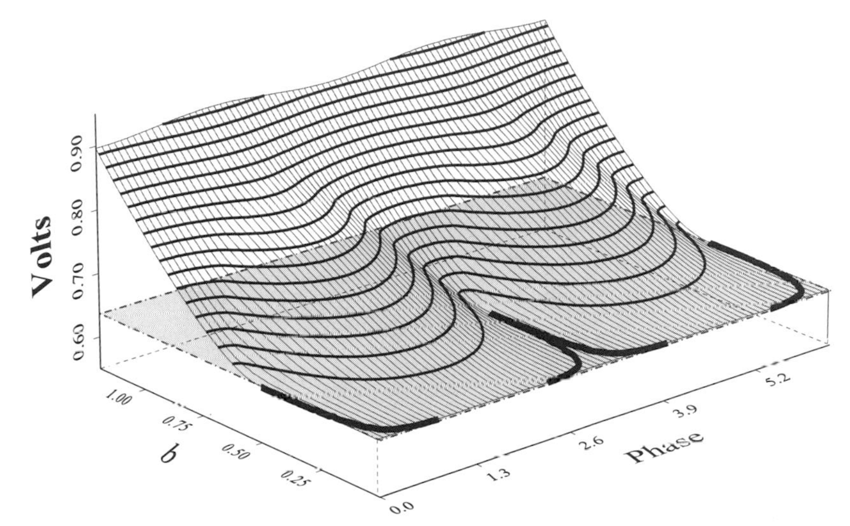

FIGURE 17.6. Average-rectified value $|f|_{av}$ as a function of b and ϕ for second-harmonic distortion.

The limiting value when $b \to 0$ agrees with the prediction $\frac{2}{\pi} = 0.637$ as given by Eq. (17.7).

For a pure sine wave, the form factor is $\frac{\pi}{2\sqrt{2}}$ and the rms value is correctly given by the product $\frac{\pi}{2\sqrt{2}} \times |f|_{av}$. When $b > 0$, this is no longer the case. If the true rms value of $f(t)$ is also obtained numerically, then

$$\epsilon = \frac{f_{rms} - \frac{\pi}{2\sqrt{2}}|f|_{av}}{\frac{\pi}{2\sqrt{2}}|f|_{av}} \tag{17.16}$$

will indicate the relative error that would be present in the displayed reading of an average-rectified meter. The resulting $\epsilon - b - \phi$ surface is illustrated in Fig. 17.7.

n = 3: third-harmonic distortion

An analysis similar to that just described for the influence of a second harmonic can be carried out for the case of third-harmonic content. The results are shown in Figs. 17.8 and 17.9.

As these examples make clear, the relationship between the reading from an average-rectified meter and the sought-after rms value of a signal is not simple. All meters of the average-rectified type presume pure sine waves and have scales which are calibrated with that assumption in mind. Nonsinusoidal signals will not be measured correctly with this approach.

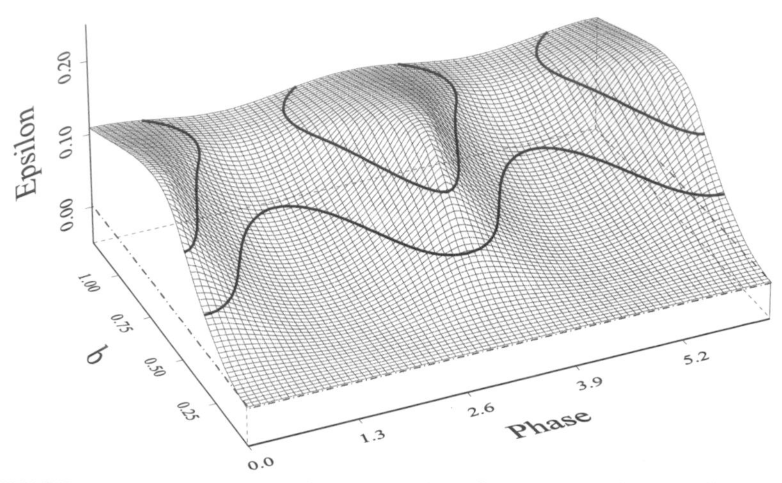

FIGURE 17.7. Relative error of "rms" readings from a meter that actually measures average-rectified values. This is plotted as a function of the second-harmonic amplitude and phase shift. Two contours at constant ϵ are shown.

17.2 TRUE RMS

There are several different basic approaches to the problem of correctly measuring the rms value of a periodic waveform of arbitrary shape. Before discussing these techniques, it is necessary first to introduce a quantity of some importance in measurements of ac signals: *crest factor*.

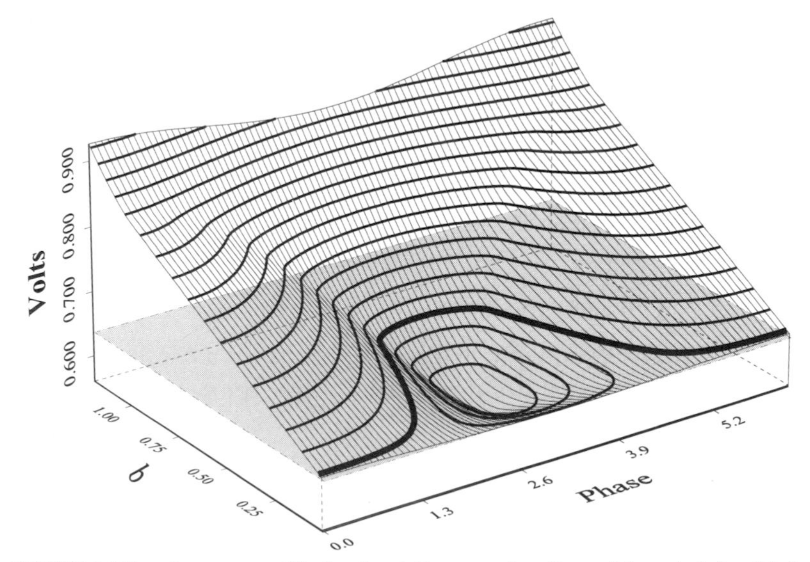

FIGURE 17.8. Average-rectified value $|f|_{av}$ as a function of b and ϕ for third-harmonic distortion.

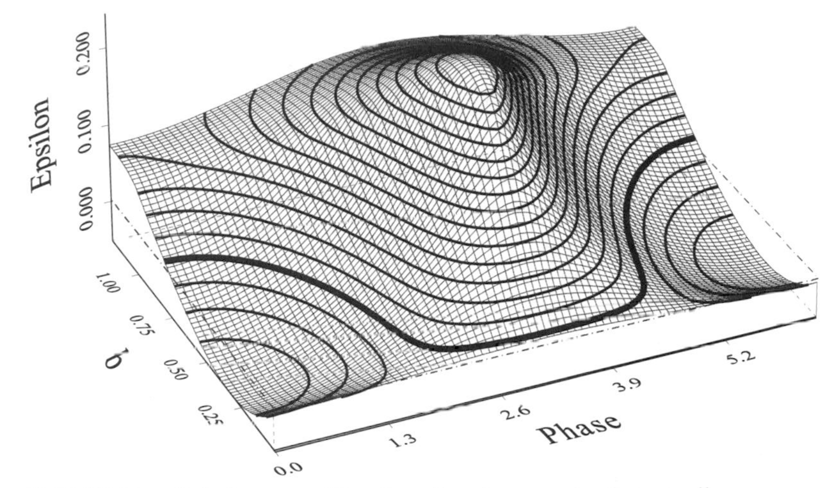

FIGURE 17.9. Relative error of "rms" readings from a meter that actually measures average-rectified values. This is plotted as a function of the third-harmonic amplitude and phase shift. Contours at constant ϵ are shown; the contour emphasized in bold is at $\epsilon = 0$.

Crest factor is defined as the ratio of the peak value of a periodic signal to the rms value.

$$\mathrm{CF} = \frac{|f|_{max}}{f_{rms}}. \tag{17.17}$$

For the waveforms discussed earlier, the crest factors are: sine wave $\sqrt{2}$; square wave 1.0; pulse train $\frac{1}{\sqrt{\eta}}$. As the case of the pulse train illustrates, the crest factor becomes larger as a waveform becomes more spiked. Generally speaking, it is increasingly difficult to make accurate measurements on signals with larger crest factors.

Thermal Technique

The root-mean-squared voltage emerged as the answer to the question of what equivalent dc voltage would have the same heating effect on a load as the applied ac. This line of thinking can be reversed to become a working procedure for determining rms values.

Consider the circuit shown in Fig. 17.10. The ac input signal is applied to a resistive heater, H_1, which is thermally anchored to, but electrically isolated from, diode D_1. The average power delivered to H_1 causes its temperature, and that of the diode D_1, to rise to an equilibrium value T_1. Similarly, the dc voltage applied to H_2 causes its temperature, and that of the diode D_2, to rise to an equilibrium value T_2.

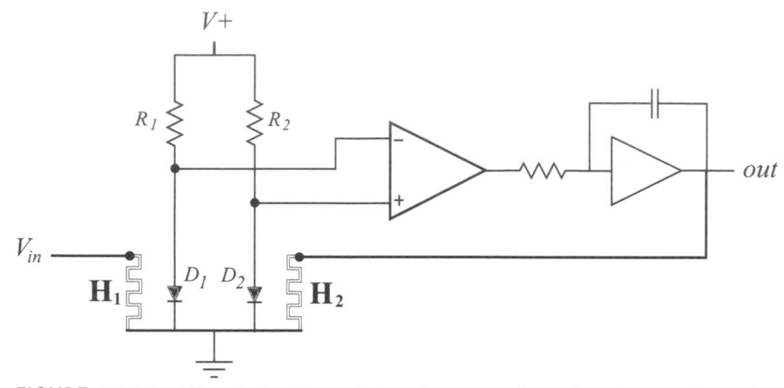

FIGURE 17.10. Circuit for determining the rms value of an input voltage by thermal balance.

Recall from the discussion of diodes in their role as temperature sensors [see Eq. (9.76) and Figure 9.20] that for silicon devices there is a shift of -2.2 mV/°C. Suppose the temperature of the heater/diode pair, T_2, is at some instant higher than T_1. Then, the voltage at the noninverting input to the comparator will be smaller than at the inverting input, so the comparator output will decrease. The dc input to heater H_2 will be reduced, and the temperature T_2 will drop toward T_1. If, on the other hand, T_2 is at some moment lower than T_1, then the comparator generated feedback drive to H_2 will rise, thus bringing T_2 back up toward T_1. In other words, this circuit provides feedback to maintain a dc voltage across heater H_2 that thermally balances the average power dissipation by the ac signal in H_1. By definition, $(V_{\text{out}})_{\text{dc}} = (V_{\text{in}})_{\text{rms}}$.

The LT1088 from Linear Technology [2] is an example of the kind of special-purpose integrated circuit available for use in this type of rms voltmeter. It includes two on-chip resistive heater/diode pairs. A circuit similar to Fig. 17.10 and using the LT1088 can provide true rms detection with an accuracy better than 1% of full scale over the range dc to 50 MHz for waveforms with crest factors as large as 50. Full-scale settling time is 0.5 sec—slow, as would be anticipated in a system limited by thermal time constants.

Analog Computation

Log and antilog amplifiers were discussed in Chapter 6. These circuits employed diodes or transistors in combination with op-amps to create output voltages that were logarthmic or exponential functions of the input voltages. Other mathematical functions can also be realized in analog electronic form. The most common operations are multiplication $(X \cdot Y)$ and division (X/Y). A typical module is the Burr–Brown DIV100 [3], which is rated at a maximum error of ±0.25% for the ratio of two input voltages. Similarly, the Burr–Brown MPY534 can perform 4-quadrant analog multiplication within a maximum error of ±0.25%. Combination multiplier/dividers with three analog inputs $[(X \cdot Y)/Z]$, such as the Analog Devices AD531 [1], are also available. These analog functions can be used to address the problem of determining the true rms value of a signal.

Explicit Analog Computation

The most straightforward analog implementation of an rms-to-dc converter is shown in Fig. 17.11. The first stage is an analog multiplier which serves to square the incoming signal. This is followed by a low-pass filter and op-amp buffer, which smooth the squared waveform. Finally, a square-root module generates the dc output. The overall sequence of operations follows precisely the prescription for rms: the square root of the average of the square. This explicit implementation suffers from at least one major drawback—limited dynamic range. This is because of the demands exacted by the initial squaring operation, which, for example, turns a 100:1 input range (e.g., 0.01 to 1.0 V) into a 10,000:1 output range. This requirement may be confounded by noise properties of the electronics. Realistically, circuits with 10:1 input ranges are practical [1].

Implicit Analog Computation

Implicit analog computation makes use of a seemingly simple variation of the definition

$$v_{\rm rms} = \sqrt{\langle v^2 \rangle}, \qquad (17.18)$$

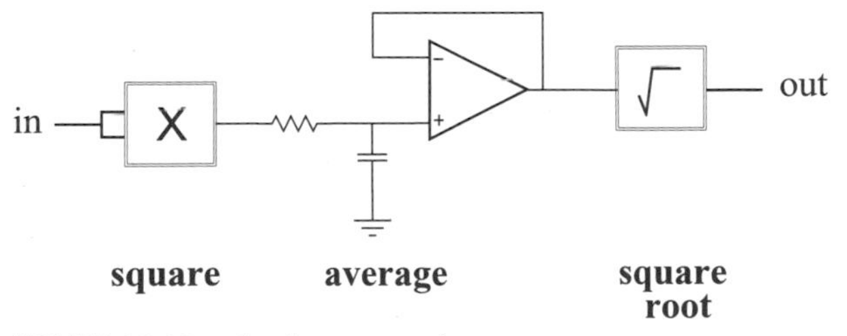

FIGURE 17.11. Analog rms-to-dc converter.

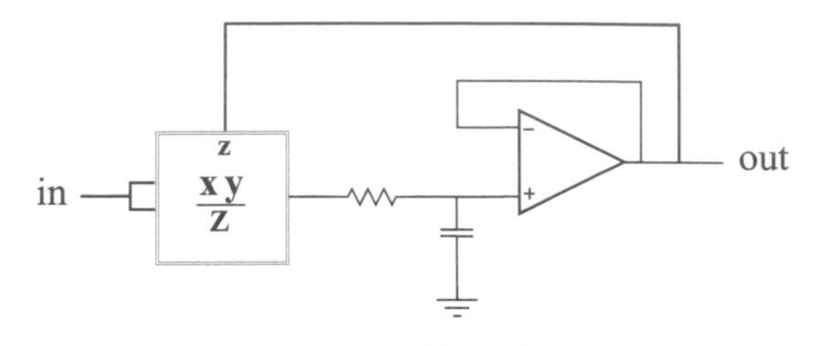

FIGURE 17.12. Schematic of a circuit that performs an implicit evaluation of the input rms.

where the brackets ⟨⟩ denote time-averaging. An alternative to Eq. (17.18) is

$$v_{\rm rms} = \frac{\langle v^2 \rangle}{v_{\rm rms}}. \tag{17.19}$$

This can be translated into a circuit implementation, as shown in Fig. 17.12. Notice that feedback to the multiply/divide module imposes the implicit relationship

$$v_{\rm out} = \left\langle \frac{v_{\rm in}^2}{v_{\rm out}} \right\rangle$$

and because the output is already time-averaged, this can be written

$$v_{\rm out} = \frac{\langle v_{\rm in}^2 \rangle}{v_{\rm out}},$$

which is just the relationship satisfied by the rms voltage, Eq. (17.19). In other words, the output voltage is the desired rms value. The Analog Devices AD536A and AD637 [4] operate on this basic principle. The AD536A can achieve conversion accuracies of less than ±0.2% for bandwidths in excess of 450 kHz. The error remains below 1% for crest factors up to 7. The AD637 is intended for lower-level inputs.

A different implicit approach is built around the structure illustrated in Fig. 17.13. The algebraic equivalent of this circuit is

$$V_{\rm out} = \langle \exp[2 \log |V_{\rm in}| - \log V_{\rm out}] \rangle$$

or

$$V_{\rm out} = \left\langle \exp\left[\log \frac{V_{\rm in}^2}{V_{\rm out}}\right] \right\rangle = \left\langle \frac{V_{\rm in}^2}{V_{\rm out}} \right\rangle = \frac{\langle V_{\rm in}^2 \rangle}{V_{\rm out}},$$

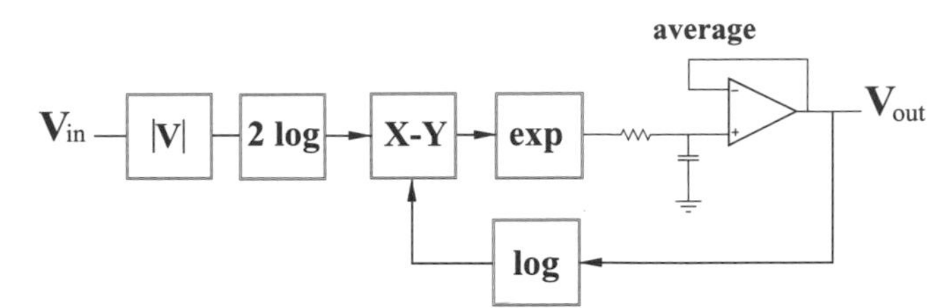

FIGURE 17.13. Implicit analog evaluation of true rms using log and antilog modules.

which is, again, just the requirement of Eq. (17.19) for V_{out} being identified with the rms value of the input signal. This method is employed in the Burr–Brown 4341 true rms-to-dc converter. This 14-pin module achieves a conversion accuracy of $\pm 0.5\%$ for sine-wave inputs in the range 0.5–5.0 V rms over the frequency interval dc–10 kHz.

Explicit Digital Computation

With the advent of sampling electronics (discussed in the next section), it has become possible to acquire the time series of a waveform and then numerically integrate the squared amplitude to determine the rms value. This entails the possible use of algorithms to interpolate additional data points and to precisely identify the period.

Some difficulties are still posed by the sampling requirements of very fast repetitive waveforms. In order to faithfully represent the signal and hence be able to compute an accurate value for its rms, there should be many data points per period. Just how many depends on the spectral content of the particular waveform— that is, its shape—or equivalently its bandwidth.

As an example, suppose a minimum of 20 sampled points per period is needed for a faithful recording of a 10 MHz waveform. This implies a sample interval of 5 nsec or a conversion rate of 200 megasamples/sec. Such a specification is not achievable with available A/D converters. However, it is possible to use a technique called *subsampling* [5] to buildup data over a number of cycles and achieve a high effective sample rate from a much slower converter.

PROBLEMS

Problem 17.1. For the waveform illustrated in Fig. 17.14:

1. Find the average-rectified value and the rms value. [Ans. 1.00 V and 1.155 V].

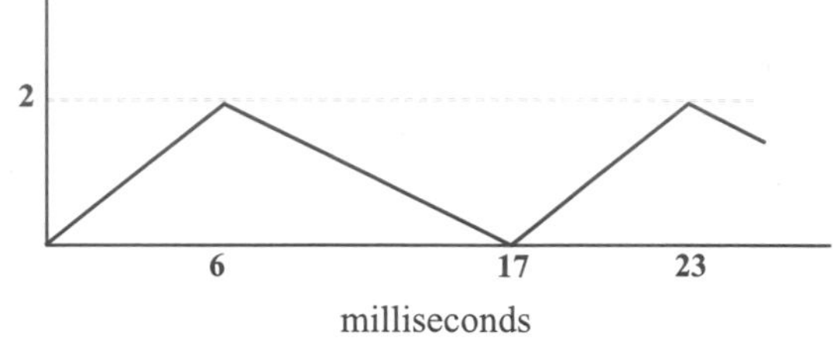

FIGURE 17.14. Problem 17.1.

2. Suppose this waveform is measured with a non-true-rms meter. What will the meter read? [Ans. 1.1107 V].

Problem 17.2. One cycle of a periodic waveform is described by $V(t) = A\ \exp(\frac{t}{T} - 1)$ over the interval $0 < t < T$. This repeats in $T < t < 2T$, $2T < t < 3T$, and so on.

1. What is the average-rectified value? [Ans. $0.632A$].
2. What is the rms value? [Ans. $0.658A$].
3. What is the crest factor? [Ans. 1.52].
4. What is the form factor? [Ans. 1.04].
5. If a non-true-rms meter gave a reading of 2.00 V, what would be the correct rms value? [Ans. 1.87 V].

BIBLIOGRAPHY

[1] Daniel H. Sheingold, editor, *Nonlinear Circuits Handbook* (Analog Devices, Inc., Norwood, MA, 1974).
[2] *Application Note 22* (Linear Technology Corporation, Milpitas, CA, 1988).
[3] Burr-Brown Corporation, Tucson, AZ.
[4] Manufactured by Analog Devices, Inc., Norwood, MA.
[5] See article by Ronald L. Swerlein in Hewlett-Packard Journal, April 1989, p. 15.

CHAPTER 18

Data Acquisition

The previous chapter dealt with periodic signals, and in particular with the measurement of characteristic values such as peak, average-rectified, and root-mean-squared levels. These are typically invoked when handheld or bench meters are placed in the ac mode.

In contrast, for many signals of interest it is the shape, not some averaged property, that is important. This is certainly the case with electrocardiograms and electroencephalograms. These may be periodic, but the details within the waveforms are significant. Also, many one-shot events (for example, accelerometer impact profiles) need to be recorded and analyzed. The demands of these sorts of measurements are not met by ac meters. Instead, transient signal acquisition is required.

Analog-to-digital conversion remains the core operation for either ac or transient analysis. However, when waveform digitization is required, sequential discrete signal quantification must be performed. In other words, the continuously evolving (and possibly nonrepetitive) signal must be observed "on the fly" at regular time intervals, so whichever type of A/D converter is utilized (e.g., successive approximation) must be able to execute a reading, wait, execute a new reading, etc., on a waveform that itself never pauses.

18.1 SAMPLE AND HOLD

In the acquisition of such a time series, a key role is played by a sample-and-hold module. The upper circuit shown in Fig. 18.1 illustrates the essential features of a S/H. Two buffers are combined with a capacitor (typical values for the hold capacitor are $\sim$1 nF) and logic-controlled switch. In the S state, the switch is

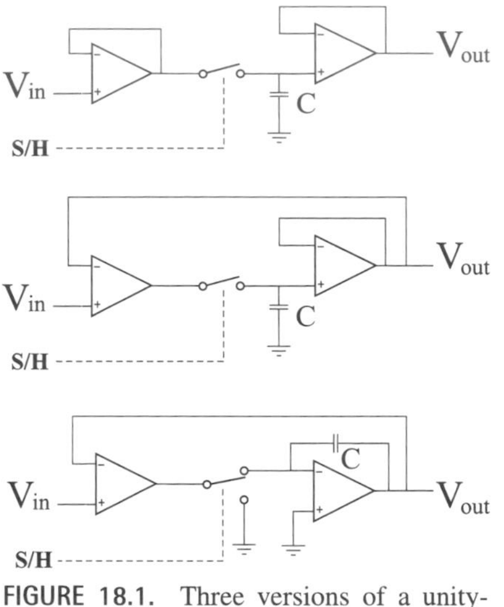

FIGURE 18.1. Three versions of a unity-gain sample-and-hold.

closed and the input signal passes through to the output. While in the sample mode, the capacitor charge $Q = CV$ continuously changes in response to the input voltage. When the hold command is invoked, the switch opens, disconnecting input from output. This freezes the capacitor charge, and hence the output voltage, at the instant of switch opening.

The middle schematic in Fig. 18.1 is a variation [1] that provides improved low-frequency tracking. The arrangement shown in the bottom schematic is also commonly used in sample-and-holds [2].

From a performance perspective, there are a number of real device limitations which must be considered. 4 distinct phases of operation are discussed as follows.

1. During Sample

While in the sample mode, the S/H is acting as a unity-gain device, so the output should exactly follow the input. But because of the finite speed of the amplifiers, the response to a full-scale step input will be a delayed output, as suggested in Fig. 18.2. Typical output slew rates range from about 5 V/μsec to several hundred volts per microsecond.

2. Sample-to-Hold Transition

This transition occurs when the switch disconnects the charge storage capacitor from the input side of the circuit. As indicated in Fig. 18.3, a finite delay exists between the mode control pulse and the physical action of the electronic switch.

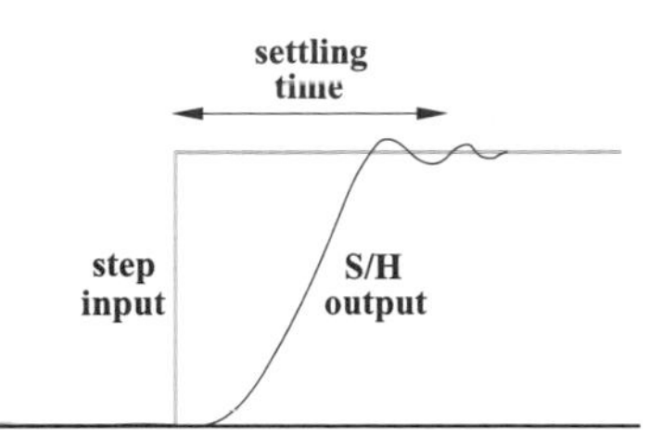

FIGURE 18.2. Transient response to a step function while in the sample mode.

This delay is known as the *aperture-delay time*. Aperture delays of 5 to 150 nsec are common.

3. During Hold

While in the hold mode, the output ideally should remain unchanging, frozen at its initial value. However, the capacitor will inevitably lose its charge, causing the output voltage to "droop." Droop rate for voltage is related to the leakage current from the capacitor

$$\frac{\Delta V}{\Delta t}(\text{volt/sec}) = \frac{I_{\text{leakage}}}{C}\frac{(pA)}{(pF)}.$$

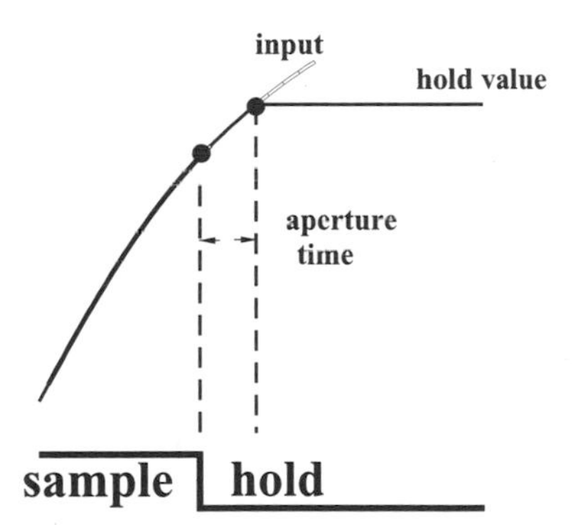

FIGURE 18.3. Illustration of aperture delay. The actual switch opening is delayed following the hold command, allowing the capacitor to continue tracking the input for this additional time. Therefore, the held value is not exactly equal to the input at the designated sample moment.

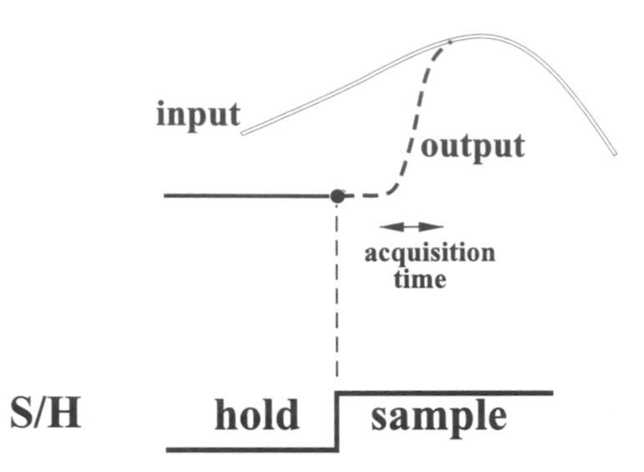

FIGURE 18.4. Illustration of acquisition time. The sample command closes the switch after a hold/sample delay and then the hold capacitor charges to the input level.

Droop rates ranging from about 0.1 V/sec to 100 V/sec are typical, with improved behavior for larger values of hold capacitance.

A second imperfection which appears during hold intervals is known as feedthrough. This is caused mainly by capacitive effects in the switch and is manifest as a small fraction of the input being present in the output. For example, with a sine-wave test input of 10 V peak-to-peak, a typical feedthrough might be 2 mV, or 0.02%.

4. Hold-to-Sample Transition

When the hold command changes to sample, the switch is closed and the S/H resumes passing the input to the output (tracking). As suggested in Fig. 18.4, this process does not occur instantaneously; the capacitor must be recharged to a new value. The relevant parameter is known as *acquisition time*; a typical value might be 5 μsec for a full-scale transition to within 0.01% of 10 V.

Notice that in situations where the sample command is quite short, the action of the device is reasonably referred to as sample-and-hold, but when the sample state is "on" for comparatively long intervals, then it might more appropriately be called track-and-hold. In other words, the action during sample is for the output to track the input.

18.2 SAMPLED WAVEFORMS

A general arrangement for a system to perform data acquisition on a time-dependent waveform is shown in Fig. 18.5. The S/H module is used to freeze the

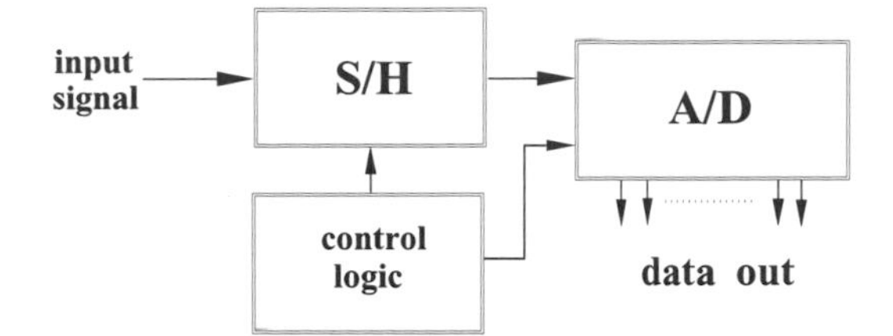

FIGURE 18.5. Simple data-acquisition system based on a sample-and-hold followed by an analog-to-digital converter. Commands from the logic module place the S/H in hold while the D/A is performing a conversion.

input to the A/D while the analog-to-digital conversion is in progress. Otherwise, the A/D would be attempting to converge to a final answer under a contradictory condition of unsteady input. The logic module orchestrates the alternation of the two phases: convert-while-hold, and not-convert-while-update. Normally, this is slaved to a system clock so that the sampling is done at equal time intervals.

A hypothetical input signal and S/H output are shown in Fig. 18.6. The conversions of the A/D are presumed to take place and be completed within the hold intervals. Thus, the input waveform is digitized at intervals spaced $\tau = H + S$ apart, as indicated in Fig. 18.7. Clearly, the sampled waveform will be a more faithful rendition of the original continuous waveform when the sampling interval is small.

Sample Rate and Aliasing

The connection between sampling rate and waveform fidelity is extremely important to the practical matter of experimental data acquisition. Sampling must be performed as frequently as possible for accuracy, but as infrequently as possible

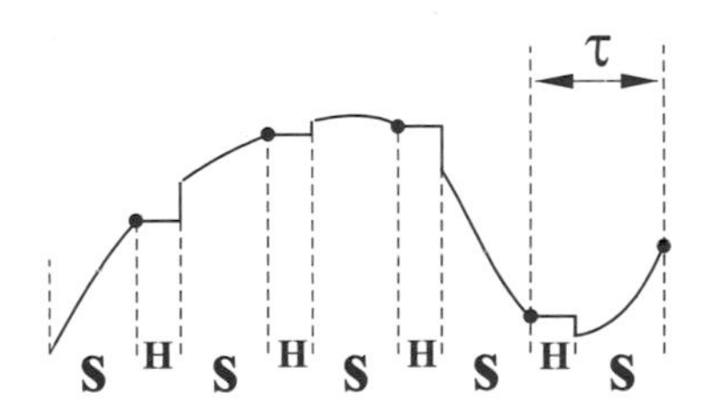

FIGURE 18.6. Input waveform and output from sample-and-hold (bold). Sample intervals (S) and hold intervals (H) are indicated. The resulting time-series points are indicated by dots.

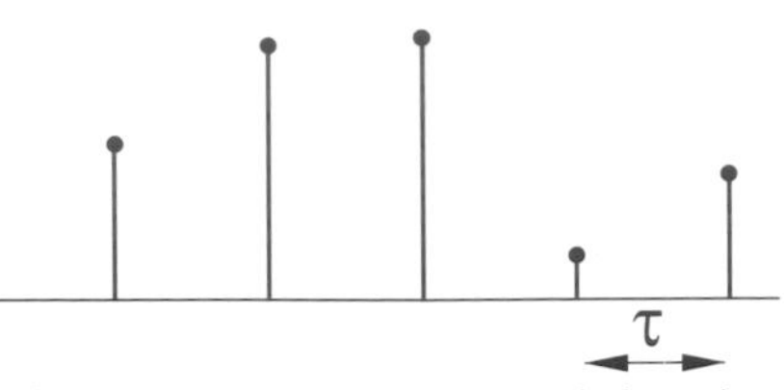

FIGURE 18.7. Discretely sampled version of original time series.

to relax demands on D/A speed and possibly on data storage. The perils of insufficient sampling are illustrated in the following discussion.

To make the discussion specific, suppose a sine wave is sampled at the low rates of 1.0472 and 0.8976 points per cycle. The results are illustrated in Fig. 18.8. The surprising revelation of these plots is that the sampled data look like sine waves of much lower apparent frequency. With 1.0472 samples per cycle, the apparent frequency is approximately 0.0476 of the true value. With 0.89760 samples per cycle, the apparent frequency is approximately 0.0997 of the true value. In other words, based only on the observation of sampled data, a false conclusion would be reached regarding the frequency of the periodic signal. This phenomenon is known as *aliasing* [3, 4].

In fact, the process of aliasing can be understood as a folding of frequencies about half the sampling frequency, $f_S/2$. To apply this idea to the example

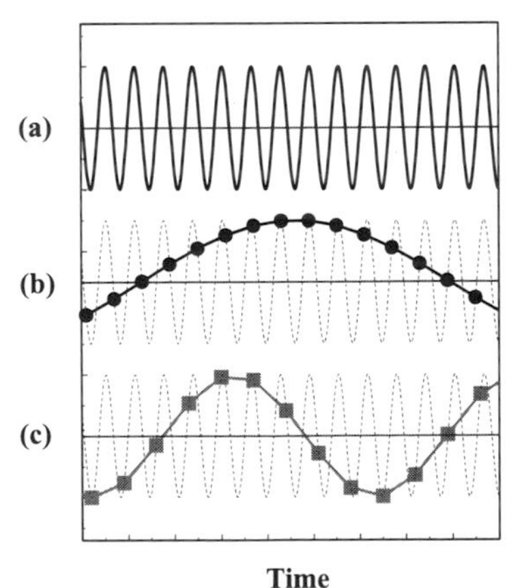

FIGURE 18.8. Illustration of aliasing caused by sampling interval. (a) original sine wave; (b) 1.0472 samples per cycle; (c) 0.89760 samples per cycle.

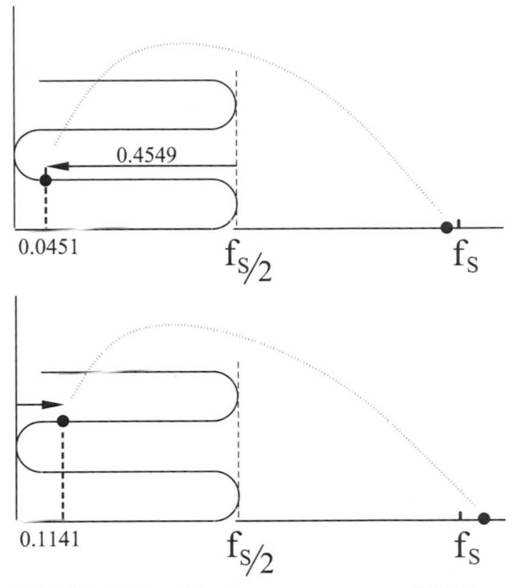

FIGURE 18.9. Aliasing as a process of folding about half the sampling frequency. The particular numerical values apply to the cases discussed in the text.

under discussion, consider Fig. 18.9. The upper part of the figure applies to case (b) in Fig. 18.8 for which there were 1.0472 samples per cycle of the sine wave. This is equivalent to saying that the sine-wave frequency was 0.9549 times the sample frequency, so the marker on the frequency axis lies just to the left of f_S. To locate the corresponding point on the folded characteristic, note that $0.9549 = 0.5000 + 0.4549$. As shown, the aliased frequency is thus $0.0451\ f_S$, or $0.0451 \times 1.0472 = 0.0472$. The period of the aliased signal would then be $0.0472^{-1} = 21.19$ sample intervals, which agrees with the data in Fig. 18.8(b).

In the second case, there were 0.8976 points per cycle, or equivalently the sine-wave frequency was 1.1141 times the sample frequency. Using folding around $f_S/2$ and the fact that $1.1141 = 1.0000 + 0.1141$, the aliased frequency is seen to be $0.1141\ f_S$, which is $0.1141 \times 0.8976 = 0.1024$. The period of this aliased signal would be $0.1024^{-1} = 9.77$ sample intervals, which again agrees with the plot in Fig. 18.8(c).

In both of these examples, the discretely sampled original signal for which $f > f_S/2$ was mapped to a lower aliased frequency within the range $0 < f < f_S/2$. Clearly, aliasing will not occur if the signal frequency lies below one-half the sample frequency. But suppose the signal has a complex shape with some

Fourier components reaching beyond $f_S/2$. These harmonics will certainly be aliased unless they are removed. In some data-acquisition systems, anti-aliasing filters are provided for this purpose. As a rule of thumb, the cutoff frequency of such filters should be about one-third of the sample frequency.

Sampling Theorem

It has just been demonstrated that sampling rates which are too small result in aliasing effects. The question which naturally arises is whether this artifact can be avoided; the answer can be found in the *sampling theorem*, which states:

Provided the function $y(t)$ contains no frequency components higher than f_M, then $y(t)$ can be completely determined by equispaced samples if the sample rate exceeds twice f_M.

This celebrated theorem has a famous history associated especially with the names Nyquist [5] and Shannon [6]. The stipulation that the signal contain no harmonics above some cutoff is also referred to as band limiting.

When adequate sampling has been performed, the procedure to retrieve the underlying continuous function $y(t)$ from the set of sampled values $y_n = y(n\tau)$, τ being the sampling interval, is

$$y(t) = \sum_{n=-\infty}^{\infty} y_n \frac{\sin[\pi(t/\tau - n)]}{\pi(t/\tau - n)}. \tag{18.1}$$

This result can be found in [4, 7, 8, 9, 10]; it is sometimes called the Cardinal Theorem of Interpolation Theory [11]. The formula requires an infinite number of observations y_n, but of course this will never be the case in a real time series containing a finite number of terms. In many situations, the reconstruction can still be quite good *within* a particular time interval, provided a large number of data points blanket that domain [10].

Preamplifiers

The specific choice of reference voltage dictates the input range of an analog-to-digital converter. The system illustrated in Fig. 18.5 functions best if the input signal occupies a large fraction of the full-scale range. However, this optimum condition will not normally be satisfied for all possible inputs. Some sensors, such as thermocouples, have relatively low-level outputs. A complete data-acquisition system thus requires an additional stage containing a programmable gain amplifier, as shown in Fig. 18.10.

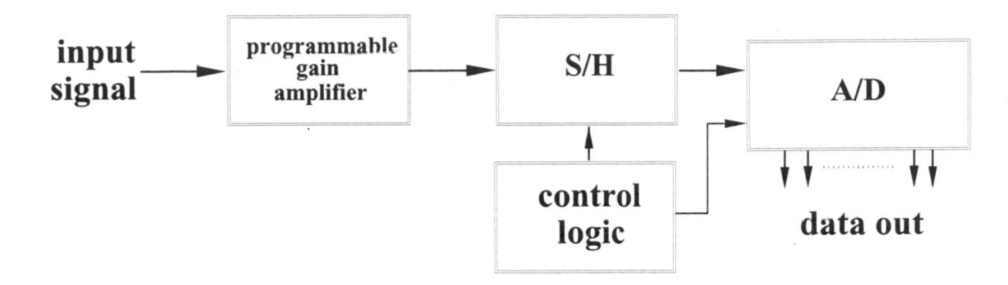

FIGURE 18.10. Data-acquisition system with programmable gain input stage.

18.3 MULTICHANNEL SYSTEMS

The previous section was concerned with aspects of a single channel devoted to data acquisition. Many applications involve more than just one signal and thus require multiple-channel acquisition capabilities. One approach would be to make as many copies of the original channel as there are signals to be measured, but this could be an expensive undertaking. Another strategy involves selectively routing the various input signals through a single acquisition module. The key element in such a design is an analog multiplexer.

Multiplexers

An analog multiplexer (MUX) is an integrated circuit realization of a single-pole, multiple-throw switch. It is completely electronic, without moving mechanical parts. The common fabrication technology is CMOS, and typical specifications include low ON resistances of a kΩ or less and settling times of just a few microseconds. Action is break-before-make, which guarantees that no switches can inadvertently be shorted together during a selection change.

Figure 18.11 illustrates the general structure of an 8-channel multiplexer. Three binary inputs are used to select one of the eight switches, thus connecting

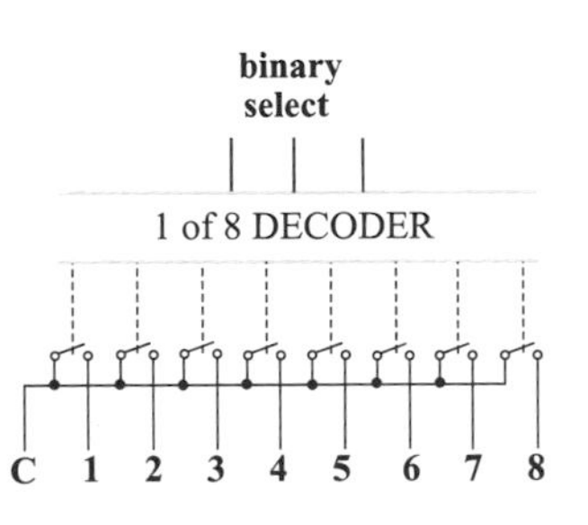

FIGURE 18.11. Structure of an 8-channel multiplexer.

common (C) to the designated output. Four-, eight-, and sixteen-channel muti-plexers are widely available.

Simultaneous S/H

For multichannel operation, each signal line is connected to one of the multiplexer inputs $(1, 2, \cdots)$. The output (C) then is passed to a single S/H and A/D combination, as depicted in Fig. 18.5 or Fig. 18.10. The finite speed of the switches causes any sequence of selected channels to become staggered in time. This channel skew means that it is not possible to measure multiple channel signals simultaneously. If simultaneous event recording is essential, then each channel must be equipped with its own sample-and-hold. A single strobe pulse can freeze all channel levels at the same moment, and these can subsequently be multiplexed into the A/D portion one at a time.

18.4 PC-BASED DATA ACQUISITION

Accompanying the dramatic advancement of the power of desktop computers have been major achievements in the capabilities of data-acquisition products designed to be installed in available internal PC I/O slots. Usually, there is provision for up to a half-dozen or more cards. With appropriate software, complex sequences of data acquisition tasks can be programmed.

Data Acquisition Cards

A multifunction data-acquisition (DAQ) card is designed to acquire analog signals on one or many channels of input, to generate analog signals on one or more output lines, and to read and write multibit digital data. Depending on the performance requirements, any of the previously discussed techniques (successive-approximation, flash, etc.) might be employed for the A/D conversions. Cards are available for all of the standard PC buses—PCI, ISA, PCMCIA (portable applications), and recently USB (Universal Serial Bus). It is also possible to obtain single-purpose cards that perform just analog output, just digital I/O, or just very high-speed analog input.

Representative features and specifications for multifunction cards include:

- up to 64 single-ended channels (32 differential) of 16-bit A/D conversion at 100 kS/sec, or over 1 MS/sec at 12-bit resolution;
- input ranges of 0.0 to $+10$ V or ± 10 V;

- programmable input gains of typically 1, 10, 100, 500;
- 16-bit analog ouput with D/A conversion at 100 kS/sec, or over 1 MS/sec at 12-bit resolution;
- output range ±10 V;
- 8 or more digital I/O lines.

When several channels are scanned in sequence, the per-channel acquisition rate is equal to the quoted speed divided by the number of channels in use. Thus, a rating of 100 kS/sec, when applied to a 10-channel sequence becomes an effective rate of 10 kS/sec for each channel.

Signal Conditioning

Various forms of signal preconditioning may be necessary, some being sensor-specific. For example, thermocouples require some form of cold junction compensation as well as preamplification. Resistive thermal detectors (RTDs) require precision current sources. Strain gages require bridge-completion circuits and excitation-voltage sources. Low-pass anti-aliasing filters and/or noise filters may be needed.

All of these types of signal modification are available. Usually, the signal conditioning hardware is in the form of one or more circuit boards which may be housed in a separate enclosure, so a complete PC-based system will appear as in Fig. 18.12. The raw transducer signals are received by the external conditioning electronics, transformed, and then passed along to a data-acquisition card situated within the PC. The connection between DAQ card(s) and conditioning module(s) may be bi-directional so that the computer can control such parameters as filter settings, amplifier gains, multiplexed channel selection, etc.

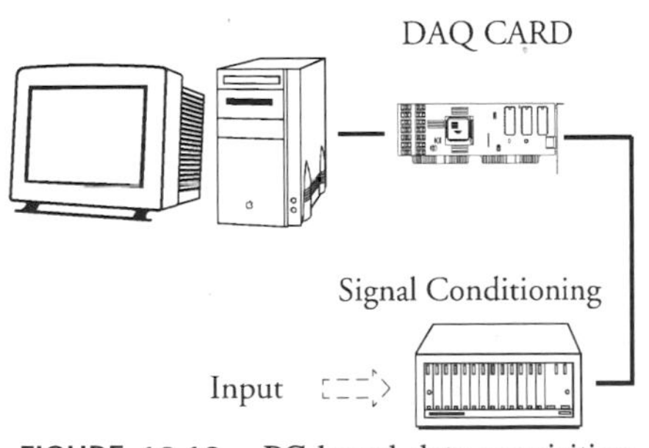

FIGURE 18.12. PC-based data-acquisition system with DAQ card and signal-conditioning modules in an external unit.

Although a PC will accept any data-acquisition card designed for one of the principal buses, the same standardization has not evolved with respect to signal-conditioning components. These often are tailored to work with particular PC-installed DAQs. This means that the user cannot "mix-and-match," but might have to choose from within the same manufacturer's product line.

At the time of writing, some typical capabilities of signal-conditioning modules include:

- precise dc voltage-excitation sources to 10 V;
- precise current-excitation sources to 1 mA;
- amplifier gains of between 1 and 1000;
- high-order elliptic, Butterworth, Bessel, Cauer, and linear phase delay low-pass filters with cutoffs of between 1 Hz and 100 kHz;
- excitation and sensing for full or half bridges incorporating 120 Ω and 350 Ω strain gages;
- multichannel thermocouple inputs for type J,K,T,E,R,S,B, and N sensors with cold junction compensation;
- multichannel RTD units for 100 Ω platinum resistors with $\alpha = 0.00385$;
- high-speed multiplexer/amplifiers with channel-scanning rates to 250 kS/sec or higher;
- relay modules with 16 or more SPDT switches for high current applications.

Software

DAQ cards and signal-conditioning modules are configured and instructed to carry out their tasks via software. This places the PC in the role of central controller for the acquisition process; hard drives in the PC are also the usual data-storage devices. Scripting the details of any particular set of measurements and/or input–output operations thus amounts to writing programs in software.

Card manufacturers provide software drivers for the Microsoft Windows operating systems [12] (and also in some cases DOS), which create linkages between the hardware itself and languages such as C++, Microsoft Visual Basic [12], as well as DOS-based Basic and Pascal. Depending on the capabilities of the multifunction card, function calls can activate data acquisition, waveform generation, digital I/O, data stream to disk, and counter/timer operations.

More powerful software packages such as the current versions of LabVIEW [13] and DASYLab [14] support boards from a number of manufacturers, including IOtech [15] and Computer Boards [16]. Another significant programming environment is Hewlett Packard's HP VEE [17]. These and other similar products provide extensive control of the acquisition process with a user-friendly graphical programming interface.

Advances in this area are rapid; the state of the art in data-acquisition software is a quickly moving target. For this reason, it is not sensible to attempt a detailed review of presently available software packages; the material would soon become out-of-date. However, it does seem safe to say that for the foreseeable future, acquisition software will be mainly Windows-based, will increasingly adopt graphical programming metaphors, and will include more and more built-in analysis capabilities, such as filtering and Fast Fourier Transforms. There also are already trends to link the data-acquisition process with other data-analysis packages, such as Microsoft Excel [12]. Capabilities for handling very large data sets are rapidly improving, aided by the availability of ever-greater disk capacity, so that sophisticated visualization and data-mining techniques are coming to the fore.

PROBLEMS

Problem 18.1. A sampled experimental waveform has an apparent frequency of 80 Hz. The sample rate was 100 Hz. What was the most likely true frequency of the signal? [Ans. 120 Hz].

BIBLIOGRAPHY

[1] For example, SHM-IC-1, manufactured by DATEL, Inc., Mansfield, MA.

[2] For example, SHM-6, manufactured by DATEL, Inc., Mansfield, MA.

[3] R.B. Blackman and J.W. Tukey, *The Measurement of Power Spectra* (Dover Publications, New York, 1959), p. 32.

[4] Robert K. Otnes and Loren Enochson, *Applied Time Series Analysis*, Volume 1 (Wiley, New York, 1978), pp. 24–29.

[5] H. Nyquist, "Certain Topics in Telegraph Transmission Theory," Trans. AIEE, 617–644 (1928).

[6] C. Shannon, "A Mathematical Theory of Communications," Bell Syst. Tech. J. 27, 379–623 (1948).

[7] J.M. Whittaker, *Interpolatory Function Theory* (Cambridge University Press, Cambridge, U.K., 1935), Chapter IV.

[8] Hwei P. Hsu, *Fourier Analysis* (Simon and Schuster, New York, 1970), pp. 153, 154.
[9] R. Marks, *Introduction to Shannon Sampling and Interpolation Theory* (Springer-Verlag, New York, 1991).
[10] Anthony J. Wheeler and Ahmad R. Ganji, *Introduction to Engineering Experimentation* (Prentice-Hall, Englewood Cliffs, NJ, 1996), p. 105.
[11] R.B. Blackman and J.W. Tukey, *The Measurement of Power Spectra* (Dover Publications, New York, 1959), p. 84.
[12] Microsoft Corp., Redmond, WA.
[13] National Instruments, Austin, TX.
[14] DASYLab is a registered trademark of DATALOG GmbH Co. KG, Moenchengladbach.
[15] IOtech, Inc., Cleveland, OH.
[16] ComputerBoards, Inc., Middleboro, MA.
[17] Hewlett-Packard Company, Palo Alto, CA.

CHAPTER 19

Data-Acquisition Systems

The previous chapter ended with a discussion of PC-based data acquisition. The context there was of acquisition boards mounted *within* the desktop computer. Obvious efficiencies and economies of this approach include the use of the computer's already available chassis and power supplies—the boards need only be plugged into available internal PC slots.

A fundamentally different way of structuring data-acquisition hardware leaves all the acquisition electronics *outside* of the PC. In such an arrangement, the external units must be provided with their own power supplies and packaging. The added cost of this approach may be offset by the possibility of greater system customization, performance, and flexibility.

Two distinct external system architectures exist at the present time. In the first, stand-alone external instruments are woven into a network by linking them together with a suitable interface bus. The resulting web of instruments is then run from a desktop computer using appropriate software.

In a second, more integrated protocol, instrument functions are performed by plug-in cards of standard dimensions, and the various units plug into slots provided in a self-powered card cage external to the PC. When inserted, the individual modules derive their required operating voltages from this host and communicate with one another via a high-speed backplane built into the card cage. The desktop computer talks to this external system via a bidirectional communication link. In a sense, this concept is rather like the original cards-in-a-PC arrangement, except that the entire cage-and-slots unit is now external to the PC. Because these cards are not located within the computer itself, there is less system noise to degrade performance. Moreover, only the acquisition hardware need be placed close to the system under test; the PC can be located elsewhere for convenience. Finally, many more slots can be provided if the cage is made large enough, so much more complex acquisition systems can be assembled with this approach.

These two system architectures will be discussed in the next sections. But first, a word of caution. The reader should be aware that system architecture for data acquisition is an area of intense development and very rapid change. Advances are driven by still-increasing capabilities of both microprocessors and analog-converter hardware. Software protocols remain in a state of flux at this time. Therefore, the bulk of this chapter is, of necessity, presented in the spirit of a present-day snapshot meant to convey what seem to be the emerging paradigms.

19.1 GPIB BUS

The General Purpose Interface Bus (GPIB) provides a means by which separate instruments such as meters and oscilloscopes may be connected to form an interactive network. Usually, a desktop computer serves as the administrator or *controller*, managing the actions of the various instruments (see Fig. 19.1).

An illustrative example of a minimal system would be a PC connected to a multimeter. Clearly, there are two essential requirements.

The first is the ability to send instructions from the PC to the meter that will place the meter in the desired measurement mode (volts, amps, dc, ac, number of digits precision, etc.), and then initiate a reading. This is a remote version of pushing buttons on the front panel of the meter.

The second requirement is the ability to transfer the meter reading to the PC. This is a remote version of manually entering the meter display into the PC via the keyboard.

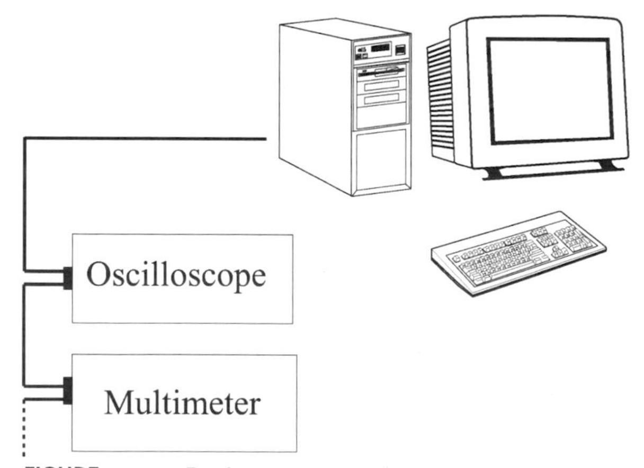

FIGURE 19.1. Desktop computer controlling external instruments via a daisy-chained bus connection.

Note that process #1 is a communication from PC-to-meter, whereas process #2 involves data transfer from meter-to-PC. In the first case, the PC talks and the meter listens, while in the second case the meter talks and the PC listens.

When many instruments are involved, some means must be provided to regulate who is talking and who is listening—in other words, the flow of information (data and commands) within the network must be controlled and specified.

All of the above-mentioned capabilities are provided by the GPIB, which originated as the HPIB from Hewlett-Packard [1]. Ultimately, this bus design was adopted throughout the instrumentation industry and, after almost three decades, is still the most widely used communication interface. Because of its adoption in 1975 as an interface standard by the Institute of Electrical and Electronic Engineers, it is also referred to as the IEEE 488 Bus [2, 3, 4, 5]. In 1987, the standard was updated and became known as IEE-488.2. The newest development in this bus is HS488 from National Instruments [6].

The GPIB standard defines cables, connectors, interface hardware (which must be resident in every instrument), and command protocols. The bus is based on negative logic and TTL levels; hence, the relationship between logic state and signal voltage is as follows: logic 0 $\Rightarrow$ high voltage (between 2.0 and 5.2 V); logic 1 $\Rightarrow$ low voltage (between 0.0 and 0.8 V).

Bus Organization

The cable is comprised of a total of 24 separate wires, which may be grouped as follows: 8 bidirectional data lines, 3 bidirectional handshake lines designated DAV, NRFD, and NDAC, 5 bus-management lines designated ATN, IFC, SRQ, REN, and EOI, a cable shield, and 7 grounds.

The connector shown in Fig. 19.2 is unique to the GPIB—it is designed to allow stacking so that instruments may be daisy-chained together. This physical stacking of connectors obviously implies that each of the bus wires electrically ties together corresponding pins from all instruments; that is, all pin 1's are joined, all pin 2's are joined, and so forth.

The eight data pins on each instrument's GPIB port are bidirectional and can function as input or output. Conflict avoidance dictates that there can be only one device placing data on the bus at any given time—the *Talker*. Up to 14 devices at any one time may read a piece of data; hence, there can be many *Listeners*.

In the handshake group, DAV is a signal controlled by the single talker, whereas NRFD and NDAC are bus-connected outputs from all the listeners. Data Valid (DAV) is "true" (low voltage) when the talker has assembled a valid byte at its eight data bus lines. The NRFD and NDAC drivers are in reality open

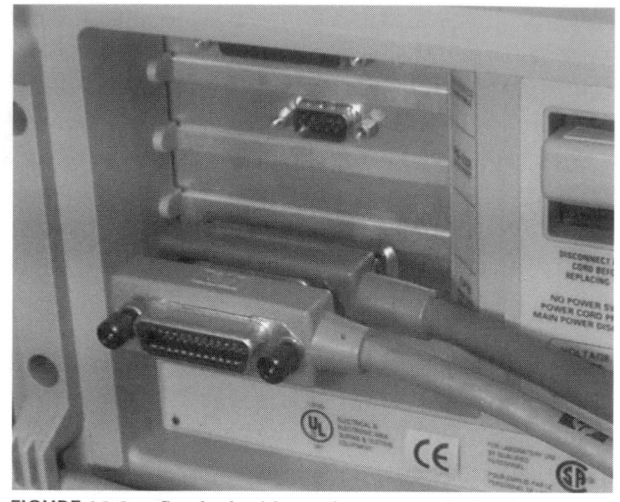

FIGURE 19.2. Stacked cables at the rear panel of an instrument with a GPIB port.

collector transistors with pull-up resistor(s). In simplified form, this arrangement is equivalent to the structure illustrated in Fig. 19.3, where each output transistor has been replaced by a switch. Note the important property that the bus voltage will go high (logic 0) only when all outputs are high (logic 0); that is, all switches must be open. This is just the truth table of a multi-input OR gate.

In the case of NRFD, the pin signifies when a device is Not Ready For Data. The quiescent condition of any listener places this output in the "true" condition (logic 1, low voltage), meaning that it is not prepared to receive data from the bus. Only when all devices move to a ready state (every switch open) will this bus line go high, indicating a condition of global readiness.

In the case of NDAC, the pin signifies when a device has not accepted data from the bus—in other words, that a read operation has not yet been completed. Again, the quiescent state of this pin for all listeners is "true" (logic 1, low voltage). If a number of instruments are meant to read a data byte, this line cannot go "false" (high voltage) until the final and slowest device has successfully completed the read operation.

Handshake

The power and flexibility of the GPIB bus arises largely from the three-wire handshake which governs the transfer of data bytes. The precise protocol for handshaking among instruments is illustrated in Fig. 19.4. The flowchart is somewhat simplified in order to highlight the key moments in the sequence.

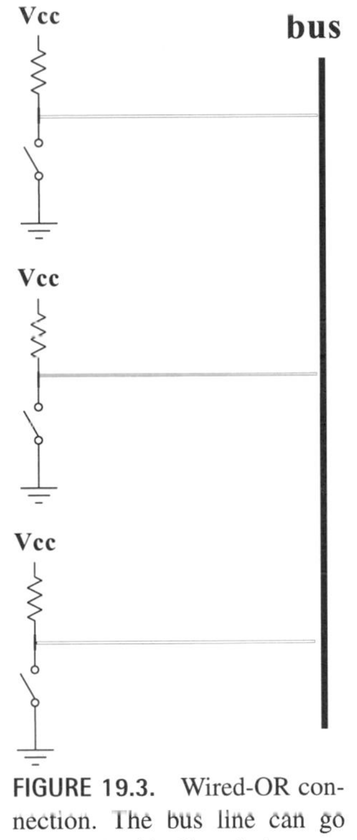

FIGURE 19.3. Wired-OR connection. The bus line can go high only if all switches are opened.

Another view of the handshake is afforded by a time series of the voltage signals on the three lines, as illustrated in Fig. 19.5. The circled numbers correspond to the matching points in the flowchart.

The bus is asynchronous, with talker and listeners independently carrying out their tasks. The handshaking protocol imposes pauses in the actions of talker and listeners to guarantee that no data are lost or that invalid data are not processed.

Checkpoint #1

The talker must wait for NRFD to become false (high-voltage state). This occurs only when all listeners have become ready to read new data.

Checkpoint #2

The listeners must wait for the talker to announce that its data byte is valid (DAV: logic 1, voltage low) and hence ready to be read.

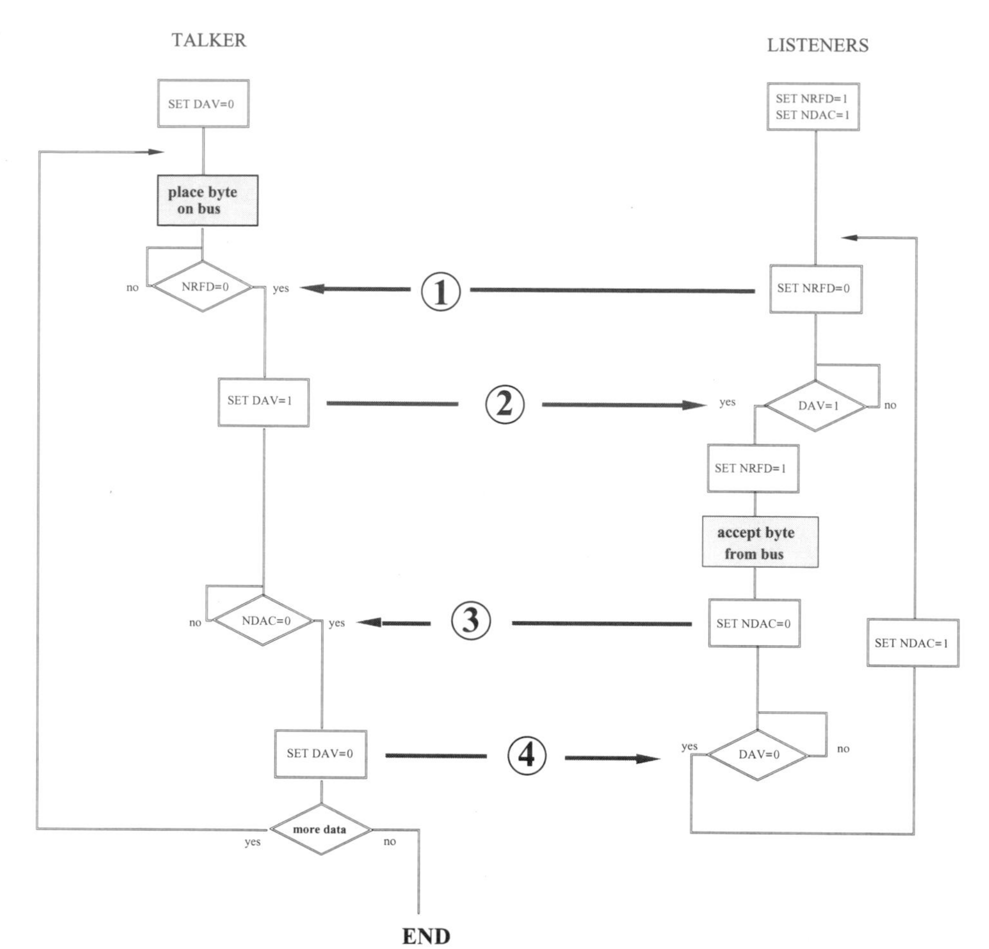

FIGURE 19.4. Flowchart for the GPIB handshake.

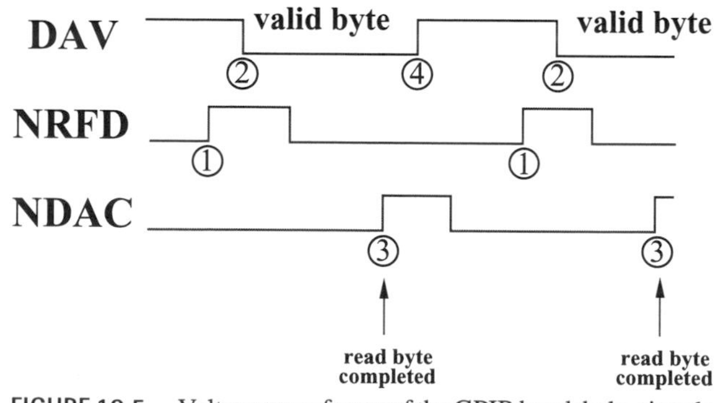

FIGURE 19.5. Voltage waveforms of the GPIB handshake signals.

Checkpoint #3

The talker must wait for NDAC to become false (high-voltage state), indicating that all listeners have successfully finished reading the data byte.

Checkpoint #4

The listeners wait for the talker to set DAV to logic 0, indicating the end of the byte-transfer cycle. When NDAC is then set to logic 1, a new handshake sequence can begin.

Talk and Listen Addresses

The assignment of talker and listeners by the controller is achieved by means of so-called talk and listen addresses. Each GPIB instrument typically has a rear-panel switch or a software menu that allows the user to choose a *device number*. No two instruments can be assigned the same device number, but otherwise the user is free to choose any value within the range 0–30. In binary, these numbers can be represented by a 5-bit code.

In order to distinguish between talker and listener functions, each 5-bit device number is augmented by two higher-order bits, as illustrated by the shaded columns in Fig. 19.6. The most significant bit in the completed byte is not used in this scheme and so is designated x (Don't Care). The byte $x01b_4b_3b_2b_1b_0$ specifies that the device whose number (in binary) is $b_4b_3b_2b_1b_0$ now must act as a listener. Similarly, the byte $x10b_4b_3b_2b_1b_0$ specifies that the device whose number (in binary) is $b_4b_3b_2b_1b_0$, is now to function as a talker.

Within the bus-management group mentioned earlier is the ATN (Attention) line. The state of ATN may be set only by the system controller. Normally, ATN is left at logic 0 (high-voltage state). Each instrument automatically monitors the ATN line and responds to its assertion (logic 1, low-voltage state) by looking to the bus for some message or command from the controller.

To issue talk and listen addresses, the controller asserts ATN and then places the appropriate byte on the data bus as suggested in Fig. 19.7. All instruments examine the byte, but only the one identified by $b_4b_3b_2b_1b_0$ responds by becoming a talker or listener, as dictated by the fifth and sixth bits.

Untalk and Unlisten

Two special commands are provided to cancel the assignment of talker and/or listeners. The specific bytes are:

$$\begin{aligned}\text{Unlisten: UNL} \quad & x0111111, \\ \text{Untalk: UNT} \quad & x1011111,\end{aligned}$$

	LISTEN ADDRESS								TALK ADDRESS							
DEVICE	b8	b7	b6	b5	b4	b3	b2	b1	b8	b7	b6	b5	b4	b3	b2	b1
0	X	0	1	0	0	0	0	0	X	1	0	0	0	0	0	0
1	X	0	1	0	0	0	0	1	X	1	0	0	0	0	0	1
2	X	0	1	0	0	0	1	0	X	1	0	0	0	0	1	0
3	X	0	1	0	0	0	1	1	X	1	0	0	0	0	1	1
4	X	0	1	0	0	1	0	0	X	1	0	0	0	1	0	0
5	X	0	1	0	0	1	0	1	X	1	0	0	0	1	0	1
6	X	0	1	0	0	1	1	0	X	1	0	0	0	1	1	0
7	X	0	1	0	0	1	1	1	X	1	0	0	0	1	1	1
8	X	0	1	0	1	0	0	0	X	1	0	0	1	0	0	0
9	X	0	1	0	1	0	0	1	X	1	0	0	1	0	0	1
10	X	0	1	0	1	0	1	0	X	1	0	0	1	0	1	0
11	X	0	1	0	1	0	1	1	X	1	0	0	1	0	1	1
12	X	0	1	0	1	1	0	0	X	1	0	0	1	1	0	0
13	X	0	1	0	1	1	0	1	X	1	0	0	1	1	0	1
14	X	0	1	0	1	1	1	0	X	1	0	0	1	1	1	0
15	X	0	1	0	1	1	1	1	X	1	0	0	1	1	1	1
16	X	0	1	1	0	0	0	0	X	1	0	1	0	0	0	0
17	X	0	1	1	0	0	0	1	X	1	0	1	0	0	0	1
18	X	0	1	1	0	0	1	0	X	1	0	1	0	0	1	0
19	X	0	1	1	0	0	1	1	X	1	0	1	0	0	1	1
20	X	0	1	1	0	1	0	0	X	1	0	1	0	1	0	0
21	X	0	1	1	0	1	0	1	X	1	0	1	0	1	0	1
22	X	0	1	1	0	1	1	0	X	1	0	1	0	1	1	0
23	X	0	1	1	0	1	1	1	X	1	0	1	0	1	1	1
24	X	0	1	1	1	0	0	0	X	1	0	1	1	0	0	0
25	X	0	1	1	1	0	0	1	X	1	0	1	1	0	0	1
26	X	0	1	1	1	0	1	0	X	1	0	1	1	0	1	0
27	X	0	1	1	1	0	1	1	X	1	0	1	1	0	1	1
28	X	0	1	1	1	1	0	0	X	1	0	1	1	1	0	0
29	X	0	1	1	1	1	0	1	X	1	0	1	1	1	0	1
30	X	0	1	1	1	1	1	0	X	1	0	1	1	1	1	0

FIGURE 19.6. Talk and Listen addresses.

ATN

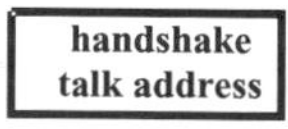

FIGURE 19.7. Use of the ATN line by the controller to issue a talk address. The same could apply to a listen address. The box is meant to signify the entire process of handshaking the data byte to the instruments on the bus.

which look like the listen and talk addresses except that the five lowest-order bits are all one's (decimal 31). Unlisten can be sent by the controller at the beginning of a command string to assure that initially there are no listeners, while the command string might end with Untalk as a final "housekeeping" statement.

Interrupts

Suppose that, while performing a series of programmed measurements, an instrument suddenly detects an internal hardware error. The occurrence of this, or possibly other, conditions should dictate that the normal flow of the acquisition process be halted so that appropriate action can be taken. Instruments thus need the capability to alert the controller that some condition requires immediate intervention. The controller, for its part, must be able to determine which instrument made the request and what condition caused it.

Two registers located in every GPIB instrument (per the IEEE-488.2 standard) play a central role in defining and reporting interrupt conditions; they are the *Standard Event Status Register* (SESR) and the *Status Byte Register* (SBR). The contents of each eight-bit register are defined as follows.

SESR:

Bit	Condition
0	operation complete
1	request control
2	query error
3	device-dependent error
4	execution error
5	command error
6	user request from front panel
7	power on

SBR:

Bit	Condition
0	
1	
2	
3	
4	MAV : message available
5	ESB : event status bit
6	RQS : request service
7	

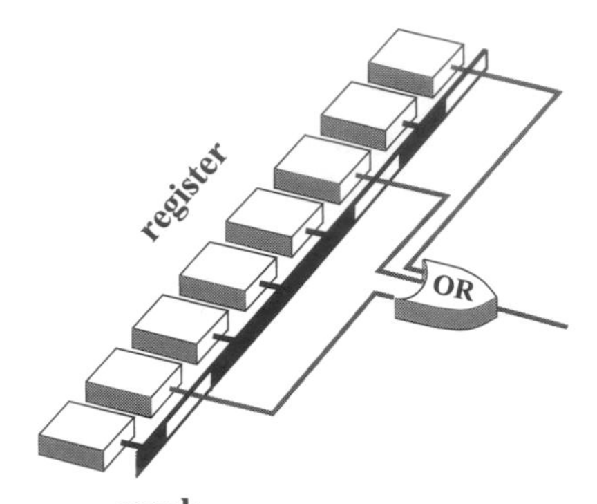

FIGURE 19.8. Basic principle of masking a status register.

The blank lines in this last table may in fact be used by some manufaturers to indicate other conditions particular to individual instruments.

Each of these registers is accompanied by a mask. The role of a mask is to control which conditions in a register are recognized, by being passed through to an associated OR gate as illustrated in Fig. 19.8. The mask is set by a user-definable byte. Thus, for example, a mask byte of 10100010 would "enable" register bits b_1, b_5, and b_7.

The standard event and status byte registers, and their user settable masks, are logically connected, as illustrated in Fig. 19.9. In this example, the standard event conditions RQC, CME, and PON are enabled by the mask (10100010) so that the occurrence of any or all such conditions will set the Event Status Bit in the Status Byte. The Status Byte mask (again, in this example) is set to $0x100000$, which means that a service request will be generated (RQS bit $= 1$) if any of these events is detected.

The Don't Care condition of the sixth bit in the Status Byte mask is built into the hardware so that b_6 in the Status Byte is *never* passed through to the OR gate (this prevents a feedback-loop condition from developing around b_6).

The Standard Event Status Register mask is defined by the *Standard Event Status Enable Register*, which can be written to with the command *ESE followed by a decimal argument in the range 0–255. The binary equivalent of this argument is the 8-bit mask. For example, "*ESE 162" would create the SESR mask mentioned earlier.

Similarly, the *Service Request Enable Register* controls the Status Byte Register mask; it is written to with the command *SRE. In the example of Fig. 19.9, the mask could be created with "*SRE 32."

As a final example, the command "*SRE 16" would set the mask so that the presence of data in the instrument's output buffer would cause $b_6 = 1$ in the SBR.

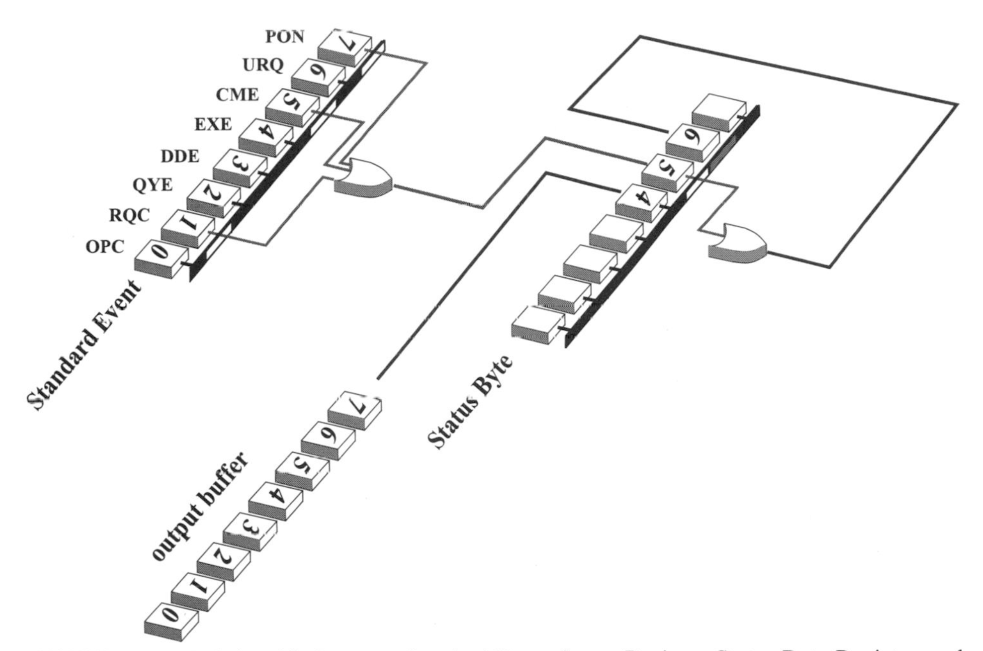

FIGURE 19.9. Relationship between Standard Event Status Register, Status Byte Register, and their masks. The particular choice of open and closed mask cells is for illustration purposes only.

From the preceding discussion, it is clear that b_6 in the SBR amounts to a global flag indicating that some user-specified event has indeed occurred. When this flag is set, the instrument will assert the SRQ (Service Request) line in the GPIB bus-management group. This particular bus line is monitored by the controller, which can be preinstructed to react to SRQ:true by conducting a *serial poll* first to determine which instrument issued the service request and then to ascertain what condition caused the interrupt.

The logic of a serial poll is depicted in Fig. 19.10. The SPE and SPD commands are

Serial Poll Enable: SPE 00011000,
Serial Poll Disable: SPD 00011001.

The SPE command is issued by the controller with ATN active (low voltage) and so is read by all instruments on the bus. It causes each instrument to prepare its status byte for reading if and when it is addressed as a talker.

In essence, the polling procedure involves addressing the instruments on the bus one at a time and asking each in turn to report its status byte back to the controller. When a status byte is encountered whose sixth bit is 1, the polling

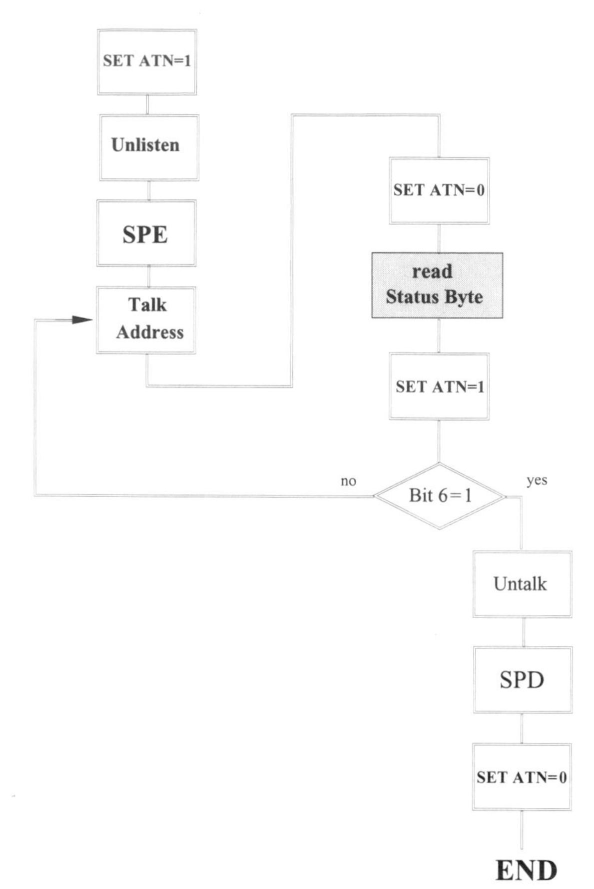

FIGURE 19.10. Logic flow diagram of a serial poll conducted by the system Controller.

terminates. As is evident, most of the actions are directed by the controller during intervals in which ATN is active (low voltage), the principal exception being the necessary release of ATN (high voltage) so that the status byte of the addressed instrument may be read.

It should be added that under the IEE-488.2 standard, commands also exist for reading from the registers, as summarized in the following table.

Register	Write to	Read from
Standard Event Status Register	internal	*ESR?
Standard Event Status Enable Register	*ESE	*ESE?
Status Byte Register	internal	*STB?
Service Request Enable Register	*SRE	*SRE?

19.2 GPIB FOR THE USER

To a scientist wishing to use the GPIB bus as a means of realizing a PC-based instrumentation system, the intricate details of the GPIB may simply be a burden. Fortunately, it is not necessary to become involved in such matters as handshake protocols. Every GPIB-capable instrument comes equipped with a built-in hardware interface terminating in the standard GPIB connector and with internal intelligence to handle all GPIB functions.

The desktop computer will need a GPIB card (these are available from a number of manufacturers), which is inserted into one of the computer's internal slots (either ISA or, more commonly, PCI). A GPIB connector on the card is mounted so that it becomes available as a rear-panel plug when the card is installed in the PC. A variety of compatible software drivers are also available to enable the user's programming language to communicate properly with the internal GPIB card.

This combination of additional hardware (card) and software (driver) is crucial to enabling the PC to function as the GPIB controller. The detailed activity of the bus is completely hidden from the user, who needs only to concentrate on writing programs suited to the data-acquisition tasks.

Instrument Commands

Suppose a multimeter is to be set in the dc volts mode and then told to make a measurement. This requires that it be addressed as a listener and then provided with the appropriate instruction sequence. For example, the Model 45 multimeter from Fluke [7] requires the character string: "*RST; VDC; RANGE 2; TRIGGER 2; *TRG; VAL?" in order to reset, go to the dc volts mode with range of 3 V, use external triggering with no delay, apply a trigger pulse, and finally take a reading from the meter. Compare this with the command string for the 3478A multimeter from Hewlett-Packard [8]: "Z1F1R0N4T3," which accomplishes the same tasks.

This illustrates the fact that command strings are quite different for different instruments and vary among manufacturers. The user must determine appropriate codes from the owner's manuals.

Of course, a command sequence is a string of ASCII characters, which in turn is just a string of bytes. These can be issued one at a time by the controller and read one at a time by the instrument (as a listener) according to the previously discussed GPIB handshake protocol. Similarly, instrument readings (e.g., a decimal number like 6.173) can be transferred to the PC in the form of a string of ASCII characters.

Software

There are at the moment two distinct paradigms for programming instrumentation systems: command-line based and graphical icon based.

An example of the first approach is provided by the following Microsoft QuickBASIC [9] program, which controls two external instruments from a PC via the GPIB bus. The driver software that established linkage to the GPIB hardware port was, in this case, NI-488.2 from National Instruments [10]. The program sends commands to the power supply, first to set the output at some desired value and then to read back the dc output as determined by the power supply's internal voltmeter. Next, a second measurement of the supply output is made, this time by the external multimeter. The three values—desired, internally measured, and externally measured—are then written to a file. These steps are repeated for a set of target voltages ranging from 0 to nearly 10 V.

The function IBFIND locates the named instrument on the bus and assigns it an alias to be used elsewhere in the program. Thus, bias% refers to the power supply, whereas meter% refers to the multimeter. The function IBWRT writes a character string to the designated instrument; IBRD reads a character string from the instrument.

```
 ' QuickBASIC control of HP 6632A power supply
 ' and HP 3478A multimeter.
 ' The power supply will be programmed
 ' to a series of output values.
 ' After each setting, the supply's internal meter
 ' will be read and this value will be compared
 ' to a measurement made by the external meter.
COMMON SHARED IBSTA%, IBERR%, IBCNT%
DEFSNG A-H, O-Z
DEFINT I-N
DELAY=5!
 ' DELAY IS NUMBER OF SECONDS TO WAIT
 ' AFTER SETTING A NEW BIAS VALUE
 ' BEFORE READING THE POWER SUPPLY OUTPUT VOLTAGE.
OPEN "READING.DAT" FOR OUTPUT AS #50
ZSPLY$=SPACE$(8)
ZMETER$=SPACE$(8)
CALL IBFIND("HPPOWER", bias%)
CALL IBFIND("HPMETER", meter%)
CALL IBWRT(meter%, "F1R1Z1N4")
```

```
CALL IBWRT(bias%, "CLR")
'
FOR ZSET=0! TO 9.6 STEP .05
' SET POWER SUPPLY OUTPUT
ZSET$="VSET" + STR$(ZSET)
CALL IBWRT(bias%, ZSET$)
' WAIT FOR DELAY SECONDS
TIME0=TIMER
DO WHILE (TIMER-TIME0)<DELAY
LOOP
' READBACK OUTPUT FROM POWER SUPPLY
CALL IBWRT(bias%, "VOUT?")
CALL IBRD(bias%, ZSPLY$)
XSPLY=VAL(ZSPLY$)
' NOW READ ACTUAL OUTPUT WITH METER
CALL IBWRT(meter%, "T3")
CALL IBRD(meter%, ZMETER$)
VOLT=VAL(ZMETER$)
' SAVE DATA TO FILE
PRINT #50, USING "######.#####"; ZSET; XSPLY; VOLT
'
NEXT ZSET
'
' NOW RESET EVERYTHING
CALL IBWRT(meter%, "F1RAT1")
ZSET=0!
ZSET$="VSET" + STR$(ZSET)
CALL IBWRT(bias%, ZSET$)
CALL IBLOC(bias%)
CALL IBLOC(meter%)
END
```

This program shows how it is possible with just a few lines of code to program a desktop computer to send commands to GPIB instruments and to read back data from those instruments.

The second programming convention is to be found in Microsoft Windows-based software such as LabVIEW [11, 12, 13] and DASYLab. Code is generated from logic flowcharts in which functional icons are interconnected in the desired manner. Samples of two icons from LabVIEW are shown in Fig. 19.11. These graphical objects represent read and write operations to a GPIB device

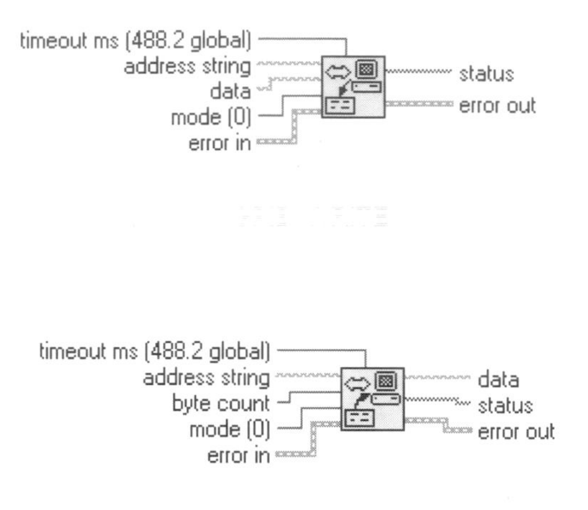

FIGURE 19.11. LabVIEW icons for GPIB Write and Read.

at a specified address. In fact, LabVIEW is provided with an extensive library of function icons covering basic operations such as instrument input/output, file input/output, data acquisition, string handling, and data analysis. Beyond these tools, instrument-specific libraries are also available that permit the user to construct compact graphical programming code.

A more complete "program" in LabVIEW is shown in the example of Fig. 19.12. Here, an HP3478A multimeter with GPIB address 23 is sent an ASCII string consisting of M01Z1N5 plus one of the measurement codes (F1, F2, ...) plus one of the range codes (R-2, R-1, R0, ...) plus one of the trigger codes (T1, T2, ...). The first code (M01) is a mask which causes an SRQ to be generated when a meter reading is completed. The next commands invoke autozero mode (Z1) and $5\frac{1}{2}$ digit display (N5). Only a small amount of additional graphical code would be needed to then read back the voltages.

Note that this graphical program accomplishes tasks similar to those performed by the QuickBASIC program. There is therefore something of a choice for the user as to which programming style is preferred.

It should be noted that in LabVIEW the logic diagram is only part of the story—there also is an associated front panel which is jointly created. The control panel for this acquisition example is shown in Fig. 19.13. Each of the sliders and switches is activated at run time by user actions on the computer screen (i.e., pointing, clicking, and dragging with a mouse).

Virtual instrument control panels can also be created in the command-line programming environment. In a language such as Microsoft Visual Basic, this is typically done with the aid of additional graphical software tool sets.

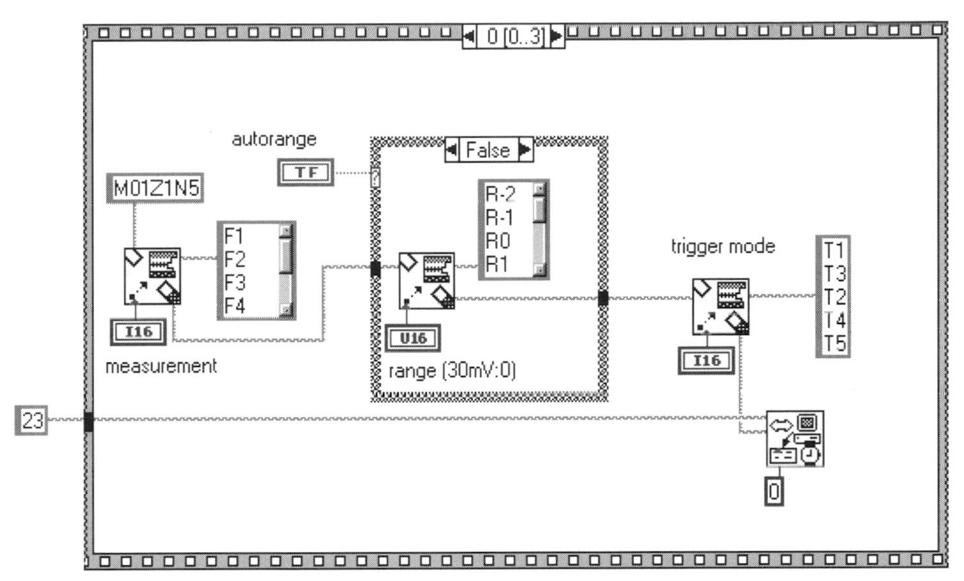

FIGURE 19.12. LabVIEW graphical program for sending command codes to a Hewlett-Packard 3478A multimeter which is at GPIB address 23. The sequence begins with the string "M01Z1N5" and then appends appropriate codes for measurement type, range, and trigger mode.

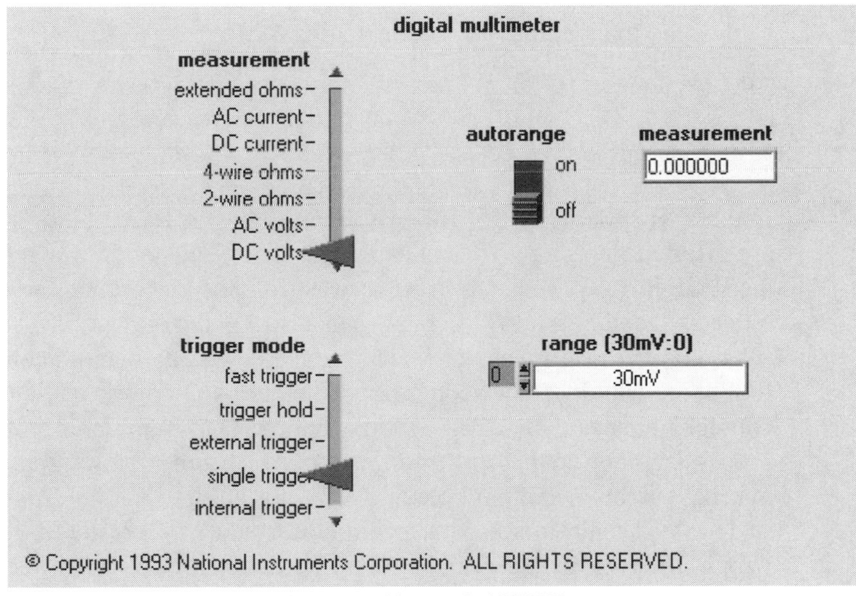

FIGURE 19.13. Front panel for the multimeter LabVIEW program.

Performance

The GPIB bus performance and specifications are as follows.

permitted device addresses	0–30
maximum number of active devices	15
maximum total cable length	15 meters
maximum data transfer rate: PCI bus	1.5 Mbyte/s
maximum data transfer rate: PC/XT	300 kbyte/s

HS488

A recent enhancement to the GPIB is the HS488 from National Instruments [6]. For this interface, an HS488 card is installed in the PC. If an HS488 equipped instrument is on the bus, it is automatically detected and the new protocol is invoked. Otherwise, standard IEEE-488 rules are followed as a default mode. The strategy adopted in HS488 is to speed up the bus by removing propagation delays inherent in the original 3-wire handshake. In HS488, the user must inform the bus-management software what total length of interface cable is in use. This information is employed to calculate optimal bus timings. At cable lengths of 15 meters, the speed is essentially the same as for IEEE-488, but with shorter cables, the transfer rate goes up, reaching 8 Mbytes/s at 2 meters.

19.3 VXI

In the VXI system [14, 15, 16], instruments take the form of plug-in cards which are inserted into slots in a host card cage. The VXI standard was formalized by the IEEE in 1992. The name VXI derives from the fact that this architecture is an outgrowth of the earlier VME bus—hence, the XI denotes an extension of V(ME). VXI modular instruments now are produced by many manufacturers. Available functions include oscilloscopes, multimeters, A/D converters, D/A converters, counter-timers, gigabyte data-storage modules, programmable amplifiers, signal processors, loop controllers, arbitrary waveform generators, signal conditioning, matrix switching, and multiplexing.

The VXI mainframe has a maximum capacity of 13 slots (see Fig. 19.14). Approximate overall dimensions of a 13-slot box are 15 inches high $\times$ 18 inches wide $\times$ 26 inches deep. The mainframe provides an integral electrical power distribution system for all plug-in cards, a controlled air-cooling capability, and excellent electrical shielding for the installed components.

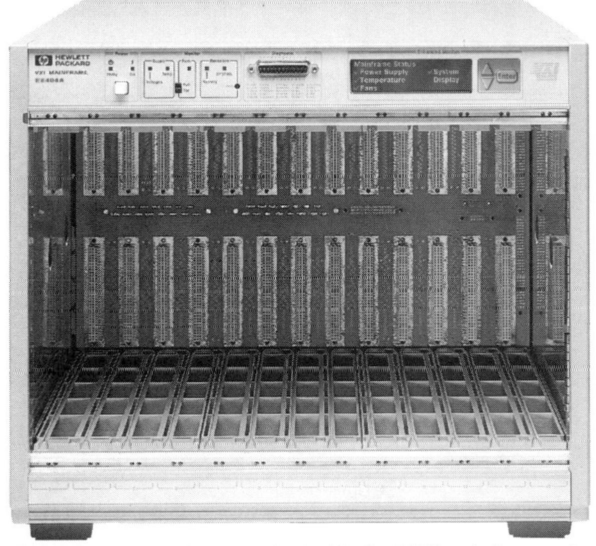

FIGURE 19.14. Photograph of a 13-slot VXI mainframe cabinet. (Courtesy of Agilent Technologies.)

VXI modules can come in any of four standard sizes, denoted A, B, C, D, with C being the most common choice. The dimensions of the C module are 9.2 inches high × 13.4 inches deep (23.4 × 34 cm). Slot spacing for C cards is 1.2 inches (3 cm). A typical VXI instrument is shown in Fig. 19.15.

At the rear of B- and C-sized cards are two 96-pin connectors denoted P1 and P2. When a module is inserted into an available mainframe slot and is latched into position, it becomes electrically connected to a backplane via these plugs. All instrument cards thus become hardwired to this internal bus, in contrast to the external cabling of the GPIB architecture.

As discussed earlier, the GPIB bus is based on a 24-wire interconnecting cable/connector, which links all external instruments in a parallel configuration. By comparison, the 192-pin combination of P1/P2 gives B- and C-sized VXI cards a very large effective backplane bus organized into the following subgroups: a 32-bit VME bus, a 10 MHz clock bus, an analog sum bus, a 12-pin local bus, a module identification bus, a TTL/ECL trigger bus, and a power distribution bus for −5.2, −2, +5, ±12, ±24 V. The inherent transfer rate of 32-bit data on the VXI backplane is 40 Mbytes/sec.

The local bus differs from the other system-wide buses in that it consists of individual wires running directly between 12 specified sets of facing pins on adjacent P2 backplane connectors. This arrangement provides a form of private communication strictly between adjacent VXI modules; that is, each module

FIGURE 19.15. Example of a C-sized VXI instrument plug-in module. (Courtesy of Agilent Technologies.)

can communicate with the modules to its immediate left and right. Such an arrangement can be well suited to situations where data from one module need to be passed quickly to another module for subsequent action (say an A/D must transfer data to a digital signal processor for real-time analysis).

The left-most position in the mainframe is special and must be occupied by a so-called "Slot 0" device. Whatever its other duties, a Slot 0 device manages the backplane, providing the system with clocks and being responsible for bus arbitration, interrupt processing, and slot and module identification.

Each VXI device has a logical address assigned from the range 0–255. The two most common device types are *register-based* and *message-based*. Register-based devices use a binary communication format and are thus fast. Message-based devices follow an easy-to-use word serial protocol (WSP), which permits information transfer in the form of ASCII strings and commands as in GPIB.

A complete VXI setup would typically consist of a desktop computer serving as the system controller and a VXI mainframe box with its installed device modules [17]. There are three standard ways of making the necessary communication link between the PC and the VXI box.

1. The link is GPIB. In this case, the PC must have a GPIB card, and the VXI mainframe must have a GPIB-VXI module installed. An additional benefit of this scheme is that a more elaborate hybrid system may be created that

combines the VXI box and its internal devices with separate daisy-chained external GPIB instruments. Data-transfer rates are around 1 Mbyte/sec.

2. The link is IEEE 1394. This high-speed six-wire serial bus also goes by the name "FireWire." The PC requires a FireWire card, and the VXI mainframe requires a 1394-VXI module. Present data-transfer rates are reported up to 15 Mbytes/sec.
3. The link is MXI. This very high-speed 62-wire bus allows the internal hardware of the PC to interact directly with the VXI backplane [15]. The PC requires an MXI interface card, and the VXI system requires an installed VXI-MXI module. Data-transfer rates of the original version of MXI introduced by National Instruments in 1989 were 10 Mbytes/sec. This has now been doubled to 20 Mbytes/sec by MXI-2 (1995).

As with the GPIB, the user of a VXI system generally would prefer not to be involved in the complex activities of the bus. The hardware design of VXI, and in particular the role of the Slot 0 device, does indeed relieve the user of most of the responsibility for detailed system management. By selecting appropriate software for the controller, it becomes possible to orchestrate complex data-acquisition tasks solely at a high-level graphical interface. Also as with GPIB, software choices range from Microsoft Visual Basic to icon-based products like LabVIEW from National Instruments and HP VEE from Hewlett-Packard [18].

19.4 PXI

The PXI system architecture is relatively new [19]. PXI adopts the mainframe/plug-in module concept, as does VXI, but is designed as an extension of the 132 Mbyte/sec PCI bus rather than the more specialized VXI backplane. The result is a system that has capabilities somewhere between VXI and a PC equipped with internal cards.

One problem that PXI seeks to alleviate is the limitation associated with the maximum of four PCI slots within a desktop computer. This places a restriction on the number of instrument functions that can be resident in a PC in the form of plug-in cards. A PXI mainframe (Fig. 19.16) has a total of eight slots, the leftmost of which is reserved for the system controller. Thus, seven slots remain available for instrument plug-ins.

Approximate overall dimensions of an 8-slot box are 7 inches high $\times$ 11 inches wide $\times$16 inches deep. The standard PXI card has a 3U form factor (see Fig. 19.17) and is 6.3 inches deep $\times$ 3.9 inches high (16 $\times$ 10 cm).

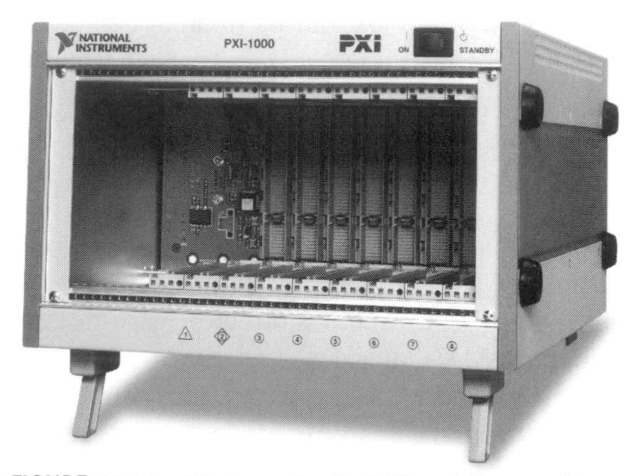

FIGURE 19.16. Photograph of a PXI mainframe cabinet. (Courtesy of National Instruments.)

When launched in 1997, PXI required a controller module containing an embedded Pentium CPU. This internal computer needed an external monitor and keyboard to become operational. In that configuration, the PXI system became equivalent to a PC tailored to instrumentation tasks and with a capacity of seven rather than four plug-in cards.

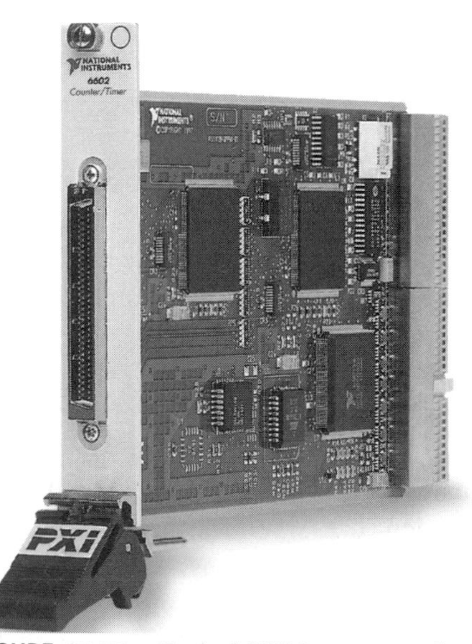

FIGURE 19.17. Typical PXI instrument plug-in module. (Courtesy of National Instruments.)

The latest version of PXI removes the need for an embedded CPU and allows control from an external PC via a new ultrafast MXI-3 serial link rated at about 90 Mbytes/sec. The advantage of this scheme is the comparative ease and economy with which conventional (nonembedded) PCs can be upgraded as technical enhancements are introduced. Because it is founded on the PCI architecture and is in essence just a customized PC, PXI is inherently compatible with Microsoft Windows software.

Time will tell if PXI can reach the level of wide acceptance and multivendor support already achieved by PC data-acquisition cards, GPIB instruments, and the VXI system.

BIBLIOGRAPHY

[1] "A practical interface system for electronic instruments," Hewlett-Packard J., October 1972.

[2] Anthony J. Caristi, *IEEE-488 General Purpose Instrumentation Bus Manual* (Academic Press, London, 1989).

[3] Richard Newrock and James R. Matey, "Catching the right bus III: IEEE-488.1," Comput. Phys. 8, 259–268 (1994).

[4] Eric D. Jones, Bryan L. Preppernau, P.G. Stein, and James R. Matey, "Catching the right bus, Part IV: IEE-488 Programming," Comput. Phys. 9, 24–33 (1995).

[5] Eric D. Jones and Bryan L. Preppernau, "Catching the right bus V: A Beginners' Guide to Programming the IEEE-488 Bus," Comput. Phys. 9, 140–147 (1995).

[6] Andrew Thomson and Amar Patel, "Using HS488 to Improve GPIB System Performance," Application Note 096 (National Instruments Corporation, Austin, TX, December 1996).

[7] John Fluke Mfg. Co., Everett, WA.

[8] Hewlett-Packard Company, Loveland, CO.

[9] Microsoft Corp., Redmond, WA.

[10] For this example, the GPIB PCIIA card and NI-488.2 software from National Instruments Corporation, Austin, TX.

[11] National Instruments, Austin, TX.

[12] John Essick, *Advanced LabVIEW Labs* (Prentice-Hall, Englewood Cliffs, NJ, 1999); see particularly Chapter 12: GPIB-Control of Instruments.

[13] Lisa K. Wells and Jeffrey Travis, *LabVIEW for Everyone* (Prentice-Hall, Englewood Cliffs, NJ, 1997), pp. 354–358.

[14] Kenneth Jessen, "VXIbus: A new interconnection standard for modular instruments," Hewlett-Packard J., April 1989, pp. 91–95.

[15] Ron Wolfe, *Short Tutorial on VXI/MXI*, Application Note 030 (National Instruments Corporation, Austin, TX, April 1996).

[16] David Haworth, *Using VXI Bus: A Guide to VXIbus Systems* (Tektronix Inc., Beaverton, OR, 1992).

[17] Yet another configuration eliminates the need for an external PC by embedding a CPU in the Slot 0 module.

[18] Robert Helsel, *Visual Programming with HP VEE* (Prentice-Hall, Englewood Cliffs, NJ, 1998).

[19] PXI Specification, Revision 1, August 1997, National Instruments Corporation, Austin, TX.

Index